RADIATION PHYSICS OF METALS AND ITS APPLICATIONS

RADIATION PHYSICS OF METALS AND ITS APPLICATIONS

L I Ivanov and Yu M Platov

A.A. Baikov Institute of Metallurgy and Materials Science, Russian Academy of Sciences, Moscow

Viva Books

New Delhi | Mumbai | Chennai | Kolkata | Bangalore | Hyderabad | Kochi

First Indian Edition 2008

VIVA BOOKS PRIVATE LIMITED

- 4737/23 Ansari Road, Daryaganj, New Delhi 110 002
  E-mail: vivadelhi@vivagroupindia.net, Tel. 42242200
- Plot No. 76, Service Industries, Shirvane, Sector 1, Nerul, Navi Mumbai 400 706
  E-mail: vivamumbai@vivagroupindia.net, Tel. 27721273, 27721274
- Jamals Fazal Chambers, 26 Greams Road, Chennai 600 006
  E-mail: vivachennai@vivagroupindia.net, Tel. 28290304, 28294241
- B-103, Jindal Towers, 21/1A/3 Darga Road, Kolkata 700 017
  E-mail: vivakolkata@vivagroupindia.net, Tel. 22836381, 22816713
- 7, Sovereign Park Apartments, 56-58, K. R. Road, Basavanagudi, Bangalore 560 004
  E-mail: vivabangalore@vivagroupindia.net, Tel. 26607409, 26607410
- 101-102 Mughal Marc Apartments, 3-4-637 to 641, Narayanguda, Hyderabad 500 029
  E-mail: vivahyderabad@vivagroupindia.net, Tel. 27564481, 27564482
- Beevi Towers, First Floor, SRM Road, Kaloor, Kochi 682 018, Kerala
  E-mail: vivakochi@vivagroupindia.net, Tel: 0484-2403055, 2403056

www.vivagroupindia.com

Published by arrangement with
Cambridge International Science Publishing
7, Meadow Walk, Great Abington
Cambridge CB1 6AZ, UK

ISBN-10: 81-309-0836-0
ISBN-13: 978-81-309-0836-6

Published by Vinod Vasishtha for Viva Books Private Limited, 4737/23, Ansari Road, New Delhi-110 002. Printed and bound by Anand Sons, Delhi-110 092.

Contents

Chapter 6
MAIN PRINCIPLES AND MECHANISMS OF RADIATION DAMAGE OF STRUCTURAL METALLIC MATERIALS 216

Lev Ivanovich IVANOV, Doctor of Physico-Mathematical Sciences, Laureate of the State Prize of the USSR, Honoured Activist in Science and Technology of Russia, Head of the Laboratory 'The effect of radiation on metals' of the A.A. Baikov Institute of Metallurgy and Materials Science, Russian Academy of Sciences, Moscow

Yurii Mikhailovich PLATOV, Doctor of Physico-Mathematical Sciences, Chief Scientist at the same Institute.

The authors of this books belong to pioneers of Russian radiation materials science who originated systematic investigations of the behaviour of solids in the conditions of reactor and cosmic irradiation and worked on the development of a number of radiation-resistant materials for atomic power engineering. In this book, they present the results of many years of experimental and theoretical investigations, carried out by themselves and other leading experts, into fundamental and applied aspects of the radiation physics of metals

INTRODUCTION

The development of a modern energy based in nuclear power engineering and further advances in the exploration of space are determined to a large degree by the development of metallic materials for various functional applications, corresponding to the requirements of service reliability, economic efficiency and minimum disruption of ecology. The successful realisation of these requirements depends mainly on the application of metals and alloys characterised by high radiation resistance and accelerated decrease of induced radioactivity.

The physical fundamentals and principles of the development of these materials are based on the systematisation and further development of the considerations regarding the mechanisms and factors controlling changes in the structure, properties and activation parameters of irradiated objects, analysis of the mechanisms and the development of methods reducing the negative effect of radiation. We have also attempted to present the main material of this book using approach to the investigated problem

Special attention in the monograph is given to the analysis of the effect of the type and concentration of impurity and alloying elements and also phase changes in the mechanism of buildup of radiation defects and radiation damage to metallic materials. The considerations regarding the interaction of radiation point defects with each other and with solutes, the diffusion of point defects and solutes, radiation-enhanced diffusion and phase transformations in irradiation were used as a basis when describing these processes. These problems are studied extensively in the book.

Detailed analysis of radiation damage and methods of suppressing this type of damage on the basis of taking into account the actual structure and the chemical and phase composition of metals and alloys has become possible to a large degree because of the advances in a number of fundamental areas of the radiation physics of solids: theory of defects in alloys, low-temperature kinetics of the buildup and annealing of radiation defects in diluted and concentrated solid solutions, radiation-enhanced diffusion, phase instability. The most important results obtained in this area include mainly the development of considerations on the interaction of interstitials with solute atoms, their diffusibility and the mechanisms of migration in diluted and concentrated solid solutions, the diffusion transfer of solid elements and their interaction with sinks. A significant role in the development of considerations regarding the radiation damage in alloys also belongs to the determination of relationships of the phase transformations and instability of the metallic materials under irradiation, including the processes of ordering, phase separation and breakdown of solid solutions, segregation, coalescence and dissolution of the phases.

A special section in the monograph is concerned with the selection of components of alloys taking into account radiation ecological requirements.

In particular, this problem is important for 'clean' fusion power engineering, because the absence in these reactors of traditional fission fuel creates, when using materials with accelerated decrease of induced radioactivity, the most suitable conditions for the efficient solution of the problem of increasing the service reliability of reactors, utilisation and processing of radioactive waste. At present, the problem of application of reduced-activation materials is becoming more and more important also in the area of conventional atomic power engineering.

Within the framework of this problem, the appropriate chapter of the monograph includes the calculation and experimental estimates of the parameters of activation of individual chemical elements and alloys, examination of a number of methods of reducing the activation of materials, analysis of general directions and problems of the development of reduced-activation radiation-resistant alloys. In the final chapter of the monograph, attention is given to the main experimental relationships and the mechanisms of radiation damage in a number of pure metals and structural metallic materials in atomic and fusion power engineering, determined by the processes of radiation hardening, embrittlement, creep and swelling.

When writing the monograph we have used to a large degree the theoretical and experimental data obtained in the laboratory 'The effect of radiation on metals' of the Institute of Metallurgy and Materials Science of the Russian Academy of Sciences, and also the results of the joint investigations with other laboratories of the Institute, domestic and foreign scientific centres. We are grateful to all co-authors of these studies.

A significant contribution to the presented material has been provided by Prof. L.N. Bystrov of the laboratory 'The effect of radiation on metals' and Profs. A.C. Damask and J.J. Dienes of the Brookhaven National Laboratory, USA.

In preparation of the monograph we were greatly helped by scientists of our laboratory S.V. Simakov, V.I. Tovtin, N.A Vinogradov, O.N. Nikitushkina and V.A. Polyakov. We are also grateful to Academician N.P. Lyakishev, Director of the Institute of Metallurgy and Materials Science of the Russian Academy of Sciences, for his considerable attention to studies in the radiation physics of solids and metals science and for his help in publishing this monograph.

Chapter 1

FORMATION OF RADIATION POINT DEFECTS AND THEIR INTERACTION WITH EACH OTHER AND WITH SOLUTE ATOMS IN METALS

1.1. INTRODUCTION

When an atom of a metal receives from a moving particle an energy equal to or higher than some threshold value T_d (of the order of 20–40 eV), the atom is displaced from its position in the crystal lattice. Depending on the energy of the primary knocked out atom T which is determined mainly by the type and energy of radiation, the nature of the irradiated target, the crystallographic direction and the angle of collision of the particle with the atom of the lattice, various configurations and structures of radiation defects may appear, starting from isolated point defects–vacancies and interstitials (Frenkel pairs) and ending with spatial zones of damage characterised by a complicated structure of defects.

Single Frenkel pairs form at energies of the primary-knocked out atoms in the range $T_d \leq T < 2.5\ T_d$ which is characteristic of, for example, irradiation of copper with the electrons with an energy of < 1 MeV. At higher energies of the primary knocked out atoms, a cascade of atoms collisions appears leading to the formation of a more complicated structure of radiation defects. The processes of interaction of radiation with the crystal lattice of metals during the formation of radiation defects and also the methods of modelling these processes have been examined in detail in [1].

The defects, formed during elementary acts of the interaction of the moving particles with matter usually belong to the primary radiation defects. The subsequent kinetic evolution of the structure of radiation damage is determined by the processes of spatial overlapping of the volumes of primary damage and/or diffusion of freely-migrating point defects and solutes.

In this chapter, attention will be given to the main relationships governing the formation of primary radiation defects. Special attention is given to the problem of formation of freely-migrating vacancies and interstitials, because the concentration of these defects plays a significant role in the processes of formation and evolution of secondary radiation defects.

In this chapter, attention is also given to the current assumptions regarding the interaction of point defects with each other and with the solute atoms. In a general case, this interaction determines the structures of defect and defect–impurity complexes and diffusion parameters of the vacancies, interstitials and solutes. The interaction mechanisms may already play a specific role in the stage formation of primary radiation defects where the athermal or thermally-activated formation of different defect and defect–impurity configurations, may modify the final structure of defects, characteristics of pure metals. At temperatures at which the point defects are mobile, their concentration and mechanisms of interaction with each other and with the solute atoms may be the main factors determining the evolution of the structure of radiation damage and the variation of the properties of materials during irradiation.

1.2. FORMATION OF PRIMARY RADIATION DEFECTS

At present, the theoretical analysis of the formation of primary radiation defects is carried out using two main approaches: calculations using the theory of simple binary collisions and the modified Kinchin–Pease model, and computer modelling using the molecular dynamics method.

Within the framework of the theory of binary collisions, the total number of the Frenkel pairs, formed during transfer of energy T to a primary-displaced atom from a particle with the initial energy E, can be determined from the equation:

$$\nu(E)=\int_{T_d}^{T_{max}} dT\frac{d\sigma(E,T)}{dT}\nu(T) \tag{1.1}$$

In equation (1.1) $d\sigma(E, T)/dT$ is the differential cross-section for the transfer by an incident particle of energy in the range from T

to $T + dT$ to the primary-displaced atom, $\nu(T)$ is the number of Frenkel pairs formed by recoil atoms of this energy range, T_d is the threshold energy of displacement, and T_{max} is the maximum energy which can be transferred by the particle to the primary-displaced atom. For the electrons, this energy is:

$$T_{max} = \frac{2E\left(E + 2m_e c^2\right)}{Mc^2} \tag{1.2}$$

For the neutrons and heavy ions, T_{max} is expressed by the equation:

$$T_{max} = \frac{4EM_1M_2}{\left(M_1 + M_2\right)^2} \tag{1.3}$$

In equations (1.2) and (1.3), E is the energy of the bombarding particle, M and M_2 is the mass of the atoms of the target, m_e is the mass of the electrons, c is the velocity of light, M_1 is the mass of the neutron or ion. The boundary conditions for the cascade function in the modified Kinchin–Pease model [2] are determined by the equations

$$\begin{aligned} &\nu(T) = 0 \,\text{at}\, T < T_d, \\ &\nu(T) = 1 \,\text{at}\, T \le T < 2.5T_d, \\ &\nu(T) = \frac{E_d}{2.5T_d} \,\text{at}\, T \ge 2.5T_d, \end{aligned} \tag{1.4}$$

where E_d is the energy of damage. This energy differs from energy T by the value of inelastic losses of the primary-displaced atom.

The primary-displaced atom with the recoil energy in the range $T_d \le T < 2.5\ T_d$ forms only one stable Frenkel pair. A cascade of atomic collisions forms at higher energies. At relatively high recoil energies, intensive atomic collisions, formation and recombination of Frenkel pairs take place in the vicinity of the trajectory of the primary-displaced atom, i.e. the core of the cascade. This process is accompanied by the formation of chains of focused atomic collisions leading to the separation of vacancies and interstitials in the Frenkel pairs and localisation of the interstitials at the periphery of the cascade. The formation of chains of atomic displacements was detected for the first time in [3] in computer modelling of the process of atomic collisions in copper, using the molecular dynamics method. The final structure of the cascade consists of a neutral zone with a higher concentration of defects of the vacancy type, and a periphery characterised by the localisation of individual interstitials or by clusters of these atoms.

Intracascade recombination, characterising cascade efficiency, for similar models may be evaluated by means of analytical expressions presented in, for example, [4,5].

At specific energies of the primary-displaced atom, branching of the cascade may take place. The mean number of sub-cascades in a cascade within the framework of the Kinchin–Pease modified model is expressed by the equation [6]:

$$\nu_{sc} = \frac{E_d}{2{,}5E_{sc}} \tag{1.5}$$

The comparison of the theoretical calculations within the framework of the examined model of simple binary collisions with the experimental data results in a large difference of the results, especially at high energies of the primary-displaced atoms [7]. With increase of recoil energy to several kiloelectronvolts in the experiments for particles of different type, the results show a large decrease of the efficiency of formation of defects with a subsequent tendency for saturation [8]. This disagreement between the theory and the experiment is evidently associated with the factor of mutual recombination of the point defects, which is very difficult to take into account within the framework of similar models, especially at the energies of formation of sub-cascades.

The main special features of the process of development of a cascade and the structure of the damaged zone, examined within the framework of the model of simple binary collisions, are in qualitatively agreement with the results of computer modelling conducted using of the molecular dynamics method. At the same time, in current experiments with computer modelling at high energies of primary recoil atoms, the results show a principal special feature of the process of development of a cascade, i.e. the formation of a thermal peak [7,9,10]. In the range of the thermal peak, temperature may be considerably higher than the melting point. The realistic nature of formation of the molten zone in the experiments with computer modelling has been confirmed by estimates of the kinetic energy of the atoms, atomic density and the parameter of the long-range order in the region of the thermal peak [7, 10].

The rapid increase of temperature in the zone of the thermal peak greatly intensifies the process of mutual recombination of defects, and the formation of a molten zone leads to the formation of cluseters of interstitials at the periphery of the cascade as a result of the effect of two mechanisms which were not previously observed: the mechanism of ballistic displacement of interstitials from the molten zone [9], and the mechanism of formation of clusters under the effect of a shockwave formed during rapid cooling of the melting zone [10].

The intensification of the mutual recombination of the point defects during the formation of a thermal peak is one of the most important consequences of computer modelling because this fact is in agreement with the actual experimental data.

It is also important to mention that the formation of subcascades which in the experiments with computer modelling was observed for the first time in [10] for a recoil energy of 25 keV, was accompanied by the merger of molten zone of two subcascades. In fact, this result contradicts of widely held opinion [4,6,11] according to which the formation of subcascades results in a decrease of the spatial correlation between the defects in the cascade, decreases their density and, consequently, suppresses the recombination processes, supporting the retention of a large part of residual defects.

One of the applied aspects of the theory of formation of primary radiation defects was associated with the need for correct calculation of the number of displacement per atom (dpa):

$$N_d = t \int_{T_d}^{T_{max}} dT \nu(T) \int_{E_{min}}^{E_{max}} \varphi_i(E) \frac{d\sigma(E,T)}{dT} dE \tag{1.6}$$

In equation (1.6) E_{min} is the minimum energy of the particle required for the displacement of the atoms, E_{max} is the maximum energy of the particles in the spectrum, $\varphi_i(E)$ is the spectral density of the flux of particles, t is radiation time.

The dpa parameter has been regarded for many years as one of the main parameters in comparison and modelling of the effect of radiation of different types and energy on materials and also in evaluation of the dose dependence of the degree of radiation damage. In reality, this parameter can be regarded only as a very rough approximation because, depending on recoil energy and the radiation dose, it does not expresses adequately the degree of radiation damage and, evidently, does not provide information on the structural composition of radiation defects. One can present a large number of examples of inadequacy of this criterion, for example, as clearly indicated by the results of [12,13].

In recent years, there has been a tendency for comparing the efficiency of structural phase changes under the effect of radiation of different type and energy on the basis of the evaluation of the concentration of freely-migrating defects avoiding recombination or merger into clusters in the process of primary radiation damage. Evidently, this criterion has an obvious advantage in comparison with the dpa parameter, especially at elevated temperatures, characterised by rapid diffusion-controlled processes of the nucleation and growth of clusters, dislocation loops and voids, and also phase changes of different type.

Theoretically, the fraction of freely-migrating point defects may be evaluated either on the basis of the theory of binary collisions in the modified Kinchin–Pease model or directly in computer modelling of the process of atomic collisions.

In a general case, within the framework of the model of binary collisions, it is possible to calculate any fraction of defects $\eta(T)$ formed by all primary-displaced atoms with an energy lower than T [11]:

$$\eta(T)=\frac{1}{\nu(E)}\int_{T_d}^{T} dT'\frac{d\sigma(E,T')}{dT'}\nu(T') \tag{1.7}$$

where quantity $\nu(E)$ is determined by expression (1.1), and the calculation of $\nu(T')$ using equation (1.7) is carried out using conventional boundary conditions (1.4)

Evidently, equation (1.7) may also be used for calculating the fraction of freely-migrating defects within the framework of the following simple model. Assuming, to a first approximation, that the freely-migrating defects are formed only by those primary-displaced atoms whose recoil energy is in the range $T_d = T < 2.5\ T_d$, their fraction may be evaluated using equation (1.7) with the appropriate integration range.

The general relationships governing the formation of primary radiation defects, including freely-migrating defects, are directly indicated by the results of calculations carried out in [11] using equation (1.7) for particles of different type and energy (Fig. 1.1).

The energy of the primary-displaced atoms $T_{1/2}$ at which $\eta(T) = 0.5$ in Fig. 1.1, characterises the 'hardness' of the primary recoil

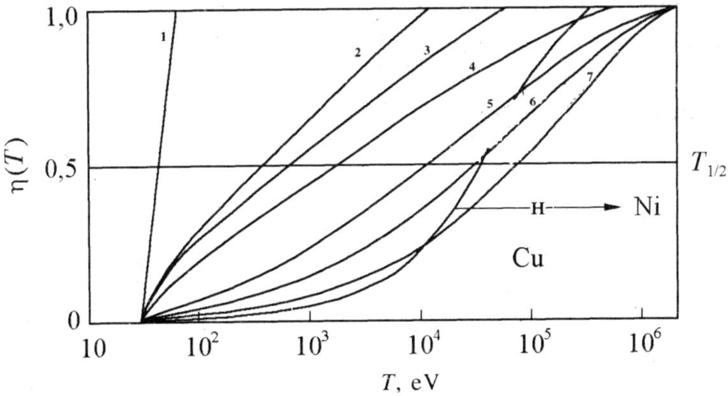

Fig. 1.1. Fraction of defects formed by primary-displaced atoms with an energy lower than T by particles of different type and energy in copper and nickel [11]. 1) 1 MeV e, 2) 200 keV H, 3) 2 MeV He, 4) 2 MeV Ne, 5) 2 MeV Ar, 6) 2 MeV Kr.

spectrum. This quantity is the mean-weighted energy of the recoil spectrum, with 50% of all radiation defects formed above and below this energy.

Figure 1.1 indicates that with an increase of the energy and mass of the particles the number of defects formed in the range of high recoil energy continuously increases. The degree of spatial correlation of defects increases in this case. It is characteristic that for the neutrons, regardless of their small mass, the spectrum of primary-displaced atoms is relatively hard. This is associated with the fact that the atomic displacements during neutron radiation takes place at very low impact parameters. Therefore, on the basis of comparison with the ions which may transfer not only high energies during direct elastic collisions but also small amounts of energy during Coulomb interaction over large distances, the fraction of the defects, formed in the energetically dense cascades, is considerably higher for neutrons. The very narrow energy spectrum of recoil (30–60 eV) is provided by electrons with an energy of 1 MeV characterised by the formation of only isolated Frenkel pairs.

The above considerations are applicable in qualitative analysis of the experimental data obtained in evaluation of the fraction of stable defects. The information on the fraction of freely-migrating defects may be obtained on the basis of analysis of the results of direct and indirect experiments.

The indirect experiments include the investigations carried out at temperatures at which the point defects are immobile. The concentration of stable Frenkel pairs is evaluated in these experiments on the basis of measurement of the variation of the properties of materials (electrical resistance, lattice parameters, *etc.*) in the process of radiation using the available values of the properties for the unit concentration of Frenkel pairs. However, it is evident that these experiments provide some averaged-out information on the total concentration of residual radiation defects because of the non-additive contribution of defects of different type and configuration to the measured properties. The direct investigations include investigations carried out at temperatures at which point defects are mobile and the variation of the properties of irradiated objects is determined directly by the concentration of the defects and the diffusion parameters.

Analysis shows that the main relationships, determined in both direct and indirect experiments, are in qualitative agreement with the results of the previously-examined theoretical calculations. As an example, Fig. 1.2 shows the relative efficiency of formation of mobile defects in relation to the mean-weighted recoil energy $T_{1/2}$ obtained on the basis of analysis of the experimental data in the examination of surface segregation in Ni–12.7at.% Si alloy at a temperature of ~800 K [14]. The efficiency, presented in Fig. 1.2, is normalised for the efficiency of formation of freely-migrating defects for protons with an energy of 1 MeV. Identical dependences were also obtained for Cu–Au and Mo–Re alloys [15,16].

Figure 1.2 shows that the fraction of freely-migrating defects

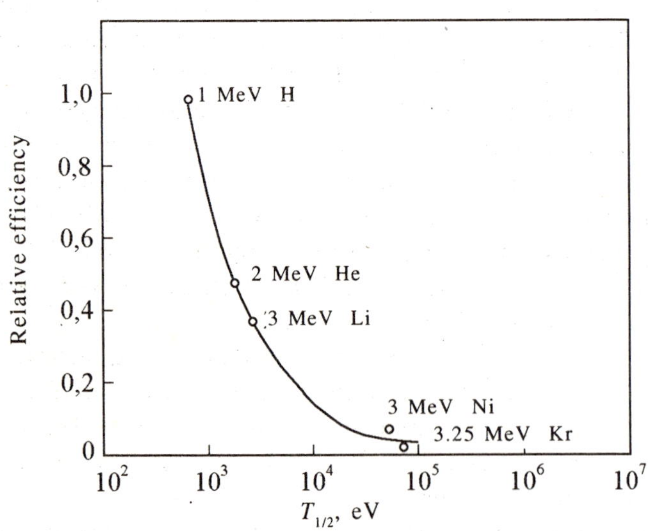

Fig. 1.2. Relative efficiency of different ions for the formation of freely-migrating defects in relation to the mean-weighted recoil energy $T_{1/2}$ [14].

decreases with the mass and energy of the ions, i.e. with an increase in the energy of primary-displaced atoms, showing a tendency to saturation. The identical nature of the dependence, as already mentioned, is also observed in low-temperature experiments [8].

As reported in [15, 16], one of the main problems in the quantitative estimates of the fraction of freely-migrating defects is associated with its temperature–energy dependence. Analysis of the experimental data in the appropriate low temperature [8] and high-temperature [14, 16] experiments shows that the fraction of freely-migrating defects, evaluated on the basis of the results of experiments carried out at elevated temperatures, decreases more rapidly with the energy of primary-knocked out atoms [15, 16]. In the low-temperature experiments, the fraction of the stable Frenkel defects decreases with increasing recoil equal energy, reaching a saturation at a concentration of ~25% of the calculated number of displacements.

For increased radiation temperatures, the concentration of freely-migrating defects for a recoil energy of > 25 keV, typical of fission and fusion neutrons, decreases to approximately 1% [15]. The qualitative interpretation of the temperature–energy dependence was carried out in [15, 16], assuming the intensification of intra-cascade recombination and of the processes of recombination and formation of clusters of defects in the region of adjacent cascades. In our view, in the interpretation of this temperature–energy dependence it is also necessary to take into account the effect of thermal oscillations of the atoms of the lattice on the dynamics of defect formation. As shown in [17, 18], in modelling the process of atomic collisions in Cu, the mean distance between the vacancy and the interstitial in the Frenkel pairs is almost halved with an increase of the irradiation temperature from zero to 293 K. This is associated with an increase of the probability of defocusing of sequences of substituting atomic collisions with an increase of the amplitude of atomic oscillations. It is evident that the given process supports an increase of the spatial correlation between defects and, consequently, increase of the intensity of mutual recombination and probability of the formation of clusters. This mechanism should be more pronounced in cascades with a higher energy density in which the total density of atomic displacements is higher.

The analysis results showing a reasonably good agreement of the experimental and calculated data in the evaluation of the fraction of freely-migrating defects have been published in [19]. The calculations were based on the model proposed in [19, 20] which, in fact, represents a modified variant of the previously examined model within which the fraction of the defects for any energy range of primary-displaced atoms is evaluated using equation (1.7). In the model [19, 20) in the evaluation of the fraction of freely-migrating defects attention is given to the formation of stable isolated Frenkel pairs not only from the primary-displaced recoil atom but also in subsequent secondary generations of atomic collisions. In this case, in addition to the energy condition of formation of stable isolated pairs $T_d \leq T < 2.5\ T_d$, of the factor of the spatial instability of individual vacancies and interstitials was also taken into account. It was assumed that the stable isolated point defects form in sequences of atomic collisions whose distance exceeds the radius of interaction of point defects for their recombination or formation of complexes.

Taking these factors into account, equations modified in relation to equation (1.7) were obtained and used as a basis for calculations of the fraction of stable Frenkel pairs in nickel in irradiation with ions of different energy and mass. The results of these calculations and the corresponding data are presented in Table 1.1.

In Table 1.1, from [19], the experimental data for [21] are presented in the form of quantities η_{exp}, normalised in relation to the absolute fractions of freely-migrating defects. On the basis of the relatively good quantitative agreement between the calculated and experimental data it is possible to assume that the main assumptions of the accepted model are evidently accurate. In the course of calculations, it was also possible to determine several relationships governing the effect of the order of generation of secondary atomic collisions on the evaluated fraction of freely-migrating defects. For light ions H^+, He^+ and Li^+ up to 95% of all free Frenkel pairs are formed by primary-displaced atoms. Their fraction may be evaluated using equation (1.7) in which the integral should be additionally multiplied by parameter $\beta < 1$, characterising the spatial instability of the point defects. For high-energy ions of Ni^+ and Kr^+, a significant

Table 1.1 Calculated and experimental values of the fraction of Frenkel defects in nickel irradiated with particles of different type and energy [19]

Irradiation	η_{cal}, %	η_{exp}, %	Literature
1 MeV H^+	8.8	20	[21]
2 MeV He^+	7.1	9.6	[21]
2 MeV Li^+	6.4	7.4	[21]
300 keV Ni^+	1.2	1.5	[22]
3 MeV Ni^+	1.6	1.6	[21]
3.25 MeV Kr^+	1.4	<0.4	[21]
2 keV O^+	1.7	1.5	[23]

contribution to the fraction of formation of freely-migrating defects is provided by generations of higher orders. For example, for Kr^+ and Ni^+ (300 keV) only 40 and 25% of the total concentration of mobile defects are formed by primary-displaced atoms. These special features are also in agreement with the calculations of formation of defects only by primary-displaced atoms (Fig. 1.1) which show that the fraction of the freely-migrating defects decreases with an increase of the energy and mass of the ions.

In conclusions, it should be mentioned that the analysis of the fraction of the freely-migrating defects in direct experiments should be carried out on the basis of radiation-enhanced diffusion in the examined materials. With the exception of [19], this approach has not been used in the analysis of experimental data because, in particular, the mechanisms of radiation-enhanced diffusion determine the dynamic concentration of freely-migrating defects and, consequently, variations of the properties of the irradiated materials.

1.3. INTERACTION OF INTERSTITIALS WITH EACH OTHER AND WITH SOLUTE ATOMS

A significant contribution to the investigation of the examined a problem has been provided by the initial studies for computer modelling of the stability of different configurations of interstitials and their complexes [27–29]. It was established that in the FCC and the BCC lattices the dumbbell configurations <100> and <110>, respectively, are most stable, Fig. 1.3. The results of subsequent theoretical calculations and experimental data are in complete agreement with these considerations. They have been published in a number of original and review studies [28–33].

Figure 1.4 shows the results of calculation of the stability of double complexes of the dumbbell configurations of interstitials in the FCC lattice, obtained by Vineyard [25]. The stability of double and more

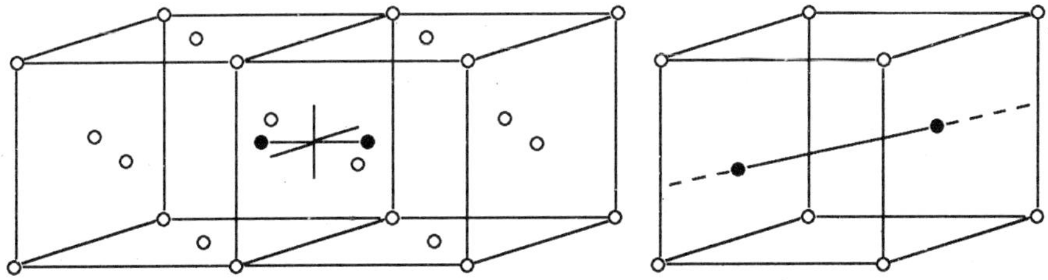

Fig. 1.3. Dumbbell configurations of interstitials in FCC (a) and BCC (b) lattices.

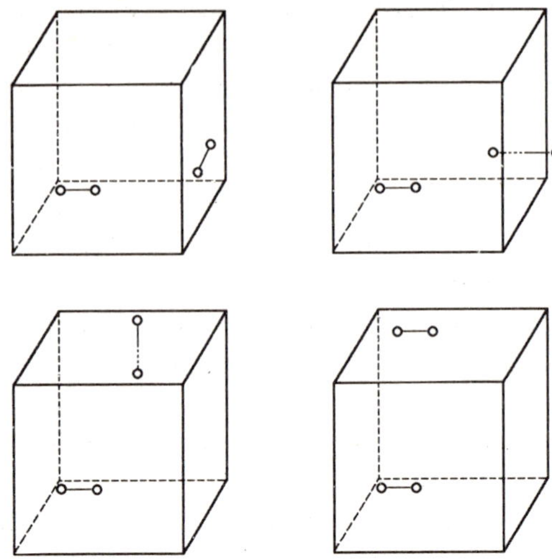

Fig. 1.4. Configuration of complexes from two interstitials in the FCC lattice [25]: a) unstable; b) binding energy B = 0. 36 eV; c) B = 0.40 eV; d) B = 0.46 eV.

complicated complexes of the interstitials in the FCC and BCC lattices has been investigated many times in a number of investigations. The results of these investigations can be found in, for example, [29, 30, 34–36]. According to the results of calculations carried out in [35, 36], the most stable double complex in the FCC lattice is represented by two parallel dumbbells in the position of the nearest lattice sites, slightly inclined (< 10°) in the {110} plane. The energy difference between the inclined configuration and the identical non-inclined complex consisting of two parallel [001] dumbbells is very small [~0.01 eV]. The evaluation of the stability of the volume clusters of interstitials in the dumbbell configuration in relation to the number of interstitials in the clusters shows [36] that when the number of interstitials in a cluster exceeds 10–13, they become unstable and transform into two-dimensional formations.

At the same time, it should be mentioned that the validity of the results of theoretical calculations in relation to the reflection of the real structure of defect configurations depends to a large degree on the selection of the calculation methods and the potential of atomic interaction. For example, in [37] when modelling by the molecular dynamics method on the basis of the pseudopotential theory of the stability of different configurations of self-interstitials in α-Fe, it was shown that the most stable configurations of the interstitials are, in contrast to the results of other identical calculations, crowdions in the close-packed direction <111>. One of the crowdions represents a configuration in which two atoms are localised symmetrically in relation to the centre of the elementary cube of the BCC lattice. In the second crowdion configuration, an interstitial atom is implanted between the nearest neighbours. Within the limits of the calculation error, the energy of formation of these crowdions is identical. Evidently, only accurate experiments can confirm the validity of different theoretical estimates.

The results of interaction of interstitials with substitutional atoms were initially analysed in [38] for the octahedral configuration in the FCC lattice of copper. The authors of [38] used the method of lattice statistics and also an approach in which the forces, acting from the side of the defect and the solute atom were calculated from the values of their dilation changes in the lattice.

The results of the calculations show that the interaction of the interstitials with the solute atoms with the dilation volumes of different sign greatly differs. The most stable positions of the interstitials for the substitutional atoms with $\Delta V > 0$ (oversized atoms) positions (1,0,0) and (1,1,1) (Fig. 1.5a) which, at the same time, for the atoms with $\Delta V < 0$ are unstable. In the vicinity of an impurity with $\Delta V < 0$ (undersized atoms) the interstitial is stable only in the position (2,1,1) (Fig. 1.5b).

Figure 1.5c shows the configuration of the most stable heterogeneous nucleus of a cluster of interstitials, constructed on the basis of the results of calculations conducted by the summation of the energies of the corresponding paired interactions.

The theoretical analysis of the stability and mechanisms of migration of the self-interstitial and mixed dumbbell configurations was carried out for the first time in [13]. The calculations were conducted by methods of the theory of perturbation and computer modelling using the Morse and Born–Mayer potentials. In the calculations, it

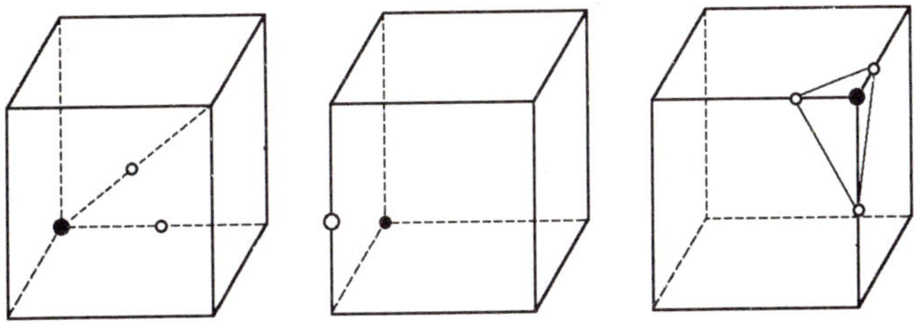

Fig. 1.5. The nearest stable positions of an octahedral interstitial in the vicinity of a substitutional atom with $\Delta V > 0$ (a), $\Delta V < 0$ (b) and the structure of a heterogeneous nucleus of a cluster of interstitials in the FCC lattice (c) [38].

was assumed that the impurities with the positive and negative dilation volumes displace the position of the minimum of the potential of atomic interaction in the pure solvent R_0 by the value $+r$ and $-r$, respectively. The binding energy of a mixed dumbbell is characterised consequently by the value and sign of the so-called mismatch parameter $\varepsilon = r_0/R_0$, associated with the relative dilation volumes of the impurity by the relationship: $\Delta V/V\Delta \approx 6\varepsilon$ [39]. The results of calculations of the binding energy of the mixed dumbbell by the method of the theory of perturbation in relation to the value of parameter ε are presented in Fig. 1.6. They show that the positive energy of binding in the mixed dumbbell <100> is possible only for $r_0 < 0$, i.e. the formation of these dumbbells is possible only for undersized impurities ($\Delta V < 0$). Calculations by the method of computer modelling using the Morse and Born–Mayer potentials show that for high negative values of r_0 when $-\varepsilon \geq 0.06$, the mixed dumbbell becomes unstable, as in the case of positive values of $\varepsilon > 0.03$.

The results of calculations of the binding energy in the <100> self-interstitial dumbbell–substitutional atom complexes, carried out also in the above study, are presented in Fig. 1.7. The values of the binding energy in Fig. 1.7 are given for undersized atoms in the units of the binding energy of the mixed dumbbell. For the atoms with positive dilation volumes, the sign of the interaction energy changes to the opposite sign together with the variation of the sign of r_0. In addition to the data for the Morse potential, the values in the brackets

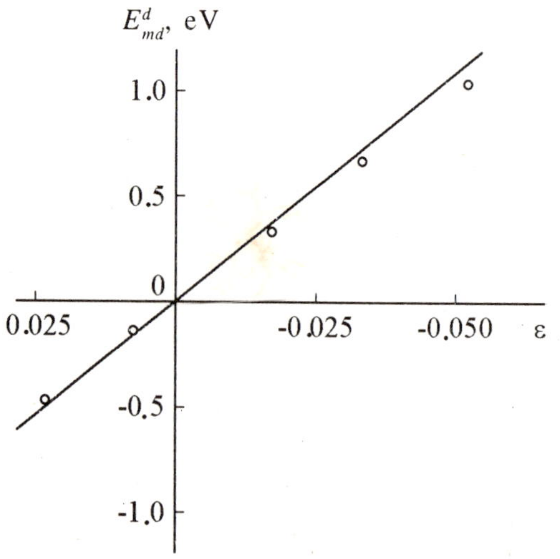

Fig. 1.6. The binding energy of a mixed dumbbell in relation to the value of the mismatch parameter ε in the FCC lattice [30].

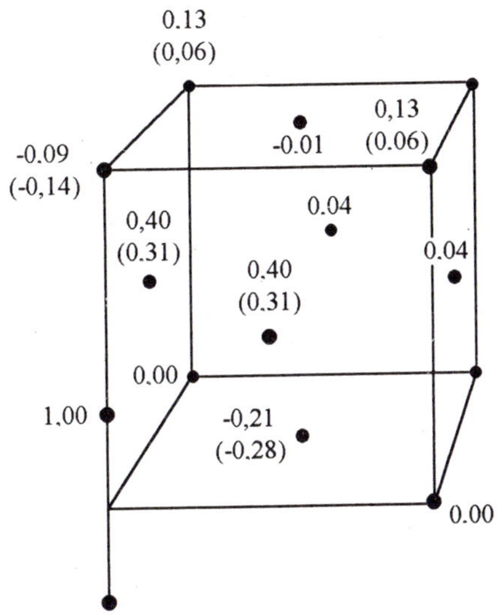

Fig. 1.7. Binding energy in different self-interstitial dumbbell <100>–impurity atom complexes in the FCC lattice [30].

are the results of calculations for the Born–Mayer potential.

The calculations indicate that in the (101) and (202) positions, the undersized interstitials are bonded relatively strongly in the complexes with the self-interstitial dumbbells. For low values of r_0, these configurations become unstable and at $-\varepsilon \geq 0.025$ they transform to the mixed dumbbells.

The binding energy of the self-interstitial dumbbells with the solutes with the positive dilation volumes is lower than that with the undersized substitutional atoms and is considerably lower than the binding energy of the mixed dumbbell. This shows that the oversized substitutional atoms are less effective traps for the self-interstitial dumbbell than the solutes with negative dilation volumes.

Subsequent theoretical calculations [40–44], confirmed one of the main conclusions of [30], according to which stable mixed dumbbells do not form for the atoms of the solutes with $\Delta V > 0$. This result is one of the most important consequences of theoretical analysis because it shows that the diffusion transfer of impurities and solutes with positive dilation volumes by the self-interstitial dumbbell mechanism in metals and alloys with the FCC lattice ineffective. In this case, the mobility of the self-interstitial dumbbells should decrease because of their periodic capture by the atoms with $\Delta V > 0$. The transfer of the solutes should in this case be controlled by the vacancy mechanism. The configuration of the most stable complexes, calculated in [40–42], for the majority of impurities in the FCC lattice of aluminium did not, however, correspond to the results obtained in [30].

In these investigations, using the molecular dynamics method and interaction potentials obtained from the first principles, the configurations of the complexes were determined for a number of substitutional impurities with $\Delta V < 0$ (Be, Li, Zn) and $\Delta V > 0$ (Mg, Ca). In the FCC lattice of Al, the configuration of the most stable complex corresponded to the calculations in [30] only for Zn and was represented by a mixed dumbbell <100>. For the oversized solute Ca, for example, the most stable configuration was the one in which the self-interstitial dumbbell was situated in the nearest lattice site in relation to the calcium atom. According to the estimates in [30], this complex is generally unstable, as indicated by Fig. 1.7.

The formation of stable mixed dumbbells <100> in a number of irradiated FCC metals for undersized atoms has been confirmed by direct channelling experiments [31]. For irradiated BCC metals, as indicated by these experiments, the interstitials with the undersized impurities form stable mixed dumbbells <100>. This result is in agreement with relatively recent theoretical calculations carried out in [33] in which the molecular dynamics method was used to estimate the stability of mixed dumbbells using the approach applied in the calculations in [30]. In this work as in [30], the potential used in the calculations (these calculations were carried out using the nonequilibrium Johnson potential) was modified taking into account the mismatch parameter. The results show that, in this case, the mixed dumbbell <110> is stable, like the mixed dumbbell <100> in the FCC lattice, only for impurities with negative dilation volumes. At $\varepsilon \leq -0.15$, it is not stable and transforms into the tetrahedron or octahedral configuration.

A large part of experimental investigations in the evaluation of the efficiency of the interaction of interstitials with the solute atoms has been carried out using the method of measurement of electrical resistance in examination of annealing or of the kinetics of buildup of defects at temperatures of the recovery stage II. In these investigations, the defect–impurity interaction is analysed on the basis of the value of the relative radii of capture of the interstitials by impurity traps, determined when processing the experimental data within the framework of specific modelling representations.

Analysis shows that it is usually not possible to accurately systematise these and other experimental data obtained in the examination of the interaction of interstitials and solutes within the framework of the model [30]. A detailed analysis was also carried out, in particular, in a review in [45] on the basis of analysis of a large number of experimental results. We believe that a highly characteristic example of this type is provided by comparison of the results obtained for diluted silver–copper and copper–silver alloys in which the relative dilation volumes are equal to –0.28 and +0.43, respectively [46]. Experiments show [47] that the oversized atoms of Ag and Cu are far more efficient traps for interstitials than the undersized atoms of copper in the silver lattice. Estimates carried out in [47] show that the relative radius of capture of interstitials by atoms of silver in copper–silver alloys is three times higher than the value for silver–copper alloys. On the whole, it is note possible to detect any specific relationships in the efficiency of capture of the interstitials by the solutes with the dilation volumes higher than and lower than zero.

One of the reasons for mismatch is associated with the simplified theoretical representations of the model in [30]. The type and stability of the complexes may be determined not only by the value and sign of the dilation volumes but also by special features of the electronic structure of the defects and impurities which are not taken into account in the model in [30]. As already mentioned, this possibility is shown in [40–42] which considered the interaction potentials obtained from the first principles.

At the same time, when processing the experimental data, it is very complicated to take into account the actual reactions of the defects and defect–impurity interaction. In irradiation or subsequent annealing, these reactions may lead to the formation of complexes of different type and size. Taking also into account the non-additive nature of the contribution from the defects of different type to the measured properties (for example, electrical resistance), the results of the investigations may provide some average values in this case.

To conclude this section, we shall examine several consequences resulting from the considerations on the non-equivalent groups of the interstitials in the FCC lattice [38, 48]. They make it possible to analyse the possible structure of the interstitial complexes within the framework of a simple geometrical approach.

Analysis carried out in [38, 48] shows that all possible positions of the interstitials during their migration through the crystal are subdivided into four characteristic groups which are such that every interstitials remains in its group during migration.

If the octahedral configuration is the stable configuration of the interstitials, and the dumbbell configuration is the saddle configuration, then in a single jump any of the three coordinates of the interstitials should change by $\pm 2a$. In relation to some separated atom of the lattice, these groups have the coordinates: $(2n_1 + 1, 2n_2, 2n_3)$, $(2n_1, 2n_2+1, 2n_3)$ $(2n_1, 2n_2, 2n_3 + 1)$, and $(2n_1, 2n_2 + 1, 2n_3 + 1)$, where n_i are integer numbers, $2a$ is the lattice parameter. All the coordinates, presented above, are expressed in the units of a.

Identical non-equivalent groups for the migration of interstitials in the dumbbell configuration <100> are characterised by the coordinates of the centre of the masses of the dumbbell (n_1, n_2, n_3) and its orientation $\xi(x, y, z)$ (Table 1.2).

During migration of a defect in an ideal infinite crystal, all these groups are equivalent because of the translational invariance of the ideal crystal. In the presence of a second defect, this equivalency is partially or completely moved and, consequently, this results in a number of interesting consequences.

On the basis of the non-equivalent groups, we shall examine initially the stability of the complexes presented in Fig. 1.4, taking into account the results of calculations by Vineyard [25].

It is assumed that one of the interstitials in the complex occupies the fixed position $(0,0,0,X)$, i.e., according to Table 1.2 it belongs to the group I. It may also be seen that an unstable configuration (Fig. 1.4a) forms only when both interstitials belong to the same group. The case in Fig. 1.4a corresponds to the position of the second interstitial $(1,1,0,Y)$.

The configuration in Fig. 1.4b corresponds to the position of the

Table 1.2 Non-equivalent groups <100> of interstitials in the dumbbell configuration [48]

Group	n_1	n_2	n_3	ξ
I	$2n_1$	$2n_2$	$2n_3$	X
	$2n_1 + 1$	$2n_2 + 1$	$2n_3$	Y
	$2n_1 + 1$	$2n_2$	$2n_3 + 1$	Z
II	$2n_1$	$2n_2$	$2n_3$	Y
	$2n_1$	$2n_2 + 1$	$2n_3$	X
	$2n_1$	$2n_2 + 1$	$2n_3 + 1$	Z
III	$2n_1$	$2n_2$	$2n_3$	Z
	$2n_1 + 1$	$2n_2$	$2n_3$	X
	$2n_1$	$2n_2 + 1$	$2n_3 + 1$	Y
IV	$2n_1 + 1$	$2n_2 + 1$	$2n_3$	Z
	$2n_1 + 1$	$2n_2$	$2n_3 + 1$	Y
	$2n_1$	$2n_2 + 1$	$2n_3 + 1$	X

second interstitials (1,1,0,X) from group II.

The identical configuration forms if the second interstitial atom belongs to group III and occupies the position of the type (1,0,1,X). The configuration in Fig. 1.4c may form if the second interstitial belongs to the groups II, III and IV with the positions of the type (0,1,1,Z), (0,1,1,Y) and (1,1,0,Z), respectively. The formation of the configuration, is shown in Fig.1.4d, is possible is the second interstitial belongs to the group IV with the position of the type (0,1,1,X).

All the stable complexes of the interstitials (Fig. 1.4b–d) are characterised by similar binding energies. Analysis shows that the probabilities of formation of the corresponding configurations are expressed by the ratio 2:3:1 and mutual transitions are possible between the configurations 1.4b–1.4c and 1.4c and 1.4d.

In conclusion, it should be stressed that in both the octahedral and dumbbell configurations of the interstitials, the binary complexes of the defects, belonging to one group, are unstable.

Within the framework of the considerations regarding the non-equivalent groups it was also shown [38,48] that the interaction of self-interstitial dumbbells with the solutes may be accompanied by the formation of complexes of two types: the self-interstitial–impurity atom complex with the possibility of a subsequent transition to a mixed dumbbell and stable 'self-interstitial dumbbell–impurity atom' complexes. Figure 1.8 a–c shows the formation of mixed dumbbells during the interaction of self-interstitial dumbbells, belonging to the groups I–III, respectively (Table 1.2) with the substitutional atom situated at the origin of the coordinates. Figures 1.8a–c reflect in this case, for each group, the results of reactions of one of the self-interstitial dumbells with the substitutional atom, localised at the origin of the coordinates. The self-interstitial dumbells, belonging to group IV (Table 1.2), cannot form a mixed dumbbell. Possible configurations of the complexes with the solute for the self-interstitial dumbells of the group IV are shown in Fig. 1.8d. It should be mentioned that in the case of the dumbells, belonging to the groups I–III, the formation of complexes of the type of 1.8d is not possible. The different binding energies in the complexes 1.8a–1.8c and 1.8d assume the possibility of existence in FCC metals of traps of at least two types. This does not contradict the experiments with the buildup and annealing of radiation defects in diluted solid solutions at temperatures of stage II of recovery [45, 49–52].

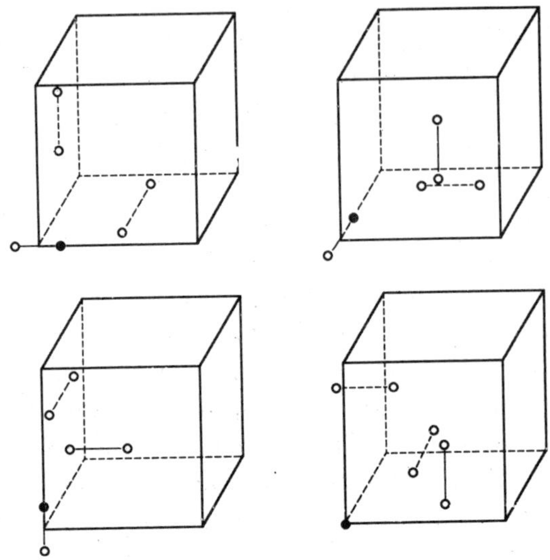

Fig. 1.8. Configuration of complexes of dumbbells with solutes for non-equivalent groups I–III (a–c, respectively) and group IV (d) in the FCC lattice [48].

1.4. INTERACTION OF VACANCIES WITH EACH OTHER AND WITH SOLUTE ATOMS

In this section, we shall not examine different models and methods of calculating the interaction of vacancies and solutes and we shall not analyse in detail the large number of experimental results obtained in the determination of the binding energy of vacancy complexes. These data have been published in detail in, for example, the proceedings of Argonne, Kyoto and Berlin conferences on defects, their properties and interaction in metals [53–55]. However, we shall discuss briefly several aspects of the examined problem.

In [56], using the method of the Green functions, the authors presented the results of calculations of interaction of vacancies with impurities of different type in copper, nickel, silver and palladium (3*d* and 4*ps* in Cu and Ni and 4*d* and 5*sp* in Ag and Pd). Calculations were carried out for the position of the nearest neighbours. The results of [56] provide convincing information on a number of general quantitative and quantitative relationships of the interaction of vacancies with the solute atoms in metals and are in a relatively good agreement with the literature experimental data.

The calculations show that for all four metals, *sp* impurities strongly interact with vacancies. Their positive binding energy is approximately proportional to the difference of the valencies of the matrix and solutes. In fact, a similar result is obtained from a simple model of interaction within in the framework of the Thomas–Fermi approximation (equation (1.8)). In contrast to this, the impurities of transition metals in the positions of the nearest neighbours are repulsed. The maximum repulsive energy is ~0.2 eV. The calculated binding energies of the divacancies in copper, nickel, silver, and palladium are 0.076, 0.067, 0.079 and 0.11 eV, respectively. The configuration and evaluation of the stability of more complicated vacancy complexes can be found in, for example, [29, 30, 57]. The results of these calculations, which have already been mentioned, are in a relatively good agreement with the experimental data presented in [56], and show that the binding energy of the vacancies in these metals with the solute atoms is in the range 0.1–0.4 eV. According to the analysis of the published experimental data, these values of the binding energy are on average characteristic of the majority of other metals.

The formation of transmutation products of nuclear reactions and the interaction with radiation defects may greatly modify the mechanisms of radiation damage and the effects of radiation. A special role in the intensification of the processes of swelling and embrittlement is displayed by transmutation gases, in particular helium and hydrogen. Therefore, the problems of interaction of the atoms of these gases with radiation defects are of considerable significance. Within the framework of this problem, attention should be given to the results of theoretical analysis of the interaction of hydrogen and also of helium and other inert gases with vacancies and interstitials, are presented in [58, 59]. Calculations in [59] were carried out for the positions of the solute atoms of inert gases: substitutional and interstitial (in the octahedral position). For the atoms of helium in the interstitial and substitutional positions, closest to the vacancy, the calculated binding energies were 0.65 and 0.47 eV, respectively. With an increase in the size of the atoms of the gases, the binding energy increases and for xenon in the interstitial position it is ≥ 3 eV. According to the calculations, the atoms of the inert gases are effective traps also for self-interstitials. For example, for helium, the binding energy in a complex with an self-interstitial dumbbell is ~0.5 eV. Calculations of the interaction of hydrogen or helium atoms with a vacancy show [59] that for the single atom of hydrogen the most stable position is directly in the vacant site of the lattice. The configurations of the stable complexes for two atoms of hydrogen or helium maybe of two types. The first configuration is a complex in which one of the atoms of the gases is in a vacancy and the atom occupies the nearest interstitial position. The second stable complex represents a dumbbell of two solutes of the gases spaced at the same distance from the vacancy. The calculations show that, being effective traps for point defects, the atoms of the gases greatly restrict their diffusibility.

Analysis of the diffusion of solutes in the alloys by the vacancy mechanism is usually carried out on the basis of the multifrequency theory of vacancy jumps [60, 61]. The theory is based on the model in which the interaction of the vacancies with the solute atoms is described within in the framework of the Thomas–Fermi approximation. The energy of interaction in this case is:

$$E = \frac{\left[Z_0 Z_1 e^2 \exp(-k_0 r)\right]}{r} \tag{1.8}$$

In equation (1.8) $Z_0 e$ is the charge of the vacancy regarded as having the valency equal to the valency of the solvent, $Z_1 e$ is the effective charge of the impurity equal to the difference $Z_1|e| = (Z_2 - Z_0)\,|e|$, where Z_2 is the valency of the dissolved element. The screening parameter k_0 has the following form:

$$k_0 = 2.95\left(\frac{a_0}{r_s}\right)^{1/2} \tag{1.9}$$

where a_0 is the Bohr radius, r_s is the radius of the sphere whose volume relates to a single conductivity electron.

An important parameter of the multifrequency theory of diffusion is the binding energy of the vacancy with the atom of the solutes which determines the degree of localisation of the solutes at the vacancy and, consequently, the efficiency of its diffusion transfer by the vacancy mechanism. This fact also indicates the importance of accurate evaluation of the binding energy on the basis of theoretical calculations and the results of experimental investigations. The fundamentals of the theory of diffusion of solutes by the vacancy and interstitial mechanisms are examined in detail in the following chapter of this book.

REFERENCES

1 Kirsanov V.V., et al, Processes of radiation defect formation in metals, Energoizdat, Moscow (1985).
2 Norgett M.J., et al, Nucl. Eng. Design, 33, 50 (1975).
3 Gibson J.B., et al, Phys. Rev., 120, 1229 (1960).
4 Thompson M., Defects and radiation damage in metals, Mir, Moscow (1971).
5 Babaev V.P., et al, Cascades of atomic displacements in metals, Preprint ITEF-110 (1982), p.40.
6 Ishino S. and Sekimura N., Ann. Chim. Fr., 16, 341 (1991).
7 De la Rubia T.D. and Phythian W.J., J. Nucl. Mater., 191/194, 108 (1992).
8 Kinney J.H., et al, J. Nucl. Mater., 121/122, 1028 (1984).
9 English C.A., et al, Mater. Sci. Forum, 97/99, 1 (1992).
10 De la Rubia T.D. and Guinan M.W., Mater. Sci. Forum, 97/99, 23 (1992).
11 Rehn L.E. and Okamoto P.R., Mater. Sci. Forum, 15/18, 985 (1987).
12 Simons R.L., J. Nucl. Mater., 141/143, 665 (1986).
13 Zinkle S.J., J. Nucl. Mater., 155/157, 1201 (1988).
14 Okamoto P.R., et al, J. Nucl. Mater., 133/134, 373 (1985).
15 Rehn. L.E., J. Nucl. Mater., 174, 144 (1990).
16 Rehn L.E. and Wiedersich H., Mater. Sci. Forum, 97/99, 43 (1992).
17 Tenenbaum A., Phil. Mag., 37, 731 (1978).
18 Tenenbaum A., Rad. Eff., 39, 119 (1978).
19 Naundorf V., et al, J. Nucl. Mater., 186, 227 (1992).
20 Naundorf V., J. Nucl. Mater., 182, 254 (1991).
21 Rehn P.R., et al, Phys. Rev., B30, 3073 (1984).
22 Miller A., et al, Appl. Phys., 64, 3445 (1988).
23 Naundorf V. and Abromeit C., Nucl. Instr. Meth., B43, 513 (1989).
24 Johnston R.A., Phys. Rev., 134, 1329 (1964).
25 Vineyard G.H., Disc. Farad. Soc., No.31, 7 (1961).
26 Johnston R.A. and Brown E., Phys. Rev., 134, 1329 (1964).
27 Jonhston R.A., Phys. Rev., 145, No.2, 423 (1966).
28 Scholz A. and Lehman C., Phys. Rev. B., 6, 1972 (1972).
29 Schroeder K, In: Point Defect Behaviour and Diffusion Processes, Metals Society, London, 1977.
30 Dederichs P.H., et al, J. Nucl. Mater., 69/70, 176 (1978).
31 Howe L.M. and Swanson M.L., In: Solute–Defect interaction: Theory and Experiment. Proc. Int. Seminar, Kingston (1985), Toronto (1986).
32 Marangos J., et al, Mater. Sci. Forum, 1, 225 (1987).
33 Kevorkyan U.R., Phys. Stat. Sol. (a), 106, 379 (1988).
34 Bullough R. and Perrin R.C., Proc. Roy. Soc. A., 305, 541 (1968).
35 Schober H.R. and Zeller R., J. Nucl. Mater., 69/70, 341 (1978).
36 Ingle K.W., et al, J. Phys. F: Metal. Phys., 11, 1161 (1981).
37 Vasil'ev A.A. and Mizandrontsev D.B., Pis'ma Zhurn. Eksper. i Teor. Fiziki, 16, No.13, 45 (1990).
38 Ivanov L.I., et al, Phys. Stat. Sol. (a), 64, 771 (1974).
39 Bartels A., et al, J. Nucl. Mater., 83, 24 (1979).
40 Lam N.Q., et al, J. Phys. F: Metal Phys., 10, 2359 (1980).
41 Lam N.Q., et al, J. Phys. F: Metal Phys., 11, 2231 (1981).
42 Doan N.V., et al, In: Point defects and defect interactions in metals, Tokyo Univ. Press, Tokyo (1982), p.372.

43 Takamura S., et al, *J. Phys: Condens. Mater.*, **1**, 4519 (1989).
44 Takamura S., et al, *J. Phys: Condens. Mater.*, **1**, No.1, 4527 (1989).
45 Robrok K.G., In: Phase transformations under radiation, Metallurgiya, Chelyabinsk (1989).
46 King H.W., *J. Mater. Sci.*, **1**, 79 (1966).
47 Mauri F., et al, In: Point defects and defect interactions in metals, Tokyo University Press, Tokyo (1982), p.383.
48 Ivanov L.I., et al, *Phys. Stat. Sol.* (a), **69**, K33 (1975).
49 Dworshak, F., et al, *J. Phys. F: Metal. Phys.*, **8**, No.7, L153 (1978).
50 Swanson M.L. and Howe L.M., *Nucl. Instr. and Meth. in Phys. Res.*, **218**, 613 (1983).
51 Bartels A., et al, *J. Phys. F: Metal. Phys.*, **12**, No.11, 2483 (1982).
52 Tokamura S. and Kobiyama M., *Phys. Stat. Sol.* (a), **90**, No.1, 269 (1985).
53 Properties of atomic defects in metals, *J. Nucl. Mater.*, **69/70**, 856 (1978).
54 Point defects and defect interactions in metals, Tokyo Univ. Press, Tokyo (1982), p.991.
55 Vacancies and interstitials in metals and alloys, *Mater. Sci. Forum*, **15/18**, 1442 (1987).
56 Klemradt U., et al, *Phys. Rev.* B., **43**, No.12, 9487 (1991).
57 Masuda K., in ref. 53, p.105.
58 Baskes M.I., et al, *J. Nucl. Mater.*, **83**, No.1, 139 (1979).
59 Whitmore M.D. and Carbotte J.P., *J. Phys. F: Metal. Phys.*, **9**, No.4, 629 (1979).
60 Le Claire A.D.: in ref.53, p.70.
61 Le Claire A.D., *Phil. Mag.*, **21**, No.172, 819 (1970).

Chapter 2

DIFFUSION PROPERTIES OF POINT DEFECTS AND SOLUTES IN PURE METALS AND ALLOYS

2.1. INTRODUCTION

The diffusion properties of point defects and solutes have a controlling effect on the special features of structural-phase changes and the efficiency of radiation damage of metallic materials. The currently diffusion considerations will be used many times in a number of chapters of this book for describing the individual mechanisms and for interpretating the experimental data. The basis for this is a brief analytical preview of the corresponding theoretical assumptions, presented in this chapter, and published in a number of monographs and original particles. The review also contains the essential experimental data. When examining this problem, special attention is given to diffusion mechanisms based on the migration of self-interstitial and mixed dumbbell configurations of interstitials because this mechanisms in comparison with, for example, vacancy mechanisms, have been studied far more extensively in the scientific literature and, basically, have been published only in the original or review articles.

2.2. DIFFUSION IN PURE METALS AND ALLOYS BY THE INTERSTITIAL DUMBBELL MECHANISM

The migration of a stable dumbbell <100> in an impurity FCC lattice is a sequence of elementary jumps with a saddle octahedral configuration. The transitions of the dumbbell from one stable position to the nearest following stable position is accompanied by a change of its initial orientation by 90°. In pure metals, the diffusion mo-

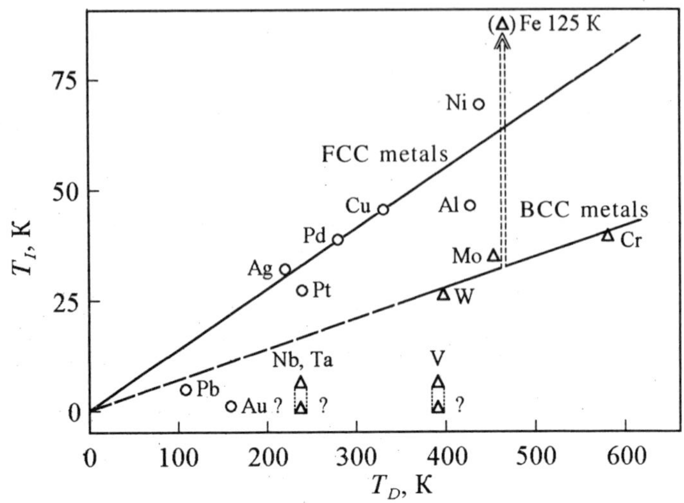

Fig. 2.1. Dependence of migration temperature of self-interstitial dumbbell configuration T_I in cubic metals on Debye temperature T_D [7].

bility of interstitials is very high and greatly exceeds the mobility of vacancies.

Theoretical estimates and experimental results for the annealing of radiation defects at temperatures of recovery stage I for the energy of migration of self-interstitial dumbbells in the FCC and BCC metals give typical values in the range 0.05–0.2 eV [1–5]. The values are considerably lower than the energy of migration of the vacancies in pure metals which, for example, for aluminium, copper and tungsten are equal to 0.62, 0.70 and 1.7 eV [6].

In [7], on the basis of analysis of the experimental data for the annealing of radiation defects in a number of metals, the authors established a correlation between the migration temperature of self-interstitials (a dumbbell consisting of two atoms of the solvent) T_I and Debye temperature T_D (Fig. 2.1). These dependences, which in [7] are analysed on the basis of special features of the phonon spectra of the examined materials and vibrational modes of defects, are approximated in the case of the FCC and BCC metals by the following equations:

$$T_I(FCC) = 0.14\, T_D\,, \qquad T_I(\mathrm{BCC}) = 0.075\, T_D \tag{2.1}$$

The solutes greatly modify the migration mechanisms of dumbbells and decrease the diffusion mobility of the interstitials both as a result of the formation of less mobile mixed dumbbells (for elements with $\Delta V < 0$) and as a result of periodic capture of self-interstitials of elements both with $\Delta V < 0$ and $\Delta V > 0$.

The model which can be used, using the results in [8], for numerical estimation of the mobility of interstitials and solutes in diluted solid solutions with FCC lattices by the dumbbell mechanism, has been proposed in [9]. The model is based on the method of Lidiard complexes [10, 11] and on assumptions regarding the complexes of the type *a* and *b* [8, 12]. The complexes of type *a* represent a configuration of an self-interstitial dumbbell with an atom of the dissolved element which may transform into a mixed dumbbell during migration of the self-interstitial dumbbell. In a complex of type *b*, the self-interstitial dumbbell prior to dissociation may only migrate in the vicinity of the impurity without forming a mixed dumbbell. Evidently, this process does not lead to the transfer of the solutes. Within the framework of the model [8], the formation of complexes of the type *a* and *b* is characteristic for undersized and oversized solutes, respectively (see section 1.3).

The authors of [9] obtained equations for calculating the coefficient of diffusion of dumbbells $D_i(c_B)$ and the coefficients of diffusibility of solutes by the interstitial mechanism in an infinitely diluted $D_i^B(0)$ and a diluted solid solution $D_i^B(c_B)$. The following characteristic frequencies were used in these calculations:

ω_0 – the frequency of jumps of a self-interstitial dumbbell configuration, including the displacement of the dumbbell with the variation of its orientation by 90°;

ω_1 and ω_1' – the frequency of rotation of the self-interstitial dumbbell around the atoms of impurities in the complexes *a* and *b*, respectively;

ω_2' – the frequency of formation of a mixed dumbbell;

ω_2 – the frequency of dissociation of a mixed dumbbell;

ω_3 and ω_3' – the frequency of dissociation of the complexes of type *a* and *b*;

ω_4 and ω_4' – the frequency of formation of complexes of type *a* and *b*;

ω_i – the frequency of retention of the mixed dumbbell;

ω_R – the frequency of variation of the orientation of the mixed dumbbell by 90°. The equation for the diffusion coefficient of the dumbbells in a diluted alloy has the following form [9]:

$$D_i(c_B)=\frac{D_i(1+\beta_i c_B)+K_{pa}\gamma_{pa}c_B}{1+K_i c_B} \tag{2.2}$$

In equation (2.2), the coefficient of diffusion of the dumbbells in a pure solvent D_i is:

$$D_i=0.44\lambda_i^2\omega_0 \tag{2.3}$$

where λ_i is the length of a elementary jump of the dumbbell.

The equations for the coefficients of diffusion mobility of element B in an infinitely diluted alloy and in the A–B alloy with the composition c_B have the following form [9]:

$$D_i^B(0)=K_{pa}D_{pa} \tag{2.4}$$

$$D_i^B(c_B)=\frac{K_{pa}D_{pa}(1+K_i c_B)+2\sigma_i D_i K_i c_B}{1+K_i c_B} \tag{2.5}$$

The values of β_i, K_{pa}, γ_{pa}, K_i and σ_i in equations (2.2), (2.4) and (2.5) are functions of frequencies ω_k and ω_k' (k = 0,1,...) [9], examined previously.

Figure 2.2 shows the resultant numerical calculations carried out in [9] of the coefficients $D_i(c_B)$ and $D_i^B(c_B)$ in a diluted alloy of copper with 1 at% of the solutes and coefficient $D_i^B(0)$ in an infinitely diluted alloy at a temperature of 300 °C in relation to the value of the parameter of dimensional mismatch ε. The frequencies ω_2 and ω_R were estimated using the dependence $E_{md}^B(\varepsilon)$ from [8], shown in Fig. 1.6. Other required frequencies were also calculated on the basis of the results of [8].

From the dependences, presented in Fig. 2.2, it follows that:

1. The mobility of interstitials in the dumbbell configuration decreases with an increase of the absolute value of the dilation volume of the dissolved element, with the largest increase recorded for the impurities with $\Delta V < 0$.

2. In an infinitely diluted alloy, the diffusion mobility of elements with $\Delta V > 0$ decreases with an increase of the dimensional mismatch in comparison with the diffusion mobility of the solvent, and in the case of $\Delta V < 0$ it increases. For a diluted alloy, the nature of the

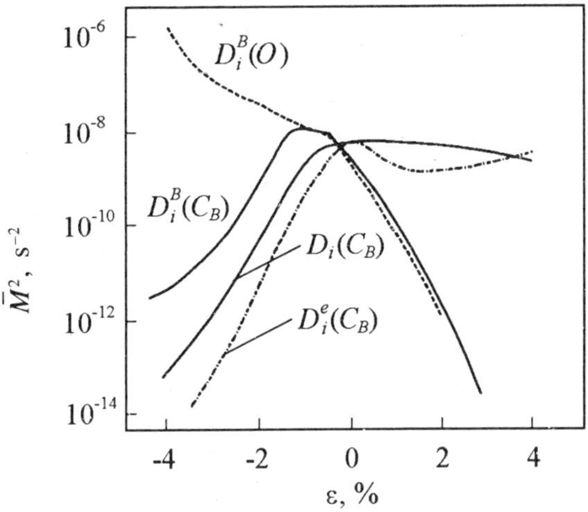

Fig. 2.2. Variation of diffusion coefficients $D_i(0)$, $D_i^B(c_B)$ [9] and $D_i^e(c_B)$ in a diluted alloy of copper in relation to the mismatch parameter ε.

dependence for $\varepsilon > 0$ does not change. For solutes with $\varepsilon < 0$, the dependence the$D_i^B(c_B)$ is non-monotonic; with increasing $|\varepsilon|$ diffusion mobility initially increases, reaches the maximum value and then decreases.

The model proposed in [9] can be used for the quantitative calculations of the diffusion parameters if we know the corresponding interaction potentials, required for estimating the frequencies ω_k and ω_k'. The approximate estimation of the diffusion mobility of the interstitials and vacancies in the diluted solid solutions is obtained usually using the expressions for the effective diffusion coefficients.

The considerations regarding the effective diffusion coefficients $D^e_{v,i}$ are based on the mechanism of periodic capture of migrating defects by impurity traps with the formation of complexes and their subsequent dissociation. It may be shown that for the case in which the recombination of the interstitials and vacancies on deffect–impurity complexes plays no significant role, the expression for $D^e_{v,i}$ has the following form (see, for example, [13]):

$$D^e_{v,i} = \frac{D_{v,i}}{1 + 4\pi r_t^{v,i} c_t^{v,i} \exp\left(E^B_{v,i}/kT\right)} \tag{2.6}$$

In contrast to the accurate expressions being the function of several frequencies, equation (2.6) contains actually only one unknown parameter, i.e. the binding energy of point defects and solutes $E^B_{v,i}$ which can be determined from the experiments.

Using equation (2.6), it is possible to determine, in relation to the mismatch parameter, the effective mobility of interstitials in a diluted copper alloy with 1 at% of solutes at 300 °C in order to compare with the results of numerical calculations carried out by A. Barbu [9]. For quantity E_i^B in equation (2.6) we shall accept, in accordance with the data in [8], the values of the binding energy in a mixed dumbbell (for impurities with $\Delta V < 0$) and the binding energy in the complexes of type b (for impurities with $\Delta V > 0$).

The comparison of $D_i(c_B)$ and $D_i^e(c_B)$ (Fig. 2.2) shows that the agreement of the results of the calculations obtained using the equations (2.2) and (2.6) is highly satisfactory. Evidently, the small quantitative difference is associated with the simplified approximation of E_i^B by the selected interaction energies.

In transition to concentrated alloys, the effective mobility of the interstitials greatly decreases in comparison with pure metals and diluted solid solutions. This is shown clearly in a number of investigations, including examination of the kinetics of buildup and and annealing of defects at temperatures of the recovery stages II and III by the method of measuring electrical resistance in silver-zinc, silver–palladium, palladium–silver [14, 15] and iron–chromium-nickel [16, 17] alloys, radiation-enhanced diffusion in silver–zinc [18] and copper–nickel [19] alloys, and structural changes in aluminium–zinc [20, 21] and silver–zinc [18, 22] alloys in electron irradiation in a high-voltage microscope.

In [14, 15] a large decrease of the mobility of interstitials results in complete suppression of the mutual recombination of defects in the process of irradiation and annealing of defects at temperatures of the second recovery stage.

For Fe–16Cr–20Ni alloy, with a composition similar to the composition of the matrix solid solution of steel 316, the energy of migration of interstitials according to the experimental data in [16] is ~0.9 eV. This value is considerably higher than the energy of migration of interstitials in pure iron and nickel (~0.3 eV [2,23] and 0.15 eV [2], respectively). The relatively high energies of migration of interstitials were also recorded for Ag–8.75 at% Zn alloy ($E_i^m > 0.46$ eV for temperatures higher than 90 °C) [18] and Cu–44 at% Ni alloy (E_i^m = 0.48 eV) [19].

High energies of the migration of interstitials were also determined for concentrated alloys Fe–18.5Mn–7.5Cr (0.7 eV) and Fe–7Mn–4.5Si–6.5Cr (1.1 eV) in examination of the nucleation of dislocation loops in the process of electron irradiation in a high-voltage microscope [24].

The authors of [14, 15] proposed a hypothesis according to which a large decrease of the mobility of interstitials in concentrated alloys is associated with a high probability of the repeated capture of the interstitials by the solute atoms after dissociation of a defect–impurity complex. Within the framework of these considerations, according to the results of theoretical investigations [24], it is possible to explain completely the main relationships of the kinetics of buildup of radiation defects in pure metals, diluted and concentrated alloys at temperatures of recovery stages II, associated with a decrease of the effective mobility of interstitials.

Attention will be also given to the main results of theoretical investigations carried out by Bocquet [26–28] which make it possible to interpret qualitatively a number of experimental special features, associated with the diffusion mobility of interstitials in alloys. The theory is based on a simple two-frequency model in which the dumbbells of the type *A–A* and *B–B* migrate with frequency ω_R. For an self-interstitial dumbbell *A–A*, this frequency is also the frequency of capture of the dumbbell by the atoms of the dissolved element *B* resulting in its transformation into a mixed dumbbell *A–B*. The mixed dumbbell may dissociate with frequency $\omega_L < \omega_R$. In the schema, the frequencies ω_L and ω_R are evidently linked by the relationship:

$$\theta = \frac{\omega_R}{\omega_L} = \exp\left(E_{AB}^{B} / kT\right) \tag{2.7}$$

where E_{AB}^{B} is the binding energy of the mixed dumbbell *A–B*.

Figure 2.3 shows the results of numerical calculations of the ratio $D_i(c_B)/D_i(0)$ for different $\theta = \omega_R/\omega_L$ in relation to c_B [27].

Figure 2.3 shows that when $\omega_L \ll \omega_R$, the mobility of interstitials in the alloy rapidly decreases. In the limit, the interstitial may be situated in a trap of atoms *B*, surrounded by the atoms *A*, and, consequently, its migration over large distances becomes almost impossible.

The concentrations of the dumbbells *A–A*, *A–B*, and *B–B* at the total concentration of the interstitials in the alloy *A–B*, being c_i, with the framework of the models [26, 28] are equal to respectively:

$$c_{AA} = c_i P_{AA} = c_i \left(\frac{c_A}{c_A + c_B\theta} \right)^2 \tag{2.8}$$

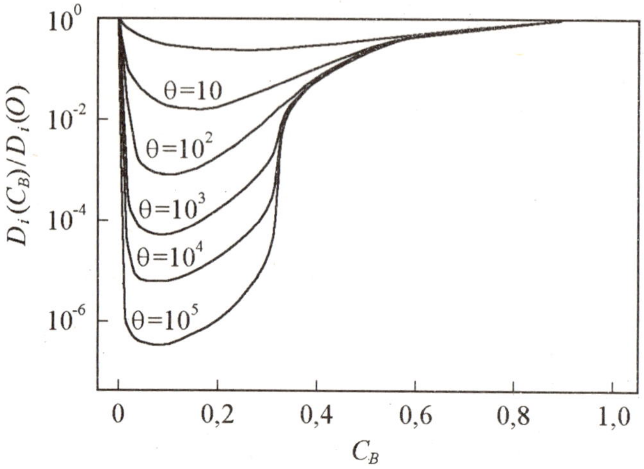

Fig. 2.3. Variation of $D_i(c_B)/D_i(0)$ ratio in relation to the concentration of solute c_B for different values of parameter θ [26–28].

$$c_{AB} = c_i P_{AB} = c_i \frac{c_B\theta}{(c_A + c_B\theta)^2} \tag{2.9}$$

$$c_{BB} = c_i P_{BB} = c_i \left(\frac{c_B\theta}{c_A + c_B\theta}\right)^2 \tag{2.10}$$

$$P_{AA} + 2P_{AB} + P_{BB} = 1 \tag{2.11}$$

The equations (2.7)–(2.10) show that at relatively high irradiation temperatures the course of different structural–phase transitions in solid solutions is controlled mainly by diffusion reactions in which the dumbbells of the type *A–A* take part. With a decrease in temperature and increasing concentration, the relative probabilities of formation of the dumbbells of the type *A–B* and *B–B* (in comparison with *A–A*) continuously increase which also increases the fraction of their participation in different diffusion reactions, including in the general flux of the interstitials to sinks. In this case, as a result of the preferential departure of the interstitials to sinks in the form of the dumbbells *A–B* and *B–B*, the tendency for the instability of the solid solution with respect to the chemical and phase composition, including the formation of segregation zones in the vicinity of the sinks, becomes stronger.

In particular, these processes are favourable for concentrated alloys in which the relative probability of formation of the dumbbells of *B–B* because of the relationships (2.8)–(2.10) increases with a decrease of temperature more rapidly than in the case of the dumbbells of type *A–B*:

$$\frac{P_{BB}}{P_{AA}} = \left\{\frac{c_B}{c_A}\exp\left(E^B_{AB}/kT\right)\right\}^2 \tag{2.12}$$

$$\frac{P_{AB}}{P_{AA}} = 2\frac{c_B}{c_A}\exp\left(E^B_{AB}/kT\right) \tag{2.13}$$

Within the framework of the Bocquet models, it is possible to explain qualitatively a number of interesting experimental special features of structural–phase changes in concentrated alloys. For example, in investigations in [20, 21], the process of formation and growth of dislocation loops of the interstitial type, typical of pure aluminium and its diluted alloys, is completely suppressed in irradiation in a high-voltage microscope in aluminium–zinc alloys with the zinc concentration higher than 1 wt%. This is accompanied by the nucleation and growth of vacancy-type dislocation loops. This result may be explained only by the dominant mobility of the vacancies in the alloys at a zinc concentration > 1 at.%.

Interesting experimental results of the same type were obtained in [18, 22]. Irradiation of Ag–9 at.% Zn alloy [22] and of a series of silver–zinc alloys with a zinc concentration in the range from 1 to 25 at.% [18] at temperatures > ~80 °C was characterised by the dominant effect of the mechanism of the nucleation and growth of dislocation loops of the interstitial type, whereas at irradiation temperatures < ~80 °C it was the mechanism of the nucleation and growth of vacancy tetrahedrons of stacking faults. This effect is also associated with the temperature transformation of the relative mobility of interstitials and vacancies in concentrated alloys.

Examination of irradiation of silver–zinc alloys (undersaturated solid solutions) with doses of ~(3–5) × 10^{26} m^{-2} in [18] also showed the effect of formation of pre-precipitates in the vicinity of tetrahedrons with subsequent formation of a spatially-oriented structure. In this case, the formation of pre-precipitates is determined by the segregation of the zinc atoms on tetrahedrons as a result of migration of Ag–Zn dumbbells to them, with the mobility lower than that of the vacancies. The phenomena was interpreted within the framework of the kinetic model [29, 30] which was also based on some assumptions of the Bocquet model [26–28].

In conclusion, it should be noted that the possibility of a large decrease of the diffusion mobility of interstitials is not always taken into account in analysis of the experimental data and in theoretical calculations. This is the basis for obtaining incorrect experimental information and inaccurate estimates when predicting and modelling different processes of radiation damage.

2.3. DIFFUSION OF SOLUTE SUBSTITUTIONAL ATOMS BY THE VACANCY MECHANISM

Special features of the diffusion of substitutional solute atoms by the vacancy mechanism in the FCC and BCC lattices have been analysed usually on the basis of frequency models of vacancy jumps. The following characteristic frequencies [31, 32] are examined in the model of diffusion of solutes in the FCC lattice (Fig. 2.4):

ω_0 – the frequency of the jumps of a vacancy in exchange with the atoms of the solvent which is not the nearest neighbour of the solute atoms;

ω_1 – the frequency of jumps of the vacancy in exchange with the atom of the solvent which, like the vacancy, is the nearest neighbour with the solute atom;

ω_2 – the frequency of jumps of the vacancy in exchange with the solute atom;

ω_3 – the frequency of exchanges of the vacancy with the atom of the solvent as a result of which the vacancy is transferred from

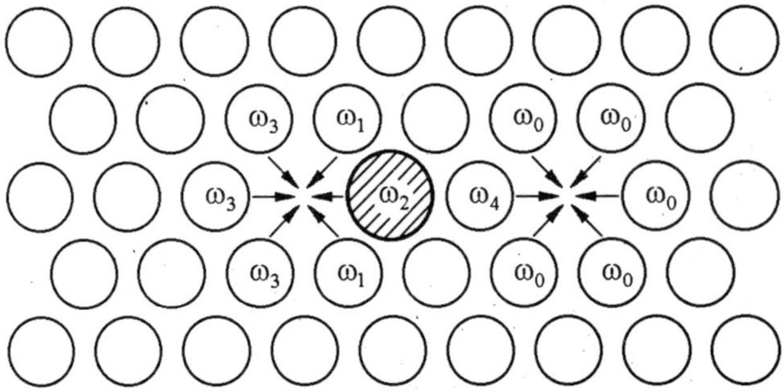

Fig. 2.4. Schema of possible jumps of vacancies in the vicinity of a solute atom in the FCC lattice.

the nearest position at the solute atom to a more remote position (dissociative jumps);

ω_4 – the frequency of exchanges of the vacancy with the atom of the solvent as a result of which the vacancy becomes the nearest neighbour of the solute atom (associative jumps).

For the BCC lattice, the characteristic vacancy frequency is identical with the exception of frequency ω_1 which is absent in this case [31, 33].

The equations for the frequencies have the following form:

$$\omega_i = \nu_i \exp\left(-E_i / kT\right) \tag{2.14}$$

where ν_i and E_i are the corresponding frequencies of atomic oscillations and the activation energy. The coefficients of diffusion of the solvent and the solute atoms are determined by the expressions:

$$D_0 = a^2 f_0 \omega_0 \exp\left(S_v^f / k\right) \exp\left(-E_v^f / kT\right) \tag{2.15}$$

$$D_2 = a^2 f_2 \omega_2 \exp\left(S_v^f / k\right) \exp\left[-\left(E_v^f - E_v^B\right) / kT\right] \tag{2.16}$$

In equations (2.15)–(2.16), a is the jump distance, f_0 and f_2 are correlation factors, S_v^f and E_v^f are the entropy and energy of formation of vacancies, E_v^B is the binding energy of the vacancy–solute atom complex.

The correlation multipliers for the FCC and BCC lattices of the solvent are equal to 0.78 and 0.73, respectively [31]. For the solute atoms, they are determined by the expression [31]:

$$f_2 = \frac{u}{\left(2\omega_2 + u\right)} \tag{2.17}$$

where u is the function of frequencies ω_0, ω_1, ω_3, and ω_4 for the FCC lattice, and ω_0, ω_3 and ω_4 for the BCC metals.

For rapidly diffusing and slowly diffusing impurities, in most cases $f_2 < f_0$ and $f_2 > f_0$,respectively.

The frequency ratio ω_4/ω_3 is an important characteristic of the diffusion mechanism,

$$\frac{\omega_4}{\omega_3} \simeq \exp\left(E_v^B / kT\right) \tag{2.18}$$

and determines the degree of localisation of the vacancy at the solute atom.

The ratio of the diffusion coefficients of the solute and the solvent is:

$$\frac{D_2}{D_0} = \frac{f_2\omega_2\omega_3}{f_0\omega_0\omega_4} = \frac{f_2\omega_2}{f_0\omega_0}\exp\left(E_v^B / kT\right) \tag{2.19}$$

Analysis of the diffusion special features of a binary alloy within the framework of the model of electrostatic interaction in the Thomas-Fermi approximation (equation (1.8)) shows [31] that for the impurities whose valency is higher than that of the solvent ($Z_2 > Z_0$), both the degree of localisation and the frequency of exchanges of vacancies with the solute atoms ω_2 is higher than for the atoms of the solvent ω_0. This conclusion is clearly indicated by equation (2.19). Correspondingly, the impurities for which the $Z_2 < Z_0$ will diffuse at a lower rate. The experimental data presented in particular in [31] are in good correlation with the examined modelling considerations.

With an increase in the concentration, the diffusion coefficient of solutes increases in accordance with the equation [31]:

$$D_2(c) = D_2(0)\left(1 + B_1 c_B + B_2 c_B^2 + \cdots\right) \tag{2.20}$$

The terms in the multiplier of the right-hand part of this equation characterise the degree of isolation of the impurity in the solid solution and the variation of exchange frequency ω_2, if another solute atom is situated alongside the atom of the impurity. The equation for B_1 has the following form:

$$B_1 = 18\exp\left\{\left(E_v^B / kT\right) - 1\right\} \tag{2.21}$$

In conclusion, attention will be given to the problem of diffusion transfer of the solute atoms to sinks by the vacancy mechanism. This problem is especially important because of the phenomenon of radiation-stimulated segregation. Here, we shall discuss only the fundamentals of this process.

In a general case, the diffusion flows of the atomic components and vacancies in the alloy are linked by the following relationship:

$$J_v = -\sum_k J_k \tag{2.22}$$

In the general form, this relationship for the dissolved element B and vacancies in the $A–B$ binary alloy is determined by the expression [34]:

$$J_B = -\frac{(L_{AB} + L_{BA}) J_v}{L_{AA} + L_{AB} + L_{BA} + L_{BB}} \tag{2.23}$$

where L_{ij} are the Onzager coefficients.

When solving this problem within the framework of the five-frequency model, Anthony [35] obtained the following expression:

$$\frac{J_B}{J_v} = \frac{c_B D_B}{\dfrac{D_A}{f} + c_B D_B} \left[\frac{\omega_1 + (13/2)\omega_3}{\omega_1 - (7/2)\omega_3} \right] \tag{2.24}$$

Analysis of equation (2.24) shows that when the vacancies are strongly bonded with the solutes ($\omega_3 \ll \omega_1$), the flux of solute atoms has the same direction as the flux of the vacancies. In the opposite case, when there is almost no interaction of the impurity with the vacancy ($\omega_3 \cong \omega_1$), the flux of solute atoms occurs in the direction opposite to the flux of the vacancies.

REFERENCES

1 Damask A. and Dines J., Point defects in metals, Mir, Moscow (1966).
2 Young F.W., J. Nucl. Mater., 69/70, 310 (1978).
3 Schilling W., J. Nucl. Mater., 69/70, 465 (1978).
4 Maury F., et al, Rad. Eff., 55, 187 (1981).
5 Mirubayashi H. and Okuda S., Rad. Eff., 54, 201 (1981).
6 Balluti R.W., J. Nucl. Mater., 69/70, 434 (1978).
7 Lucasson P., et al, Mater. Sci. Forum, 15/18, 231 (1987).
8 Dederichs P.H., et al, J. Nucl. Mater., 69/70, 176 (1978).
9 Barbu A., Acta Met., 28, 499 (1980).
10 Lidiard A.B., Phil. Mag., 46, 1218 (1955).
11 Howard R.E. and Lidiard A.B., Rep. Progr. Phys., 27, 161 (1964).
12 Jonhston R. and Lam N.Q., Phys. Rev. B., 13, 4364 (1976).
13 Schilling W. and Schroeder K., In: Physics of irradiation produced voids. Harwell, 1972. AERE-R (1994), p.213.
14 Sadykhov S.I.O. and Platov Yu.M., Fiz. Khim. Obrab. Mater., No.2, 98 (1974).
15 Platov Yu.M., et al, Fizika Met. Metalloved., 39, No.6, 1290 (1975).
16 Dimitrov O. and Dimitrov C., Point defects and defect interactions in metals, Tokyo Univ. Press, Tokyo (1982), p.656.
17 Dimitrov C., et al, Points defects and defect interactions in metals, Tokyo Univ. Press, Tokyo (1982), p.660.
18 Platov Y.M. and Simakov S.V., Fiz. Khim. Obrab. Mater., No.6, 5 (1988).
19 Ivanov L.I., et al, Fiz. Khim. Obrab. Mater., No.1, 28 (1988).
20 Kiritani M., et al, In: Proc. 8th Int. Congr. Electron. Microscopy, Vol.1 (1974), p.622.
21 Kiritani M. and Takata H., J. Nucl. Mater., 69/70, 277 (1978).
22 Halbachs M. and Yoshida N., Phil. Mag., 43, No.5, 1125 (1981).
23 Wells J.M. and Russell K., Rad. Eff., 28, 157 (1976).
24 Inone H., et al, J. Nucl. Mater., 191/194, 1342 (1992).
25 Dienes G.J., et al, Rad. Eff., 33, 59 (1977).
26 Bocquet J.L., Rapport CEA, R-5112, Saclay (1981), p.45.
27 Bocquet J.L., Point defects and defect interactions in metals, Tokyo Univ. Press, Tokyo (1982), p.644.
28 Bocquet J.L., Acta Met., 34, No.4, 571 (1986).
29 Platov Yu.M., et al, Fiz. Khim. Obrab. Mater., No.3, 20 (1990).
30 Platov Yu.M., et al, Mater. Sci. Forum, 97/99, 253 (1992).
31 Le Claire A.D., J. Nucl. Mater., 69/70, 70 (1978).
32 Le Claire A.D., Phil. Mag., 7, 141 (1962).
33 Le Claire A.D., Phil. Mag., 21, No.172, 819 (1970).
34 Allnatt A.R. and Lidiard A.B., Rept. Progr. Phys., 50, 373 (1987).
35 Anthony T.R., In: Diffusion in Solids, Acad. Press (1975), p.353.

Chapter 3

BUILDUP AND ANNEALING OF RADIATION DEFECTS IN PURE METALS AND ALLOYS

3.1. INTRODUCTION

In this chapter, attention is given to the mechanisms determining the kinetics of the buildup of radiation point defects, the formation of complexes of these defects, clusters, dislocation loops and voids. At present, these mechanisms for pure metals have been examined in a considerable detail, although in a number of cases they require further clarification. Far less attention has been paid to these relationships for solid solutions and more complicated metallic systems, although it is evident that these systems are very important for practical application. At the same time, it should be mentioned that the priority and volume of these investigations continuously increase, and the results produce a more accurate basis for the development of radiation-resistant alloys, analysis and prediction of their radiation damage.

Taking these aspects into account, in the presentation of the material of this chapter it will be attempted to place the main stress on special features of buildup of radiation defects in alloys, associated with the presence of impurities and alloying elements of different type and concentration.

The nature of the structure of the present chapter reflects to a certain degree the temperature and dose hierarchy of the formation and evolution of the structure of radiation defects. The examination of the main kinetic mechanisms of the buildup of radiation defects is preceded by a brief analysis of the stages of recovery in irradiated metals, diluted and concentrated solid solutions. These representations form a significant basis for the interpretation of the kinetic relationships governing different structure–phase processes. This is followed by the analysis of the kinetics of buildup of radiation defects during irradiation in the vicinity of the absolute zero where the evolution of the structure of defects is not determined by their thermally activated migration. Further analysis is concerned with the kinetic mechanisms of thermally activated processes. Basic equations are given for the introduction and annealing of non-correlated point defects used normally for the evaluation of the concentration of point defects during irradiation. The main sinks for point defects and their efficiency are examined.

The material presented in this chapter also includes the modelling considerations and analysis of the experimental results for the kinetics of buildup of radiation defects at temperatures of recovery stage II in pure metals, diluted and concentrated alloys, based on the mechanisms of formation of defects–impurity complexes. A large amount of material is concerned with the processes of the nucleation and growth of dislocation loops and voids. Special attention is also given here to the effect of solutes.

3.2. THE STAGES OF RECOVERY OF STRUCTURE-SENSITIVE PROPERTIES IN IRRADIATED METALLIC MATERIALS

At present, considerations regarding the nature of the stages of recovery in irradiated metallic materials have been developed most efficiently for FCC metals. They can be reduced to the following [1–5]: radiation defects formed at temperatures in the vicinity of absolute zero (usually ~4 K), are immobile, and during subsequent heating they are annealed in different temperature ranges – the stages of recovery of the structure-sensitive properties.

The first recovery stage has a discrete structure consisting of five substages. The substages I_A, I_B and I_C are characterised by the recombination of close Frenkel pairs, and the substages I_D and I_E are characterised by the correlated and non-correlated annealing of interstitials and their capture by impurity traps.

At the temperatures of the second recovery stage, the interstitials may be released from impurity traps, recombine with vacancies, migrate to other sinks and form clusters. The subject of long-term discussions has been the interpretation of the nature of the recovery stages III and IV. Within in the framework of the model of two types of interstitials it has been assumed [1,4,6–10] that at temperatures of the substage I_E and in the recovery stage III various configurations of interstitials migrate, and recovery at temperatures of stage IV is determined by annealing of vacancies. However, it should be stressed that the great majority of experimental investigations have confirmed the validity of the model of the one type of interstitials (mobility at temperatures of the substage I_E and higher) and the vacancy interpretation of recovery stage III (see, for example [1,11–14]). At higher temperatures, the recovery of the properties is linked with the annealing of vacancies captured by impurities, and annealing of the clusters of radiation defects, and voids.

The main relationships governing the stages of recovery of radiation defects in metals with other types of lattice are more or less identical: in pure metals, the mobility of interstitials is higher than that of the vacancies and they migrate at temperatures of the stages I and III, respectively [1,4,15–24]. At the same time, a number of the BCC metals have also special features. In metals such as Nb, Ta, Mo and W, the interstitials have, in comparison with the FCC metals, anomalously high mobility and characteristic special features of the reactions of interaction with vacancies [18–20]. Consequently, it is not possible to interpret accurately the stages of recovery in these metals on the basis of the fundamental model of Corbett, Smith and Walker [2].

In relatively pure metals, solutes in small concentrations may partially suppress or displace substages and the recovery stages to higher temperatures [18-31], restricting the mobility of self-interstitial defects during the formation of different defects-impurity complexes.

In concentrated alloys, as shown in [32, 33], the effective mobility of freely-migrating interstitials may decrease so much (section 2.2] that the recovery of the properties, associated with the annealing of non-correlated defects in contrast to pure metals and diluted solid

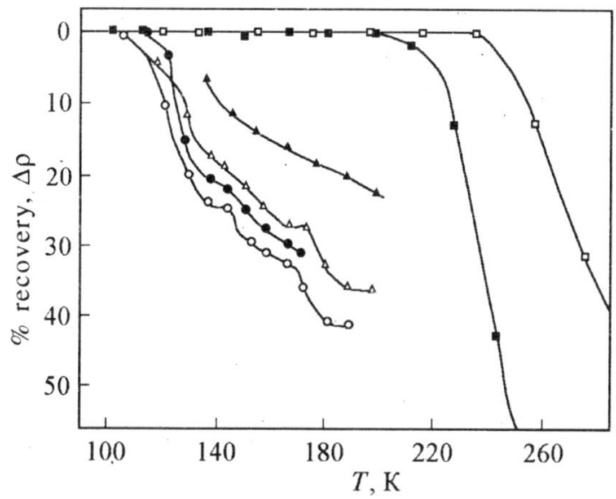

Fig. 3.1. Curves of isochronous annealing of resistivity of silver with a purity of 99.999% (Δ) and 99.99% (▲), Ag–0.038 at% Cd (•), Ag–0.15 at% Zn (○), Ag–13.6 at% Zn (■), and Ag–20 at% Pd (□) [32–34], irradiated with electrons with an energy of 2.2–2.3 MeV at 86–100 K.

solutions starts only from the temperatures of stage III at which the vacancies become mobile. Figure 3.1 shows this on the basis of the results obtained for the recovery of resistivity in silver and its alloys during annealing after irradiation with electrons with an energy of 2.2–2.3 MeV at temperatures of 86–100 K [32–34].

It should be mentioned that the investigations into the annealing of defects in concentrated alloys represented a small part in comparison with the investigations into the annealing of radiation defects in pure metals and diluted solid solutions. At the same time, it is evident that these investigations are most important for practical applications.

3.3. THE BUILDUP OF RADIATION DEFECTS DURING IRRADIATION IN THE VICINITY OF ABSOLUTE ZERO

The main special feature of radiation damage of metals at temperatures of irradiation close to absolute zero is that the process of buildup of defects is not determined here by their thermally activated migration. With an increase of the concentration of defects, the rate of buildup of the defects decreases mainly as a result of two processes referred to as 'radiation annealing': spontaneous recombination and sub-threshold collisions.

Annealing as a result of sub-threshold collisions takes place when an incident particle may transfer, to the atoms in the vicinity of the stable Frenkel pairs, an energy insufficient for displacement but sufficient for the recombination of the pair. This process is characterised by the corresponding section σ_r. Spontaneous recombination of the pair is determined by the volume of the crystal V_0 in which the Frenkel pair is mechanically unstable as a result of interaction between defects of the opposite type. According to different theoretical and experimental evaluations, the volume of spontaneous recombination is in the range from 15 to 600 atomic volumes [35, 36] which for the radius of spontaneous recombination r_0 is equal to 2–5 atomic spacings.

The kinetics of buildup of the Frenkel pairs C_F taking into account the processes of radiation annealing, is described by the equation [35, 37]:

$$\frac{dc_F}{d\Phi} = \sigma_d\left(1 - 2V_0 c_F\right) - \sigma_r c_F \tag{3.1}$$

where Φ is the radiation dose, σ_d is the displacement cross-section.

Accepting for V_0 the value of 50–600 atomic volumes from the solution of equation (3.1) without taking sub-threshold collisions into account, it may be seen that the atomic concentration of the Frenkel pairs at saturation does not exceed 10^{-3}–10^{-2}. The kinetics and extent of saturation may also be affected by the processes of formation of defect clusters [38].

Investigations of the kinetics of variation of electrical resistance are of theoretical and practical interest in the case of low-temperature irradiation. These experiments make it possible to determine important parameters such as V_0, ρ_s (electrical resistance corresponding to saturation) and estimate the individual contribution to the decrease in the rate of buildup of defects from spontaneous recombination and sub-threshold collisions. The concentrations of radiation defects at saturation, estimated on the basis of these experiments for FCC metals, are $(1–3)\times10^{-3}$. For BCC metals, such as iron, molybdenum, tungsten and zirconium, the values are slightly higher: $(4–5)\times10^{-3}$ [35].

The results of measurements of the cross-section of radiation annealing at sub-threshold collisions in Cu and Ni have been presented in [39] for irradiation with iodine ions with an energy of 100 MeV. The sub-threshold cross-sections were determined on the basis of analysis of the kinetic dependences of the measurement of electrical resistance of pre-irradiated and non-irradiated specimens on the basis of the model which makes it possible to take into account the presence of different types of radiation defects, both unstable and stable in relation to sub-threshold collisions.

For three types of defects configurations, for example, where two configurations are unstable (indexes 1 and 2), and the third configuration is stable in relation to annealing at sub-threshold collisions, the corresponding system of the equations has the following form [39]:

$$\begin{aligned} \frac{dc_1}{d\Phi} &= \sigma_{d1}\left(1-2V_0c\right)-\sigma_{r1}c_1 \\ \frac{dc_2}{d\Phi} &= \sigma_{d2}\left(1-2V_0c\right)-\sigma_{r2}c_2 \quad (3.2) \\ \frac{dc_3}{d\Phi} &= \sigma_{d3}\left(1-2V_0c\right) \end{aligned}$$

(3.2)

where $c=\sum_i c_i$.

The radiation annealing cross sections at sub-threshold collisions in nickel are 6.5×10^{-2} and 1.4×10^{-2} cm^2. As mentioned in [39], these defects belong to defects annealed thermally at temperatures of sub-stages I_B–I_C and I_D–I_E, respectively. The radiation annealing section at sub-threshold collisions, corresponding to defects annealed in Cu at temperatures of recovery stage I, was determined to be equal to 4.6×10^{-18} cm^2.

In addition to tasks associated with the necessity for determining the concentration of residual defects in primary processes of radiation damage, the examination of kinetic relationships of the variation of electrical resistance and determination of values of ρ_s is essential in particular for selecting and evaluating the efficiency of stabilising materials in the system of powerful superconducting magnets of accelerators and TOKAMAK-type fusion reactors.

With an increase of irradiation temperature up to the temperatures of sub-stage I_C, the kinetics of buildup of defects, determined by equation (3.1), will also be valid in this case (at least for FCC metals). The difference will be associated only with an increase of the effective volume of recombination whose value changes as a result of additional thermal activation. This decreases the concentration of radiation defects and the values of the variation of electrical resistance at saturation.

Impurities should not have any significant effect on the kinetics of buildup of defects, especially in the vicinity of the liquid helium temperature. The experiments show that their effect on the thermally activated recombination of close Frenkel pairs in the sub-stages I_A–I_C is extremely small [11,40].

3.4. MAIN EQUATIONS OF FORMATION AND THERMAL ANNEALING OF POINT DEFECTS DURING IRRADIATION. SINKS AND THEIR EFFICIENCY

Many currently available kinetic models, describing the processes of the formation and evolution of the structure radiation defects, radiation-enhanced diffusion and phase changes during irradiation are based on the theory of the rate of chemical reactions. In most cases, the concentration of free interstitials and vacancies in pure metals is calculated using the following system of differential equations [41–44]:

$$\frac{dc_i}{dt} = G - Rc_v c_i - c_i \sum_j K_i^j \tag{3.3}$$

$$\frac{dc_v}{dt} = G - Rc_v c_i - c_v \sum K_v^j \tag{3.4}$$

Equations (3.3) and (3.4) are valid for relatively large defects where the migration of defects to the free surfaces may be ignored.

In equations (3.3) and (3.4), $G = \sigma_f\ \varphi$ is the rate of introduction of freely-migrating defects (the problems of formation of freely migrating defects and the methods of evaluation of the concentration are examined in chapter I), σ_f is the formation cross-section of these defects, φ is the rate of irradiation, c_i and c_v is the concentration of the interstitials and vacancies. The recombination constant is expressed by the equation $R = 4\pi r_{vi}\ (D_i + D_v)/\Omega$, where r_{vi} is the radius of mutual recombination, $D_{i,v} = D^0_{i,v} \exp(-E^m_{i,v}\ /\ kT)$ at the coefficients of diffusion of interstitials and vacancies, k is the Boltzmann constant, and T is absolute temperature.

The values $K^j_{i,v} = S^j_{i,v} D_{i,v}$ represent the constants of absorption of point defects by the sinks of the j type which, in a general case, may be represented by clusters of defects, dislocations, voids, grain boundaries, etc.

The quantities S_i^j and S_v^j characterise the power of sinks of type j for interstitials and vacancies, and for edge dislocations, spherical sinks and loops they have the form:

$$S^d_{i,v} = Z^d_{i,v}\rho_d \tag{3.5}$$

$$S^s_{i,v} = 4\pi\int rZ^s_{i,v}(r)n(r)dr \tag{3.6}$$

$$S^l_{i,v} = 2\pi\int rZ^l_{i,v}(r)n(r)dr \tag{3.7}$$

where r is the geometrical radius of the sinks, $n(r)dr$ is the number of sinks with a size from r to $r+dr$ in the unit volume; $Z_{i,v}{}^j$ is the force efficiency of the sink for the capture of a point defect or the factor of the force interaction of a point defect with a sink.

The quantity r_{vi}, included in the expression for the recombination constant R in the equations (3.3) and (3.4), is the effective radius of recombination of the interstitials and vacancy or the capture radius.

If the interaction of two point defects is centrally symmetric, the effective radius of capture is determined from the following equation [45]:

$$r_c = \frac{1}{\int_{r_0}^{\infty} e^{U(r)/kT} \left(\frac{dr}{r^2} \right)} \tag{3.8}$$

where r_0 is the radius of spontaneous recombination, and $U(r)$ is the energy of interaction of the defects with the sink. For the case of elastic dilation interaction between two point defects, the general expression for U has the following form [46]:

$$U = -\left(\frac{A}{r^3} \right) \Gamma(\theta, \varphi) \Delta V_1 \Delta V \tag{3.9}$$

where A is the function of elastic constants of the lattice C_{11}, C_{12}, C_{44}, and r is the distance between the defects, $\Gamma(\theta,\varphi)$ is the function describing the orientation crystallographic dependence of the interaction energy, ΔV_1 and ΔV_2 are the dilation volumes of the defects.

In a general case, the effective radius of capture of two defects r_c, where one of them is regarded as a sink, is the distance at which the long-range potential of interaction of the defects U is higher than the thermal energy of the mobile defect kT. But the expression for r_c has the form [47]:

$$r_c = \frac{1}{4\pi d \left(c_\infty - c_0 \right)} \int_s J ds \tag{3.10}$$

In equation (3.10), c_0 and c_∞ are the concentratione of mobile defects on the surface of the sink and at a large distance from it, respectively.

The density of the flux of point defects to the sink j below the integral of equation (3.10) is expressed as follows:

$$J_{i,v}(U) = D_{i,v} \nabla c_{i,v} + \left(\frac{D_{i,v} c_{i,v}}{kT} \right) \nabla U_{i,v} \tag{3.11}$$

The first member of equation (3.11) is the diffusion flux of defects to the sink, and the second one is the drift flux superimposed on the random diffusion flux by the interaction potential U.

In the final analysis, the procedure for calculating the effective radius of the sink r_c is reduced to solving the equation (3.11) for the equilibrium conditions:

$$\nabla J = D\nabla^2 c + \left(\frac{D}{kT}\right)\nabla c \nabla U \tag{3.12}$$

and the corresponding boundary conditions for the given interaction potential U.

The force efficiency of the sink for defects of the given type in the equations (3.5)–(3.7) may be determined directly from the equation:

$$Z^j(r) = \frac{r_c}{r'_c} \tag{3.13}$$

where r'_c is the radius of capture also calculated using equations (3.10) and (3.12) and for the same boundary conditions as r_c but without the drift member in equation (3.12). In fact, quantity $Z^j(r)$ is equal to the ratio of the fluxes of point defects to the actual sink in the field of its stresses and to the sink not generating stresses, with the same geometrical configuration.

If there is no force interaction of the defects with the sink, i.e. $Z^j(r) = 1$ (as, for example, in the case of a pure void which in the first approximation is usually examined as a sink not interacting neither with the interstitials nor with the vacancy [48]), then the effective radius of capture of the spherical sink will be equal simply to the geometrical radius of the sink r: $r_c = rZ^j(r) = r$. In this case, the non-spherical sink may be characterised by the effective geometrical radius of an equivalent sphere.

In absorption of a point defect by a dislocation, their force interaction in the first approximation is usually described using the potential U [49]:

$$U = \rho \Delta V_{i,v} = \frac{(1-\nu)\mu b \sin\theta}{3\pi(1+\nu)r} \Delta V_{i,v} \tag{3.14}$$

where p is the hydrostatic field of stresses of the dislocations, ν is the Poisson coefficient, and $\Delta V_{i,v}$ is the dilation volume of the point defect, μ is the shear modulus, b is the Burgers vector.

Since the dilation volume of the interstitials ΔV_i = 1–2 atomic volumes, and for the vacancies it is $\Delta V_v = -(0.25–0.5)$ [48], equation (3.14) shows directly that, with other conditions being equal, the interstitials are absorbed by the dislocations more efficiently than the vacancies.

Preferential absorption by a sink of type j of, for example, an interstitial in comparison with a vacancy, is usually characterised by the quantities:

$$\frac{\left(Z_i^j - Z_v^j\right)}{Z_v^j} \quad \text{or} \quad \frac{Z_i^j}{Z_v^j} \tag{3.15}$$

where Z_v^j are the corresponding factors of the force interaction. For the edge dislocations, for example, the theoretical estimates of the quantity Z_i^d/Z_v^d are in the range from 1.01 to 1.5 [44].

The calculations of the efficiency of the force interaction for sinks of different type on the basis of solving diffusion problems in the given force field within the framework of the schema, described previously, have been published in [36, 49–53].

If a sink is a good absorber of point defects, the reaction of absorption of defects by a sink is controlled by diffusion. Otherwise, the process is controlled by the rate of overcoming of a potential barrier at the sink by the defect. General criteria, characterising the type of reaction of absorption of defects by sinks, can be found in [45].

On the whole, the direction of theoretical and experimental examination of the efficiency of sinks of different type in relation to the capture of point defects plays a significant role in the general problem of the radiation damage of metallic materials. The possibility of taking into account and regulating their efficiency determines mainly the solution of the problem of development of radiation-resisting materials.

3.5. CHARACTERISTIC TEMPERATURE RANGES OF RADIATION DAMAGE IN FORMATION AND THERMALLY ACTIVATED ANNEALING OF NON-CORRELATED POINT DEFECTS

From the stationary solutions of equations (3.3) and (3.4) for c_v:

$$c_v = \frac{1}{2}\frac{S_i D_i}{R}\left\{\left(1+\frac{4RG}{D_v D_i S_v S_i}\right)^{1/2} - 1\right\} \tag{3.16}$$

Since in equilibrium $c_i S_i D_i = c_v S_v D_v$, for the dynamically equilibrium concentration of the interstitials:

$$c_i = \frac{c_v S_v D_v}{S_i D_i} \tag{3.17}$$

In equation (3.16) the expression:

$$\eta = \frac{4RG}{D_v D_i S_v S_i} \tag{3.18}$$

is the parameter of mutual recombination of point defects. The function of this parameter $F(\eta)$ [43]:

$$F(\eta) = \frac{2}{\eta}\left[(1+\eta)^{1/2} - 1\right] \tag{3.19}$$

is associated with the dynamically equilibrium concentration of free vacancies c_v by the dependence:

$$c_v = \frac{F(\eta)G}{S_v D_v} \tag{3.20}$$

Consequently:

$$D_v S_V c_v = F(\eta)G \tag{3.21}$$

Equation (3.21) shows that the function $F(\eta)$ determines the number of defects absorbed by sinks, in relation to the total rate of formation of the defects G. If $\eta \to 0$, $F(\eta)$ tends to unity, i.e. all the introduced interstitials and vacancies are displaced to the sinks and their mutual recombination does not play any role.

If the value of η is high, then $F(\eta) \approx 2/(\eta)^{1/2} \to 0$, and the mutual recombination of the defects plays the controlling role. When $F(\eta) = 1/2$ ($\eta=8$), the fraction of the defects, absorbed by the sinks, is equal to the fraction of the defects annihilated by mutual recombination, i.e. at:

$$\eta = 8 = \frac{4RG}{D_v D_i S_v S_i} \tag{3.22}$$

$$Rc_v c_i = D_v S_v c_v = D_i S_i c_i = \frac{G}{2} \tag{3.23}$$

Equation (3.22) can be used to determine the relative transition temperature T_t/T_m below which the mechanism of mutual recombination will be dominant, whereas the mechanism of migration of the defects to the sink is dominant above this temperature:

$$\frac{T_t}{T_m} = \frac{E_v^m + E_i^m}{kT_m \ln\left[\dfrac{2D_v^0 D_i^0 S_v S_i}{RG}\right]} \tag{3.24}$$

Table 3.1 gives the results of evaluation of the relative transition temperature for nickel and aluminium using the values of the parameters of point defects from [54].

The estimates show that even for a power of the sinks of 10^{15} m^{-2}, characteristic of greatly deformed or irradiated metallic materials, the temperature of transition from the range of preferential recombination to the range of preferential movement of the defects of the sinks is relatively high. At temperatures of preferential recombination there are also suitable conditions for the formation of complexes of point defects, the nucleation and growth of clusters, mainly of the interstitial type. The controlling process of radiation damage of structural materials are in this case hardening and embrittlement. With a decrease of the concentration of the sinks, processes of mutual recombination and nucleation will also be controlling at high temperatures. Equation (3.24) also shows directly that the restriction of diffusivity of point defects in alloys during their capture by the solute atoms displaces efficiently the transion temperature to higher temperatures.

The temperature range $T > T_t$ is characterised by the preferen-

Table 3.1 Values of parameters for numerical calculations of the relative transition temperature in nickel and aluminium

Parameter	Symbol	Nickel	Aluminium
Melting point, K	T_m	1720	930
Vacancy formation energy, eV	E_v^f	1.39	0.70
Energy of formation of interstitials, eV	E_i^f	4.08	3.20
Energy of vacancy movement, eV	E_v^m	1.38	0.57
Energy of movement of interstitials, eV	E_i^m	0.15	0.10
Pre-exponential factor of the diffusion coefficient of vacancies, m^2 s^{-1}	D_v^0	6×10^{-6}	4.5×10^{-6}
Pre-exponential factor of the diffusion coefficient of interstitials, m^2 s^{-1}	D_i^0	1.2×10^{-5}	8×10^{-6}
Power of sinks for interstitials and vacancies, m^{-2}	$S_{i,v}$	10^{15}	10^{14}
Rate of introduction of defects, s^{-1}	G	10^{-6}	10^{-6}
Relative transition temperature	T_t/T_m	0.32	0.28

tial departure of point defects to sinks, i.e. by the favourable conditions for the growth of clusters, dislocation loops and voids, swelling. In practice, in neutron irradiation, swelling is usually detected at temperatures of $\geq 0.3\ T_m$ at which the fluxes of the vacancies to the voids are relatively strong. In solid solutions and alloys, different phase processes also take place at high intensity.

At the the same time, with increasing temperature, the dynamically equilibrium concentration of radiation point defects continuously decreases and the conditions for the nucleation of flat and volume clusters become less favourable. This reason, together with the effect of the processes of breakdown of the clusters of defects and voids, and also the increase of the thermodynamically equilibrium concentration of the vacancies, results in the restriction of radiation damage at temperatures of the order of 0.6–0.7 T_m.

3.6. KINETICS OF BUILDUP OF RADIATION DEFECTS DURING THE FORMATION OF COMPLEXES WITH SOLUTE ATOMS

One of the characteristic ranges of radiation damage of pure metals – the range of preferential recombination – includes the temperature range from the sub-stage I_E to recovery stage III (from ~0.04 T_m to 0.2–0.3 T_m), the vacancies are immobile, the interstitials freely migrate and the process of the buildup of defects is determined by the reactions in which they take place.

In efficiently annealed metals, the role of departure of the interstitials to the dislocations is not large, because even in the case of high-purity metals (for example, 99.999 and 99.9999 %), the concentration of impurity traps is ~10^{-5}–10^{-6}, which is 2–3 orders of magnitude higher than the concentration of dislocation sinks (~10^{-9}–10^{-8}). At relatively high doses of irradiation of materials, the process of the growth of clusters and dislocation loops starts to play a significant role. The special features of this process are examined in the following section. In the case of low irradiation doses, the kinetic relationships governing the buildup of radiation defects even in very pure metals should be determined by the competing mechanisms of mutual recombination and the capture of interstitials by the solute atoms.

Examination of the given problem is of great theoretical and applied value. This is determined mainly by the possibility of obtaining information on the properties of radiation defects in alloys and the efficiency of interaction of these defects with solutes. The given kinetic method significantly supplements and widens the possibilities of the method of examination of the properties of radiation defects on the basis of the analysis of recovery processes during annealing of the irradiated material. In addition, for a number of metals, especially materials in electronics, degradation of the properties is directly controlled by the mechanisms of radiation damage examined here.

The modelling representations of the kinetics of the buildup of radiation defects at the temperatures of recovery stage II were developed for the first time in a study by Walker [55], who examined the capture of interstitials by traps: non-saturated, saturated, and traps whose efficiency increases with continuing capture. In subsequent stages, the model proposed by Walker was improved in [56-58]. The main modification of the model is associated with the introduction of the radii of recombination of the interstitial atom r_v and capture of the atom by the impurity r_t so that it was possible to use the model also as a method for evaluating the efficiency of different solutes in relation to the capture of interstitials.

On the basis of analysis of a system of equations for the case in which the atoms of the impurity are regarded as non-saturated in traps, and annealing of interstitials at constantly acting sinks can be ignored:

$$\frac{dc_i}{dt} = G - 4\pi D_i c_i \left(r_t c_t + r_v c_v \right)$$
$$\frac{dc_v}{dt} = G - 4\pi D_i c_i c_v \tag{3.25}$$

The rate of variation of the concentration of vacancies within the framework of quasistationary approximation may be represented in the form [56]:

$$\frac{dc_v}{dt} = G \frac{r_t c_t}{r_t c_t + r_v c_v} \tag{3.26}$$

In order to analyse the experimental results for the measurement of electrical resistance in irradiation, equation (3.26) was presented in the following form [56]:

$$\frac{d\Delta\Phi}{d\Delta\rho} = \frac{1}{\rho_F f_c \sigma_d} \left[1 + \frac{r_v}{r_t c_t} \Delta\rho \right] \tag{3.27}$$

where Φ is the irradiation dose; ρ_f is the electrical resistivity of the unit concentration of Frenkel pairs; f_c is the fraction of interstitials avoiding recombination with a self-interstitial vacancy. The multiplier in front of the bracket in the right-hand part of equation (3.27) is the inverse value of the initial rate of variation of electrical resistivity.

On the basis of equation (3.27) a number of authors determined the relative efficiencies of capture of interstitials by impurity traps r_t/r_v in alloys based on copper, aluminium and silver [8, 34, 56–61]. The values of these efficiencies are usually in the range from hundredths to tenths of fractions of the unity. A number of the experimental values of r_t/r_v, including the results of later measurements, were presented in a review in [62].

In [34, 63], a simpler model of non-saturating traps was modified, taking into account the processes of instantaneous capture of an interstitial atom by an impurity trap and of its recombination with the vacancy, and also correlated recombination in the sub stage I_D. The expression for the initial rate of variation of electrical resistivity, obtained in [34, 63], taking these processes into account, has the following form:

$$\left.\frac{d\Delta\rho}{d\Phi}\right|_0 = \rho_F f_c \sigma_d \left\{ \left(1-\frac{r_v}{r_p}\right) + \frac{2}{3}\pi\frac{r_v}{r_p}\left[2r_t^2 + \left(r_p - r_v\right)^2 r_t c_t\right] \right\} \tag{3.28}$$

Direct comparison of this equation with equation (3.27) shows that this equation, like the general equation for the rate of variation of electrical resistance in [34, 63], greatly differs from equation (3.27) for the simpler model of non-saturating traps. Equation (3.28) shows that for the modified model, the initial rate of radiation damage depends on both the type and concentration of impurity atoms and also on the mean the distance r_p in the Frenkel pairs in the substage I_D, which is also in agreement with the experimental results.

The characteristic experimental relationships of the kinetics of the buildup of radiation defects in pure FCC metals and alloys at temperatures of recovery stage II are presented in Fig. 3.2 and 3.3 on an example of a series of experiments with electron irradiation of pure silver and its diluted and concentrated solid solutions [32-34, 64].

The following conclusions may be drawn on the basis of analysis of these dependences:

1. The rate of buildup of defects during isothermal irradiation increases with increasing concentration of the solutes (Fig. 3.2).

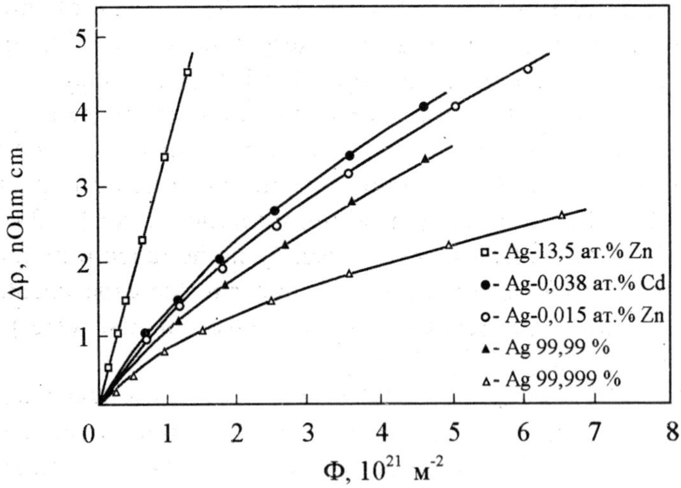

Fig. 3.2. Kinetics of variation of electrical resistivity in silver and its solid solutions in irradiation with electrons with an energy of 2.2–2.3 MeV [32–34, 64]. Irradiation temperature 86 K (●,○,▲,△) and 105 K (□). Radiation intensity 1.4×10^{17} m^{-2} s^{-1} (●,○,▲,△) and 3×10^{18} m^{-2} s^{-1} (□).

2. The rate of buildup of defects in pure silver and diluted alloys decreases with the dose and temperature of irradiation (Fig. 3.2 and 3.3a).

3. The rate of buildup of defects in the concentrated solid solutions is almost independent of irradiation temperature and dose (Fig. 3.3b).

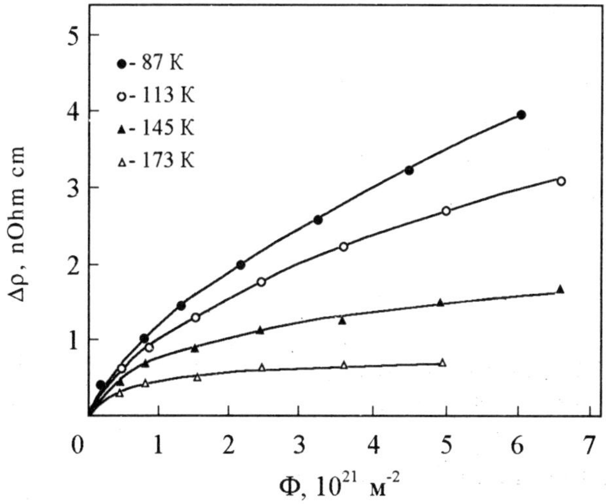

Fig. 3.3a. Temperature dependences of the variation of the electrical resistivity of silver with a purity of 99.99% in irradiation with electrons with an energy of 2.2 MeV [32–34]. Radiation intensity 3 × 10^{16} m^{-2} s^{-1}.

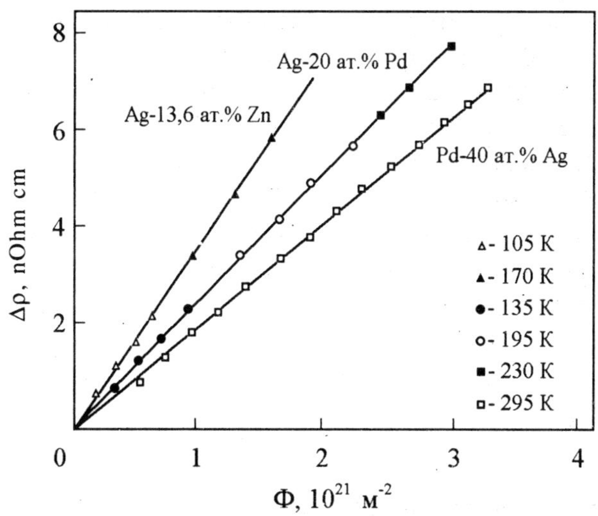

Fig. 3.3b. Temperature dependences of the variation of the electrical resistivity of concentrated solid solutions in irradiation with electrons with an energy of 2.2 MeV [32,33,64]. Radiation intensity 3 × 10^{16} m^{-2} s^{-1}.

The temperature dependence of the rate of buildup of radiation defects in diluted alloys in the recovery stage II in certain investigations (for example [58,65,66]) was associated with the variation of the radii of recombination r_v and capture of interstitials by impurity traps r_t or, more acurately, with the temperature dependence of the r_t/r relationship determined in processing the experimental data in accordance with equation (3.27). However, it is very difficult to accept this interpretation.

In fact, the determination of the radii of recombination and of capture (section 3.4) and equation (3.9) for the energy of interaction of two points defects indicates that the radii of both recombination and capture should depend in the same manner on temperature:

$$r_{t,v} = A_{t,v} T^{-1/3} \tag{3.29}$$

and the ratio r_t/r should consequently be temperature-independent.

In equation (3.29) $A_{t,v}$ are the appropriate constants from equation (3.9).

To describe the kinetics of the buildup of radiation defects at temperatures of recovery stage II, the authors of [32, 32, 67, 68] presented considerations regarding the dissociation of the interstitial atom–impurity trap complexes. The effect of the concentration of solutes and irradiation temperature on the kinetics of buildup of the defects was examined in [68] within the framework of the model which took into account the following reactions: 1. The recombination of the interstitials and vacancies in correlated Frenkel pairs; 2. The dissociation of correlated pairs; 3. The capture of interstitials by impurity traps in correlated Frenkel pairs; 4. The recombination of free interstitials and vacancies; 5. Capture of free interstitials by impurity traps; 6. The dissociation of the interstitial atom-impurity trap complexes.

The equations describing the variation of the concentration of the correlated Frenkel pairs c_c and the concentration of the free interstitials and vacancies have the following form [68]:

$$\frac{dc_c}{dt} = f_1 G - c_c\left(R_c + K_{cd} + K_{ct} c_t\right) \tag{3.30}$$

$$\frac{dc_i}{dt} = f_2 G - K_{cd} c_c - R c_v c_i - K_t c_t c_i - K_d c_{ti} \tag{3.31}$$

$$\frac{dc_v}{dt} = f_2 G - c_c\left(K_{cd} - K_{ct}c_t\right) - Rc_v c_i \tag{3.32}$$

In equations (3.30)–(3.32), G is the total rate of introduction of correlated and non-correlated Frenkel pairs, f_1 and f_2 and the fractions of the correlated and non-correlated Frenkel pairs, respectively; R_c, K_{ct} and K_{cd} are the constants of the reactions of recombination of the correlated pairs, the capture of interstitials of the correlated pairs by the impurity traps and the dissociation of correlated pairs with the formation of free defects, respectively. The quantities R, K_t and K_d are the constants of the reactions of recombination of non-correlated defects, the capture of free interstitials by impurity traps and the dissociation of the complexes 'interstitial atom–impurity trap', respectively. The dissociation constant K_d is:

$$K_d = K_d^0 \exp\left[-\left(E_i^m + E_i^B\right)/kT\right] \tag{3.33}$$

where E_i^B is the energy of the bond between the interstitial and the impurity trap, c_{ti} is the concentration of the interstitials, captured by impurity traps.

Within the framework of the quasistationary approximation, the following equation of obtained for the mean rate of radiation damage:

$$\overline{\left(\frac{dc_v}{dt}\right)} = \frac{RG}{1+\left(\frac{RW}{K_t c_t}\right)}\left[1+\frac{WRK_{ct}}{FK_t\left(R_c + K_{cd} + K_{ct}c_t\right)} - \frac{W^2 K_d R}{FGK_t c_t}\right] \tag{3.34}$$

$$W = \frac{1}{2}\left[\frac{FGt}{1+\left(FGt/2\right)}\right] \tag{3.35}$$

$$F = f_2 + \left(1 - P_c\right) f_1 \tag{3.36}$$

The probability of correlated recombination of the interstitials and vacancies P_c in equation (3.36) is:

$$P_c = \frac{R_c}{(R_c + K_{cd} + K_{ct}c_t)} \tag{3.37}$$

According to equation (3.34), the temperature dependence of the mean rate of radiation damage is described by the third member of the right-hand part, which depends on the rate of breakdown of the complexes. This term is inversely proportional to the concentration of the impurity traps, i.e., as the concentration of the impurity traps increases, the temperature dependence becomes weaker and the rate of buildup of the defects increases.

In order to clarify the results, the system of equations (3.33.32) was also solved numerically [68].

Figure 3.4 shows the dependences of the buildup of radiation defects, calculated for different concentrations of the impurity traps, their binding energy with the interstitials and irradiation temperature. The value of the dissociation constant K_d in Fig. 3.4 characterises both the binding energy of the interstitials with the impurity traps and irradiation temperature.

The following main conclusions can be drawn on the basis of Fig. 3.4 and equation (3.34):

1. The rate of buildup of defects in isothermal irradiation increases with increasing concentration of the solutes and their binding energy with the interstitials.

2. As the concentration of impurity traps decreases, the dose and temperature dependence of the rate of buildup of defects become stronger.

3. With increasing concentration of the solutes, the dose and temperature dependence of the rate of buildup of the defects weakens.

4. At high concentrations of solutes, characteristic of concentrated alloys, the dose and temperature dependence of the rate of buildup of defects may be fully suppressed as a result of a large decrease of the effective mobility of interstitials.

The comparison of the experimental results, presented previously, with the theoretical data indicates that the main relationships governing the buildup of radiation defects in both diluted and concentrated FCC solid solutions are completely explained on the basis of the examined modelling representations, reflecting the strong effect of the concentration and type of solutes. This is clearly confirmed in direct comparison of Fig. 3.2–3.3 and 3.4. Since the parameters, reflecting the type of lattice, are not included in the model in [68], the main assumptions of the model can also be used for metals and alloys with other types of crystal structure.

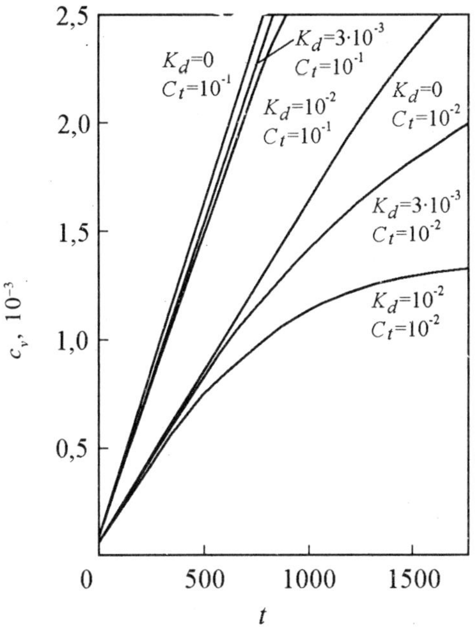

Fig. 3.4. Calculated kinetic dependences of the variation of the concentration of vacancies in diluted and concentrated solid solutions for different irradiation temperatures [68].

3.7. THE FORMATION AND GROWTH OF CLUSTERS AND DISLOCATION LOOPS IN PURE METALS AND SOLID SOLUTIONS IN IRRADIATION

In the previous section, attention was given to the kinetic mechanisms of the buildup of radiation defects which dominate at a relatively low radiation doses. With increasing radiation dose, the probability of homogeneous and heterogeneous formation of complexes from two or more interstitials increases. Consequently, the kinetics of radiation damage of pure metals and diluted alloys in the temperature range of recovery stage II and at higher temperatures starts to be controlled by the growth of clusters and dislocation loops of the interstitial type and by the buildup of vacancies, avoiding recombination with the interstitials. In the alloys where the mobility of vacancies can be, as shown previously, higher than that of the interstitials, the preferential nucleation and growth of vacancy-type clusters is possible. In a general case, the controlling type of clusters and the dislocation loops in the given time period may be determined by the crystal structure and the composition of the material, temperature, intensity and other parameters and by irradiation conditions (see, for example [69–76]).

Examination of the mechanisms of formation and growth of clusters and loops during irradiation is of considerable importance for both the theory of radiation damage and for the development of radiation-resistant alloys. In irradiated metallic materials, the clusters and dislocation loops are effective obstacles to the movement of dislocations, controlling the processes of hardening and embrittlement up to temperatures of 0.3–0.4 T_m [77–80]. At higher irradiation temperatures, the processes of formation and evolution of loops in both annealed and deformed metals, determine the formation of the dislocation structure which usually preced the period of stationary swelling and creep. This is shown in Fig. 3.5–3.8 which show the main special features of the variation of the structure of metallic materials in the period preceding swelling [81–84].

Figures 3.5–3.7 show that the main stages of the process of nucleation and growth of the loops are the same as in the conventional breakdown of supersaturated solid solutions: nucleation (nonstationary and stationary), growth with retention of the approximately constant density, and coalescence. Qualitative differences in the mechanisms are also important: two types of opposite particles take part in the nucleation in the growth of loops–vacancies and interstitials, and in addition to this, the coalescence stage is accompanied by the interaction of dislocation loops with the dislocation structure. The result of this process is the formation of a dislocation network with sufficient stability in subsequent irradiation (Fig. 3.7 and 3.8).

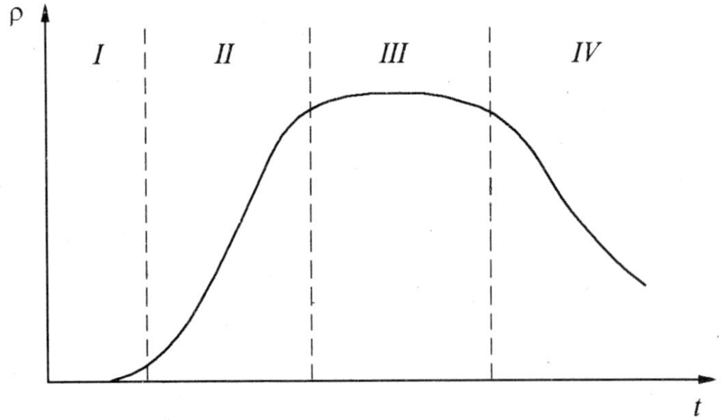

Fig. 3.5. Schematic dependence of the variation of the density of loops (in a general case precipitates in the process of phase transformation) during radiation [81,82]. I) incubation period; II) period of stationary nucleation; III) period of growth at constant density; IV) coalescence period.

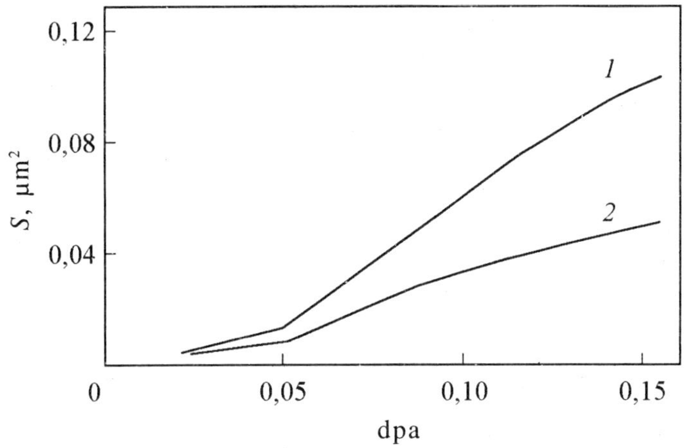

Fig. 3.6. Variation of the area of loops in irradiation of 316-type steel in a high-voltage electron microscope [83]: 1) Frank loop; 2) rhomboidal loops.

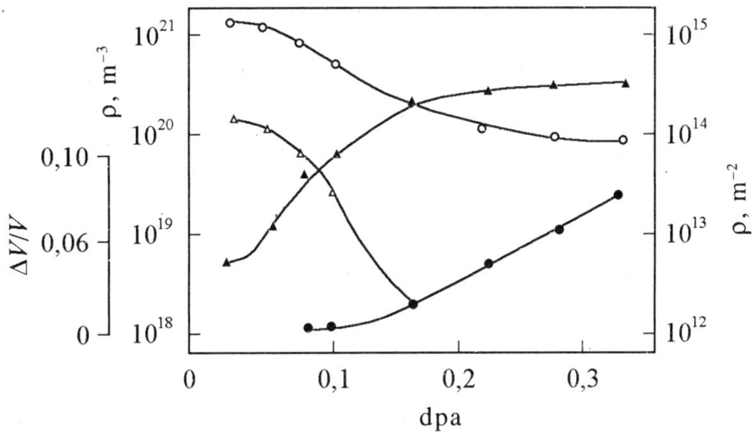

Fig. 3.7. Variation of the density of dislocations (ρ_d), loops (ρ_l), voids (ρ_v) and also of swelling ($\Delta V/V$) in irradiation of 316-type steel in a high-voltage electron microscope [83].

The theoretical and practical importance of the examination of the mechanisms of the nucleation and growth of clusters and dislocation loops is reflected in a large number of experimental studies concerned with the problem. A large part of these studies have been carried out using the informative, convenient and rapid method of high-voltage electron microscopy.

The large majority of investigations of the formation and growth of clusters of interstitials shows that even in pure metals, at relatively low temperatures ($\sim\leq 0.3\ T_m$), the heterogeneous mechanism

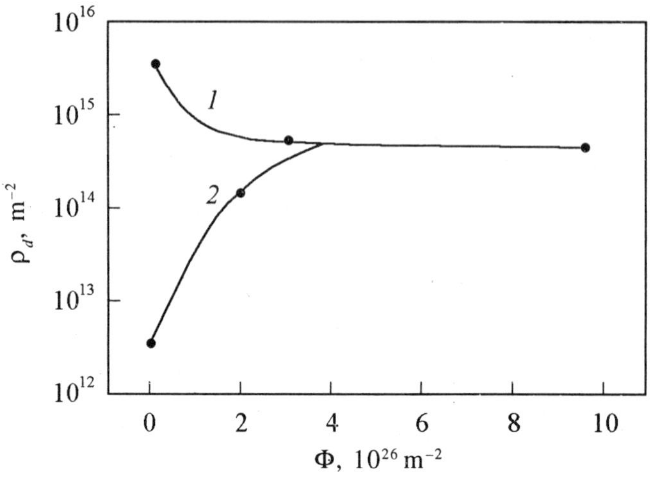

Fig. 3.8. Variation of the density of dislocations during neutron irradiation of steel 316 in cold-worked (1) and annealed (2) conditions at ~500 °C [84].

of nucleation of clusters dominate, i.e., on the atoms of the impurities and alloying elements. Experimentally, this is shown clearly, in particular, in the study by Shimomura [85], where the density of loops in electron-irradiated Au with a purity of 99.999 % was an order of magnitude higher than in Au with a purity of 99.9999 %. Convincing confirmation of the heterogeneity of the nucleation of loops has also been presented in a number of investigations concerned with the examination of the dependence of the concentration of loops on intensity and irradiation dose [86–89].

Simple theoretical estimates are in a very good agreement with the results of these experiments. For example, for aluminium with an atomic concentration of the impurities of $\sim 10^{-5}$–10^{-6} at the volume of capture of 50 atomic volumes by every impurity, the energy of migration of the interstitials is 0.1 eV and dpa = 10^{-4} s^{-1}, the probability of capture of the interstitial by the impurity trap at 200 K is more than four orders of magnitude higher than the probability of meeting of two interstitials. In addition, the mechanism of heterogeneous formation of nuclei of clusters and loops is also favourable from the energy viewpoint, both as a result of a decrease of the energy barrier for nucleation and as a result of the stabilisation of complexes of defects of different type. In a general case, the relative role of the homogeneous mechanism should increases with increasing temperature, reflecting the tendency for a decrease of the efficiency of the defect–impurity interaction. This is confirmed by the experimental results from the examination of the nucleation and growth of loops in Nb–Zr alloys with additions of oxygen and nitrogen [90, 91].

The theory of the homogeneous mechanism of the nucleation of dislocation loops within the framework of the modified classic considerations was proposed for the first time in [81]. It is identical with the modified stationary theory of the nucleation of voids in pure metals during irradiation [92,93]. All these theories are based on the stationary solution of discrete equations of the evolution of clusters in the space of the dimensions taking into account the fluxes of vacancies and interstitials and emission of particles (for vacancy formations) from the clusters. Equations identical with the equations of the classic theory of nucleation were derived for the rate of nucleation and nucleation time [81, 82]:

$$I_c = z'\beta_c\rho'_c \tag{3.38}$$

$$\tau = \left(2\beta_c z'^2\right)^{-1} \tag{3.39}$$

where z' is the Zeldovich factor, inversely proportional to the width of the dimension distribution of the free energy of formation of nuclei $\Delta G'_n$, corresponding to the height $\Delta G'_n$ kT units lower than the distribution maximum (Fig. 3.9), β_c is the speed of absorption of interstitials by a loop of the critical size.

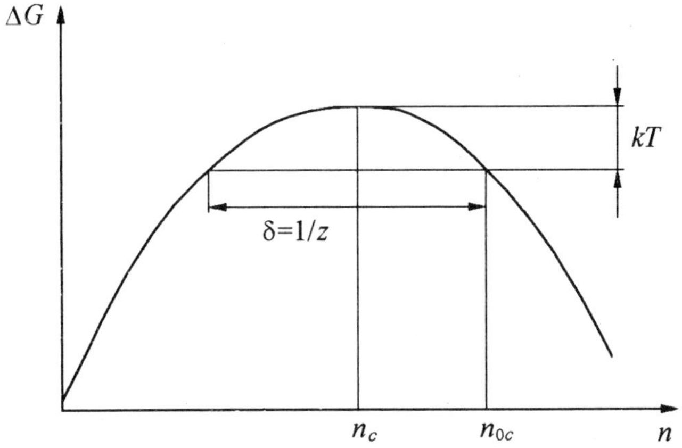

Fig. 3.9. Schematic dependence of the free energy of nucleation of dislocation loops on the number of interstitials in a loop, n_c and n_{oc} is the number of interstitials in loops of critital and supercritical sizes, respectively.

$$\rho_c' = \alpha N_0 \exp\left(-\Delta G_c' / kT\right) \tag{3.40}$$

In the equation for the concentration of the nuclei of the critical size (3.40) α is the geometrical constant, N_0 is the number of atoms in the unit volume, $\Delta G_c'$ is the maximum value of the distribution of free energy of formation of loops $\Delta G_n'$.

In contrast to the classic value, the modified value $\Delta G_n'$includes also the kinetic parameter which takes into account the ratio of the fluxes of interstitials and vacancies to a dislocation (in a general case, a cluster, a void).

To describe the evolution of dimensional distributions of dislocation loops and voids, the authors of [94–97] used for the first time the Fokker–Planck type equations. This approach provides clear information on the kinetics of the process and for cases in which analytical solutions are possible it enables considerable simplification of the procedure of the calculations, associated with the numerical solution of the system of a large number of paired differential equations. In subsequent investigations, the application of numerical or analytical solutions of the Fokker–Planck type equation in the examination of the processes of the nucleation and growth of clusters, dislocation loops and voids has been quite frequent [91,98–102].

Analysis and interpretation of the experimental data for the nucleation and kinetics of the growth of loops are carried out either using the previously mentioned equations of the Fokker–Planck type or of models based on the numerical solution of systems with a large number of equations for consecutive variations of the concentration of defects in clusters [102–107], or using simple probability analytical models for the variation of the mean size of the loops, identical to [108-111]. The simplification, accepted in similar analytical models (dominant type of sinks, relative mobility of vacancies and interstitials, the size of the loops, the temperature, dose and intensity of irradiation, etc) make it possible in a number of cases to approximate the experimental dependences by exponential functions of the type $\bar{d} - t^n$ (see, for example, [90, 91,110–113), where $\bar{d}$ is the mean diameter of the loops, t is the irradiation time. For example, when examining the growth of loops in aluminium and its alloys Al–0.1% In and Al–0.4% Si [113], the experimental results are approximated by the functions $\bar{d} - t^n$ with exponents n for the given materials being 0.7; 1 and 0.4, respectively. In some cases, the kinetics of growth of loops even in the same material but in different time and temperature ranges is represented by the authors of investigations in the form of two functions of the type $\bar{d}-t^n$ with different exponents (for example [111, 112]). As shown later, this variety in the kinetic representation of the experimental data is determined to a certain degree by different modelling approximations used in analysis of the results.

In many early investigations, the nucleation time of loops in accordance with the model [108] is regarded as synonymous with the time of meeting of two interstitials:

$$\tau_{ii} = \left(Gr_{ii}D_i\right)^{-1} \tag{3.41}$$

which, as can easily be seen, is extremely short.

A similar interpretation also contradicts the considerations of the modified classic theory of nucleation [81,82] and a large number of experimental data, for example [69,114]. In [114] the authors noted a discrepancy between the experimental results and the assumptions made in [108] and show that it is necessary to interpret the nucleation time of the loops on the basis of the modified classic theory of nucleation [81]. This time corresponds to the time periods I and II in Fig. 3.5.

A significant role in the nucleation and growth of dislocation loops is played by the solute atoms. They can not only modify the mechanism of the process but also lead to completely new effects. This problem, together with the problems discussed previously, is the subject of special attention in further investigation taking into account the results of theoretical and experimental investigations obtained in [67, 69, 94, 114–118).

3.7.1 Theory of the size distribution of clusters and dislocation loops of the interstitial type and its application for analysis of the experimental data

For the case of immobile vacancies (for example, in stage II of annealing of defects), the evolution of the size distribution of the interstitials in accordance with [89, 94] is described by the following infinite system of differential equations:

$$\frac{dn(0,\bar{m})}{d\bar{m}} = -\frac{f(0)n(0,\bar{m})}{\sum_m f(m)n(m,\bar{m})} \tag{3.42}$$

$$\frac{dn(m,\bar{m})}{d\bar{m}} = \frac{f(m-1)n(m-1,\bar{m}) - f(m)n(m,\bar{m})}{\sum_{m} f(m)n(m,\bar{m})}$$

where $f(m)$ is the efficiency of capture of the interstitials by the trap filled m times, $\bar{m}$ is the mean number of interstitials in a cluster, and $n(m, \bar{m}) = N(m, \bar{m})/N_0$ is the relative number of clusters containing m interstitials, N_0 is the total number of clusters.

The transition from discrete m to a continual variable results in an equation for the density of distribution $\rho(x, \bar{x})$ of the Fokker–Planck type [89, 94]:

$$\frac{\partial\rho(x,\bar{x})}{\partial\bar{x}} = \frac{1}{g(\bar{x})}\left\{\frac{1}{2}\frac{\partial^2 x}{\partial x^2}\left[f(x)\rho(x,\bar{x})\right] - \frac{\partial}{\partial x}\left[f(x)\rho(x,\bar{x})\right]\right\} \qquad (3.43)$$

where $g(\bar{x}) = \int_0^\infty f(x)\rho(x,\bar{x})dx$

$$g(\bar{x}) = \int_0^\infty f(x)\rho(x,\bar{x})dx \qquad (3.44)$$

with the initial condition $\rho(x,0) = \delta(x)$ (3.45)

and with the boundary condition $\rho(\infty,\bar{x}) = 0$ (3.46)

In [89, 94], the authors published analytical solutions of the differential equation (3.43) for the case $f(x) = \alpha_0$:

$$\rho(x,\bar{x}) = \frac{1}{\sqrt{2\pi\bar{x}}}\exp\left[-\frac{(x-\bar{x})^2}{2\bar{x}}\right] \qquad (3.47)$$

and $f(x) = \alpha_0 + \alpha x$:

$$\rho(x,\bar{x}) = \frac{x^{\frac{\alpha}{\alpha_0}-1}}{\Gamma\left(\frac{\alpha}{\alpha_0}\right)\left(\frac{\alpha}{\alpha_0}\bar{x}\right)^{\alpha/\alpha_0}}\exp\left(-\frac{x}{\frac{\alpha}{\alpha_0}\bar{x}}\right) \qquad (3.48)$$

For the usually examined variation of the efficiency of capture of loops $f(x)\ \alpha x^{1/2}$, the authors of [89] obtained the analytical expression for the variance of distribution:

$$D_2 = (\bar{x}) = 4\bar{x} \tag{3.49}$$

The transition from the number of atoms in the loop to its diameter gives the following equation:

$$x = \frac{\pi d^2}{4S_0} \tag{3.50}$$

where S_0 is the area per atom in the loop, determined by the orientation of the plane in which the loop is situated in relation to the surface. In accordance with the equation for the variance $D = x^2 - \bar{x}^2$ and expressions (3.49) and (3.15), the following characteristic parameter σ_2 was obtained [89]:

$$\sigma_2 = \frac{\langle d^4\rangle - \langle d^2\rangle^2}{\langle d^2\rangle} \tag{3.51}$$

which in the process of evolution of the distributions should remain constant for the efficiency of capture $\alpha x^{1/2}$.

For the variance of the distribution in the case of a linear increase of the efficiency of capture of the loop $f(x) = \alpha_0 + \alpha x$, the authors of [89] obtained the following equation:

$$D_1(x) = \frac{\alpha \bar{x}^2}{\alpha_0} \tag{3.52}$$

In the case of the linear law of the variation of the efficiency of capture, the value of σ_1 can be determined from the following equation:

$$\sigma_1 = \frac{D(x)}{\bar{x}^2} = \frac{\langle d^4\rangle - \langle d^2\rangle^2}{\langle d^2\rangle^2} \tag{3.53}$$

Parameter σ_1 (like σ_2) should remain constant in the process of irradiation, if the linear law of the variation of the efficiency of capture is valid.

The determined parameters make it possible to greatly simplify the analysis of experimental distributions.

For the case of mobile vacancies, the authors of [89] obtained the following expression instead of the system of equations (3.42):

$$\frac{\partial n(m,\bar{m})}{\partial \bar{m}}=\left[\lambda(\bar{m})+1\right]\frac{f(m-1)n(m-1)-f(m)n(m,\bar{m})}{\sum f(m)n(m,\bar{m})}+$$

$$+\lambda(\bar{m})\frac{f(m+1)n(m+1,\bar{m})-f(m)n(m,\bar{m})}{\sum_m f(m)n(m,\bar{m})} \tag{3.54}$$

In this case, it was assumed that for the capture by the loop on average of λ interstitials it is necessary to absorb vacancies and (λ + 1) interstitials, and the efficiency of capture by the loops of vacancies and interstitials in accordance with [36, 48] has the same functional dependence, and for relatively large loops they differ only in the proportionality coefficients.

The transition to a continuous approximation results in the following equation [89]:

$$g(\bar{x})\frac{\partial \rho(x,\bar{x})}{\partial x}=-\frac{\partial}{\partial x}\left[f(x)\rho(x,\bar{x})\right]+\frac{2\lambda(\bar{x})+1}{2}\times$$

$$\times\frac{\partial^2}{\partial x^2}\left[f(x)\rho(x,\bar{x})\right] \tag{3.55}$$

Using equation (3.55) for the variance of the distributions in the case with $f(x)=\alpha_0+\alpha x$ and $f(x)=\alpha x^{1/2}$:

$$D_1(x)=\frac{\alpha(2\lambda+1)}{\alpha_0}\langle x^2\rangle \tag{3.56}$$

$$D_2(x)=(4+\lambda)x \tag{3.57}$$

Direct comparison of the equations (3.49), (3.52), (3.56) and (3.57) shows that the inflow of vacancies to the interstitial loops results in widening of the distribution.

The equations for the values of σ_1 and σ_2 have the following form [89]:

$$\sigma_1 = \frac{\alpha(2\lambda - 1)}{\alpha_0} \tag{3.58}$$

$$\sigma_2 = \frac{4S_0(4+\lambda)}{\pi} \tag{3.59}$$

Figure 3.10 shows the experimental distribution of the dislocation loops of the interstitial type in Au (dotted lines) [85] and the distribution determined by equation (3.48) for the case $f(x) = \alpha_0 + \alpha x = 1 + 0.285x$ (solid line) [94]. It may be seen that the agreement of the theoretical and experimental dependences, especially in the range of low doses, is sufficiently good. The small difference in the distributions in the case of large doses is possibly associated with the variation of the nature of the efficiency of capture with an increase of the size of the loops.

Figure 3.11 shows the characteristic distributions of the loops with respect to the square of the diameter in aluminium, irradiated with

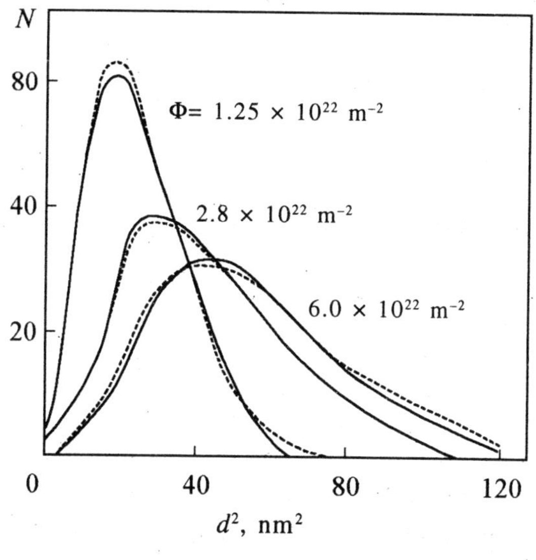

Fig. 3.10. Experimental (broken lines) [85] and calculated from equation (3.48) for $\alpha = 0.285$ (solid lines) [94] size distributions of dislocation loops of the interstitial type in electron-irradiated gold.

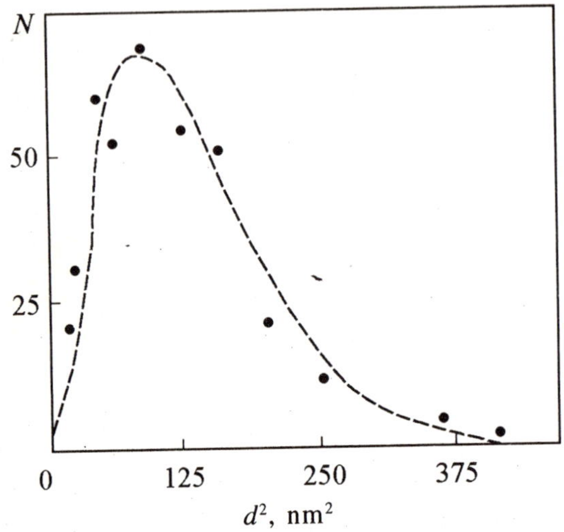

Fig. 3.11. Distributions of interstitial dislocation loops with respect to the squares of their diameter d^2 in Al [89]. Intensity of irradiation with electrons with an energy of 1 MeV was 2×10^{22} m^{-2} s^{-1}, dose 3.6×10^{25} m^{-2}.

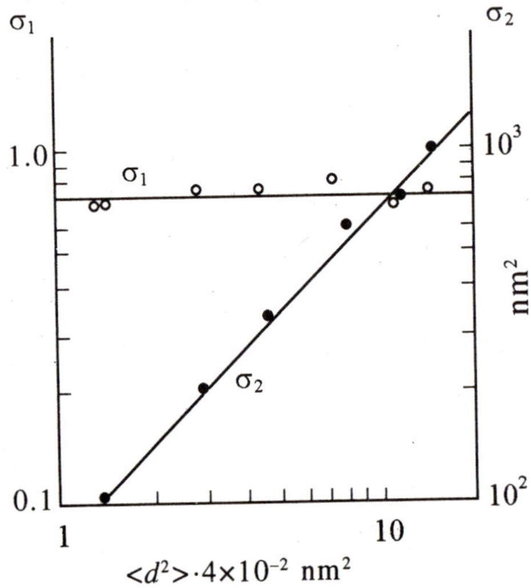

Fig. 3.12. Parameters of size distributions of dislocation interstitial loops σ_1 and σ_2 in electron-irradiated Al–0.06 at% Mg alloy.

electrons with an energy of 1 MeV in a high-voltage microscope at room temperature [89]. The form of the distribution of the loops also corresponds to the gamma distribution, described by equation (3.48).

The analysis of the size distributions of the loops in the Al–0.06% Mg alloy on the basis of parameters σ_1 and σ_2 is presented in Fig. 3.12 [89]. It may be seen that in irradiation the parameter

σ_1 remains constant (equations (3.53) and (3.58)) which in accordance with the examined theory characterises the variation of the efficiency of capture by the loops in accordance with the linear law. Identical dependences for the values of σ_1 and σ_2 were recorded in analysis of the kinetics of growth of the loops in the irradiated Al–0.02 at% Zn [116].

The results of analysis of the efficiency of loops in relation to the capture of interstitials, obtained in [89, 94], differ from the general considerations according to which this parameter is usually assumed to be proportional to the square root of the number of the interstitials in the loops. At the same time, it should be noted that this problem has not been analysed in detail using experimental data prior to the publication of [89, 94].

3.7.2. The kinetics of buildup of interstitials and vacancies in pure metals and solid solutions during the formation and growth of dislocation loops

The variation of the concentration of free interstitials c_i^f and the total concentration of vacancies c_v (i.e. the sum of the free single vacancies and single vacancies in clusters of different size) in relatively thick irradiated objects has been described in [114] by the following system of the differential equations:

$$\frac{dc_i^f}{dt} = G - Rc_v c_i^f - K_i c_v c_i^f \tag{3.60}$$

$$\frac{dc_v}{dt} = G - Rc_v c_i^f - \frac{1}{2} K_v c_v^2 \tag{3.61}$$

When writing the equations (3.60), (3.61), it was assumed that the concentration of the interstitials in the loops c_i^l is equal to the concentration of vacancies in the matrix with the accuracy to the concentration of free interstitials c_i^f, i.e. $c_i^l + c_i^f \cong c_i^l = c_v$. In accordance with the results of analysis of the evolution of the previously examined dimensional distributions, the linear law of variation of the efficiency of capture of interstitials by the loops has also been accepted.

The second term in these equations takes into account the recombination of interstitials with vacancy-type defects (in a general case, with clusters with multiplicity i), where R is the recombination constant.

The third terms in the right-hand part of the equations (3.16) and (3.61) determine respectively the rate of capture of free interstitials and vacancies by the dislocation loops. K_i and K_v are the constants of absorption of the interstitials and vacancies by the interstitial loops. Multiplier 1/2 in equation (3.61) reflects the fact that only half of the reaction acts $K_v c_v^2$ results in the disappearance of vacancies from the system as a result of their annihilation on the interstitial loops. Their second part includes the reactions of formation and growth of the vacancy-type clusters, which evidently cannot change the general concentration of the vacancies in the matrix.

Integration of the system (3.62)–(3.61) using the quasistationary approximation and subsequent transformations lead to:

$$c_i^l = c_{i0}^l + athb(t-\tau) \tag{3.62}$$

The quantity c_{i0}^l in equation (3.62) is interpreted in [114] as the general concentration of the interstitials in the nuclei of overcritical sizes, and τ is the nucleation time, determined by equation (3.39). The nuclei of the overcritical sizes are the loops which in contrast to the nuclei of critical sizes grow with a probability equal to unity. The parameters a and b in equation (3.62) are equal to respectively:

$$a = \left[\frac{2GK_i}{(R+K_i)K_v}\right]^{1/2} \tag{3.63}$$

$$b = \left[\frac{GK_iK_v}{2(R+K_i)}\right]^{1/2} \tag{3.64}$$

Since $c_{i0}^l \ll a$, the value of a in equation (3.62) corresponds actually to the general concentration of the interstitials in the loops at saturation, i.e. $a \cong c_{is}^l$. Parameter b characterises the rate of absorption of interstitials (or vacancies) by loops at saturation.

It should be stressed that the dependence (3.62) in different time periods, whose length is determined mainly by the intensity and temperature of irradiation, may be approximated by different exponential functions of the type $c_i^l - t^n$ ($0 \le n \le 1$). Therefore, taking into account, as mentioned previously, the existing difference in the interpretation of the experimental results in the form $c_i^l - t^n(d_i^l - t^n)$, it may be assumed that the examined kinetic model is of a higher universal nature.

Equations (3.62)–(3.64) show that:

$$\frac{b}{a} = \frac{K_v}{2} \tag{3.65}$$

$$\frac{R}{K_i} = \frac{G}{ab} - 1 \cong \frac{G}{ab} \tag{3.66}$$

$$t_s = \frac{\pi}{b} + \tau \tag{3.67}$$

where t_s is the time required by the concentration of the interstitials in the loops to reach saturation.

The initial rate of variation of the concentration of the interstitials in the loops during their growth or the initial rate of absorption is equal to:

$$\left.\frac{dc_i^l}{dt}\right|_0 = ab = G\frac{K_i}{R + K_i} \tag{3.68}$$

The initial rate of absorption of the interstitials by a single loop of overcritical sizes β_{0c} is:

$$\beta_{0c} = \frac{abN}{\rho} \tag{3.69}$$

where N is the number of atoms in the unit volume, ρ is the density of the loops.

Within the framework of the accepted linear law of the variation of the efficiency of capture

$$\beta_{0c} = \frac{n_{0c}\beta_c}{n_c} \tag{3.72}$$

The equations (3.39), (3.69) and (3.72) show that:

$$\left(\frac{n_{0c}}{n_c}\right)^{1/2}\delta=(2\tau\beta_{0c})^{1/2}=\left(\frac{2\tau abN}{\rho}\right)^{1/2}=W \tag{3.71}$$

Within the framework of the classic representations, quantity $(n_{0c}/n_c)^{1/2}\delta$ can be regarded as the parameter characterising the effective width of the energy distribution of the nuclei of the loops with respect to the size, i.e. the susceptibility of loops to nucleation. In equations (3.70) and (3.71), n_c and n_{0c} is the number of interstitials in the nuclei of the critical and overcritical size, respectively, β_c is the rate of absorption of the interstitials by a loop of the critical size.

These equations show that the model, proposed in [114], in contrast to other available models, makes it possible in analysis of the experimental results to obtain considerably more detailed information on the mechanism and parameters of the growth of loops and has also been used for the first time to evaluate the parameters of their nucleation τ and $(n_{0c}/n_c)^{1/2}\delta$.

In [69, 116–118], the previously examined modelling considerations for pure metals were modified directly for solid solutions. Attention was given to a *A–B* binary alloy in which it was assumed that the formation of types of dumbbell configurations of interstitials *A–A*, *A–B* and *B–B*, is possible. Taking into account the relationships for the concentrations of these dumbbells, the Boquet model [190, 120] (equations (2.7)–(2.11)) was used to obtain systems of equations for two cases: with the mobility of interstitials or mobility or vacancies being dominant.

When the mobility of the interstitials is higher, the corresponding system of the equations has the following form [69, 115]:

$$\begin{aligned}\frac{dc_i^f}{dt}&=G-\left[(R_{AA}+K_{AA})P_{AA}+(R_{BB}+K_{BB})P_{BB}+\right.\\&\left.+2(R_{AB}+K_{AB})P_{AB}\right]c_vc_i^f\end{aligned} \tag{3.72}$$

$$\frac{dc_v}{dt}=G-(R_{AA}P_{AA}+R_{BB}P_{BB}+2R_{AB}P_{AB})c_vc_i^f-\frac{K_vc_v^2}{2} \tag{3.73}$$

$$\frac{dc_B}{dt} = -\left(2K_{AB}P_{AB} + K_{BB}P_{BB}\right)c_v c_i^f \tag{3.74}$$

As in the model for pure metals, c_i and c_v are the concentrations of the free interstitials and the general concentration of the vacancies in the matrix (both in the form of single vacancies and clusters of different multiplicity), c_B is the concentration of component B in the alloy A–B, R_{ij} and K_{ij} are the constants of recombination and absorption of the dumbbells by the loops for dumbbells of different type. Equation (3.74) of the given system describes the variation of the concentration of component B in the solid solution as a result of its departure to the sinks–loops in the form of dumbbells A–B and B–B.

When solving the system (3.72)–(3.74) it was possible to use the quasistationary approximation, as in the case of pure metals. For the variation of the total concentration of the interstitials in the loops, the resultant equation has the form identical with that of the dependence for pure metals:

$$c_i^l = c_{i0}^l + a\,\mathrm{th}\,b\left(t-\tau\right) \tag{3.75}$$

where

$$a = \left(\frac{2Gm}{K_v n}\right)^{1/2}, \qquad b = \left(\frac{GK_v m}{2n}\right)^{1/2} \tag{3.76}$$

$$n = \left(R_{AA} + K_{AA}\right)P_{AA} + \left(R_{BB} + K_{BB}\right)P_{BB} + 2\left(R_{AB} + K_{AB}\right)P_{AB} \tag{3.77}$$

$$m = K_{AA}P_{AA} + K_{BB}P_{BB} + 2K_{AB}P_{AB} \tag{3.78}$$

For the mean values of the constants R_{ij} and K_{ij} (R_m and K_{im}), the equations were obtained for a and b, identical with the equations (3.63) and (3.64) for pure metals:

$$a = \left[\frac{2GK_{im}}{\left(R_m + K_{im}\right)K_v}\right]^{1/2}, \quad b = \left[\frac{GK_{im}K_v}{2\left(R_m + K_{im}\right)}\right]^{1/2} \tag{3.79}$$

The interpretation of the quantities a, b, τ, c_{i0} and other parameters of the nucleation and growth of the loops within the framework of the given model is identical with the definition for pure metals. The functional expressions of these parameters [69,114,115] are identical with the equations for pure metals (3.65)–(3.71).

The general solution of equation (3.74) for the variation of the concentration of component B in the solid solution is presented in [116]. For the case in which the flux of the dumbells B–B to the loops can be ignored, the analytical solution of equation (3.74) has the form:

$$c_B^m = c_B^0\left[1-\gamma G(t-\tau)\right], \quad c_B^l = c_B^0 \gamma G(t-\tau) \tag{3.80}$$

where the value of γ is expressed from:

$$\gamma = \frac{2K_{AB}c_B^0\theta}{\left[(R_{AA}+K_{AA})c_A + 2c_B^0(R_{AB}+K_{AB})\theta\right]} \tag{3.81}$$

c_B^0 in equations (3.80) and (3.81) is the initial concentration of component B in the solid solution, and parameter θ in equation (3.81) is determined by equation (2.7).

One of the important consequences of [115] is the fact that in real materials (like in relatively pure metals and alloys), the dislocation loops may represent loops with segregations or loops with stoichiometry of the phases. This theoretical result is in complete agreement with the experimental data (see, for example [121,123]).

For the case in which the mobility of vacancies is dominant, the appropriate system of the differential equations for the A–B binary alloy has the form [69, 117]:

$$\frac{dc_v^f}{dt} = G - Rc_i c_v^f - K_v c_i c_v^f \tag{3.82}$$

$$\frac{dc_i}{dt} = G - Rc_i c_v^f - \frac{(K_{AA}P_{AA} + 2K_{AB}P_{AB} + K_{BB}P_{BB})c_i^2}{2} \tag{3.83}$$

$$\frac{dc_B}{dt} = -(2K_{AB}P_{AB} + K_{BB}P_{BB})c_i^2 \tag{3.84}$$

The solution of the system (3.82)–(3.83) in the quasistationary approximation for the variation of the total concentration of vacancies in the loops (or tetrahedrons) in irradiation gives the dependence:

$$c_v^l = c_{v0}^l + athb(t-\tau) \tag{3.85}$$

The parameters a and b in equation (3.85) are equal to:

$$a = \left[\frac{2GK_v}{m(R+K_v)}\right]^{1/2}, \quad b = \left[\frac{mGK_v}{2(R+K_v)}\right]^{1/2} \tag{3.86}$$

The dependence of the variation of the concentration of component B in the solid solution as a result of its precipitation on the vacancy clusters due the flux of the dumbbells of the type A–B has the form [69, 116]:

$$c_B^m = \left\{c_{B0}^2 - \frac{4\alpha^2 K_{AB}\left[b(t-\tau) - thb(t-\tau)\right]}{b\theta}\right\}^{1/2} \tag{3.87}$$

Equation (3.87) shows directly that for the given mechanism of breakdown of the solid solution, there is an incubation period of the precipitation of component B in the zone of the vacancy loop or tetrahedron. This duration is approximately determined by the start of fulfilling the inequality: $b(t-\tau)thb(t-\tau)>0$. This is in complete agreement with the experimental data on the breakdown of non-saturated silver–zinc solid solutions [17]. As an example of the experimental dependences of the growth of the loops, Fig. 3.13 shows the results of calculation of the variation of the general concentration of the interstitial in the dislocation loops during irradiation of aluminium and a series of its binary alloys with magnesium,and zinc by electrons with an energy of 1 MeV in a high-voltage microscope at room temperature published in [69,114,116,118]. In this case, the selection of the type of alloying elements for investigations was also determined by the fact that the magnesium atoms are characterised by positive dilation volumes in the aluminium lattice [124], and the zinc atoms by negative volumes [124].

The experimental data and the relationships published in Fig. 3.13

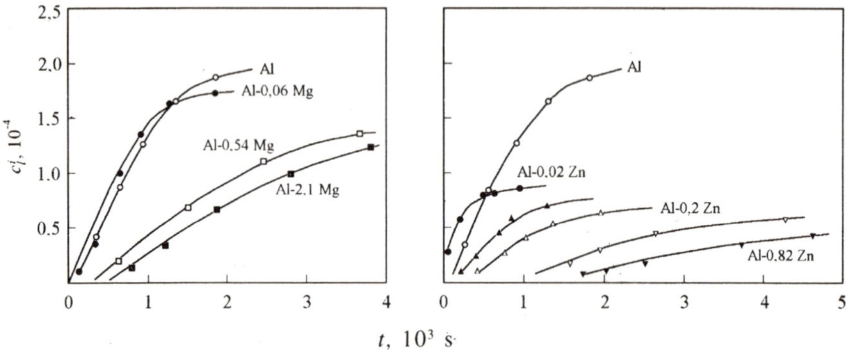

Fig. 3.13. Kinetic dependences of the variation of the toal concentration of interstitials in dislocation loops in electron irradiation of aluminium and its binary alloys with magnesium and zinc [69,114,116].

give the following main relationships:

1. All the alloys are characterised by the incubation period τ which is interpreted in [69,114,116] as the time of nucleation of dislocations within the framework of the modified classic theory of nucleation [81] (equation (3.39)).

2. In most cases, the solutes increase the density of the loops and decrease the diameter of the loops and the total concentration of the interstitials in the dislocation loops at saturation.

3. In the stage of growth of the loops, the density of the loops remains almost constant.

4. In the case of long irradiation times, the total concentration of the interstitials in the loops shows a tendency for saturation. This resulted in complete agreement with the dependences (3.62) and (3.75).

The continuous lines, drawn through the experimental points in Fig. 3.13 represent the results of processing the experimental data in accordance with the dependences (3.62) and (3.75). The concentration dependences of a number of parameters obtained by processing the experimental results are presented in Table 3.2 and Fig. 3.14. The parameter η' in Fig. 3.14 and in Table 3.2 is equal to [69,115,117]:

$$\eta' = \frac{GR_m}{K_{im}K_m\alpha^2} \tag{3.88}$$

Like parameter η [43], parameter η' characterises the degree of mutual recombination of the interstitials and vacancies.

Figure 3.13 shows that for all examined materials, the experimental

Table 3.2. Density of loops and parameters obtained in processing the experimental data in accordance with the dependence $c_i^l = c_{i0}^l + athb\,(t - \tau)$ [69,116]

Alloy, at.%	ρ_{sc}, m^{-3} (10^{21})	τ, s	a, (10^{-4})	b, s^{-1} (10^{-4})	I, m^{-2} $s^{-1}(10^{18})$	W
Al–99.99	0.66	145	2.03	10.1	4.5	74
Al–0.06Mg	1.50	113	1.72	12.8	13.2	45
Al–0.54Mg	2.00	470	1.48	5.3	4.2	47
Al–2.10Mg	2.00	565	1.44	4.0	3.5	44
Al–0.02Zn	1.50	35	0.83	43.5	42.8	32
Al–0.10Zn	1.60	166	0.77	13.9	9.6	36
Al–0.20Zn	2.00	360	0.68	9.3	5.5	37
Al–0.50Zn	3.10	1070	0.60	5.5	2.0	36
Al–0.82Zn	5.00	1620	0.57	4.4	3.0	31

Alloy, at.%	β_{0c} s^{-1}	K_v s^{-1}	η', (10^4)	t_s, s	$\bar{d}_S$ nm
Al–99.99	18.7	10.0	6	3255	41.0
Al–0.06Mg	8.8	15.0	5	2565	26.4
Al–0.54Mg	2.4	7.2	39	6400	21.2
Al–2.10Mg	1.7	5.6	73	8420	20.9
Al–0.02Zn	14.5	105.0	2	760	17.5
Al–0.10Zn	4.0	36.0	21	2426	16.3
Al–0.20Zn	1.9	27.0	62	3738	13.7
Al–0.50Zn	0.6	18.0	227	6834	10.3
Al–0.82Zn	0.3	15.2	396	8842	7.9

points are efficiently approximated by the dependences (3.62) and (3.75). Similar results were obtained [69,114,196] in processing a number of experimental literature data for the kinetics of growth of dislocation loops in aluminium [86], titanium [111], magnesium [125], Nb–Zr [19] and also 316-type stainless steel [83,126]. It is characteristic that the dependences of type (3.62) and (3.75) in the entire time period

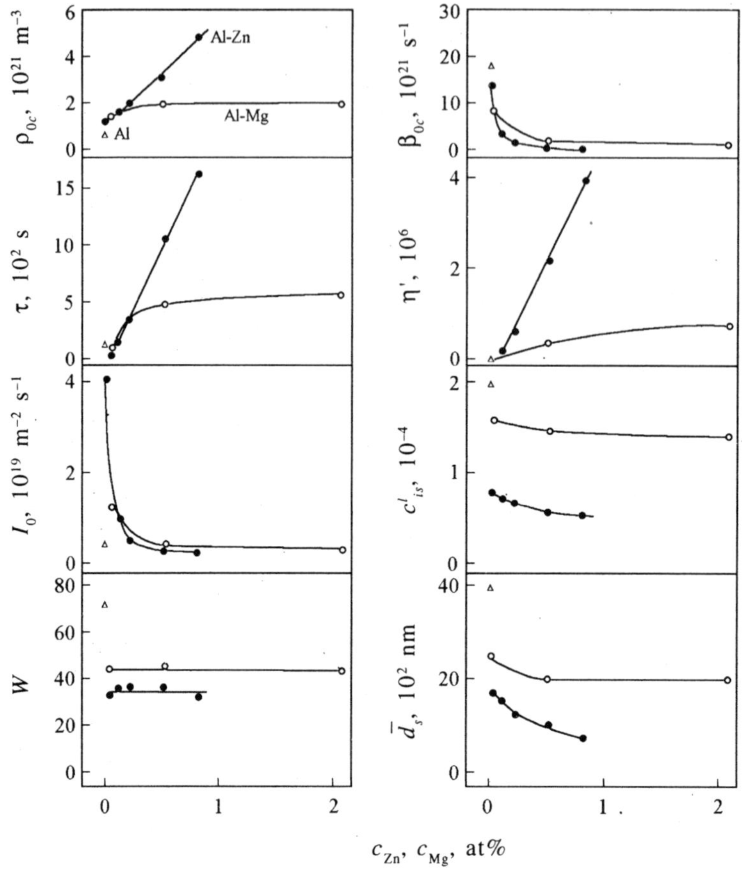

Fig. 3.14. Concentration dependences of the parameters of nucleation and growth of interstitial dislocation loops in aluminium and its binary alloys with magnesium and zinc [69,116].

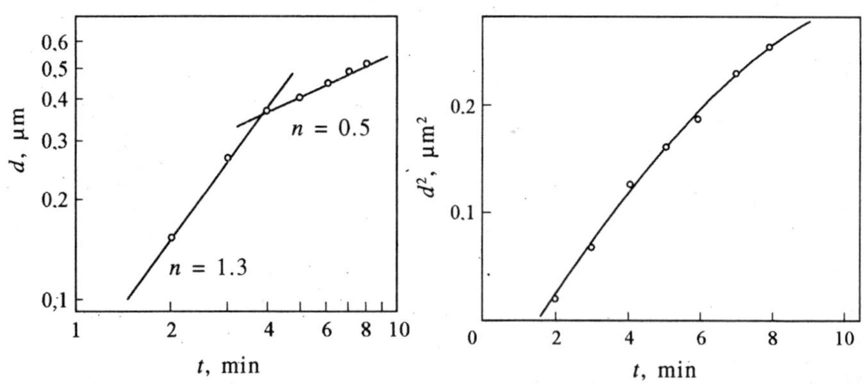

Fig. 3.15. Variation of the diameter of loops in Ti in electron irradiation [111] (a) and the results of processing these experimental data in accordance with the dependence $d^2 = d_0^2 + a' \, thb \, (t - \tau)$ (b).

also describe the experimental data for titanium (111) which, as already noted for the two time periods, were approximated in [111] by two dependences $d \sim t^{1.3}$ and $d \sim t^{0.5}$ (Fig. 3.15a). This is shown in Fig. 3.15b, where the experimental data from [111], presented in the form $d^2 \sim t$, were used to plot the most probable curve obtained in processing of experimental points in accordance with the dependence:

$$d^2 = d_0^2 + a' thb(t - \tau) \tag{3.89}$$

where $a' = \frac{4S_0}{\pi \Omega \rho}$, S_0 is the area per atom in the loop.

Equation (3.89) follows from (3.62) and (3.75) in transition from the variation of the concentration of interstitials during irradiation to the variation of the diameter of the loops. The resultant parameters a', b and τ in dependence (3.89) are equal to 3.45×10^{-1} μm^2, 2.5×10^{-3}s^{-1} and 99 s, respectively.

The equations (3.62) and (3.75) can, consequently, be examined as functions accurately describing the kinetics of buildup of defects in the process of growth of the loops during irradiation.

On the basis of these results, it is possible to draw the following conclusions:

1. The nucleation and growth of dislocation loops of the interstitial type during irradiation in all alloys take place mainly on the solute atoms, i.e. nucleation is heterogeneous. Solutes decrease the effective width of the energy barrier for nucleation W. In transition from pure aluminium to alloys, parameter W is independent of the concentration of the solutes. This indicates the controlling role of single atoms of the solutes in the process of nucleation of the loops.

2. The dependence of the rate of nucleation of the dislocation loops on the concentration of solutes is non-monotonic. In the case of very low concentrations of the alloying elements, the rate of nucleation increases and subsequently decreases, showing a tendency for saturation with further alloying.

3. In most cases, in all alloys, regardless of the decrease of the width of the energy barrier for nucleation, the incubation period increases. This is associated with an increase of the parameter of recombination mainly as a result of a decrease of the mobility of the interstitials during capture of these atoms by the solutes or the formation of less mobile mixed dumbbells.

4. The ambiguous nature of the variation of constant K_v is associated evidently with the concentration dependence of the diffusion coefficient

of the vacancies. The localisation of vacancies at the single solute atoms at low concentrations of these atoms increase the mobility of vacancies. A further increase in the concentration increases the probability of capture of the vacancies by two or more atoms of the solutes. This is accompanied by a decrease in their mobility.

5. All the atoms show an unambiguous tendency for the decrease of the mobility of the interstitials with increasing concentration of the solutes (parameter β_{0c}).

6. The tendency for the decrease in the rate of growth of the loops and decrease in the total concentration of the defects at saturation with increasing content of solutes are associated mainly with an increase of the degree of mutual recombination of the defects and with a decrease of the intensity of fluxes of the interstitials to the loops as a result of a decrease in the mobility of the interstitials.

Almost all parameters characterising the processes of nucleation and growth of the loops in solid solutions in irradiation in relation to the concentration of the solutes show a tendency for saturation. Consequently, in the development of radiation-resistant alloys, the alloying of solid solutions aimed at suppressing radiation damage may be efficient only up to specific concentrations of the solutes.

In conclusion, it is useful to mention an example confirming the correctness of the examined modelling considerations. It also describes the kinetics of growth of the dislocation loops (tetrahedrons) of the vacancy type.

Figure 3.16 shows the total concentrations of the vacancies in tetrahedrons of stacking faults (calculated from the experimental data in [70]) in the process of their growth during irradiation of an annealed

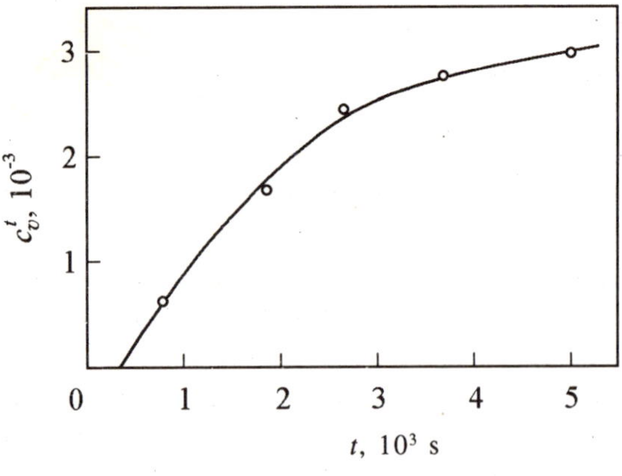

Fig. 3.16. Kinetics of variation of the total concentration of vacancies in tetrahedrons during electron irradiation of Ag–16 at% Zn alloy [69,117].

16 at% Zn alloy with electrons with an energy of 1 MeV at a temperature of ~20 °C (open points) [69,117]. The intensity of irradiation was 6.15×10^{22} m^{-2} s^{-1}. In the process of growth, the density of the tetrahedrons remains almost constant (~1.8×10^{22} m^{-3}).

The solid lines, drawn in Fig. 3.16 through the experimental points, represent theoretical dependence (3.85) with trial parameters for the given alloy τ = 550 s, $a = 3\times10^{-3}$ and $b = 5.7\times10^{-4}$ s^{-1}. Figure 3.16 shows that as in the case of interstitial loops, the modelling considerations [69,117] are also valid for describing the kinetics of growth of clusters of the vacancy type.

3.8. FORMATION AND GROWTH OF VOIDS IN PURE METALS AND ALLOYS UNDER IRRADIATION

The processes of formation and growth of voids in irradiated metallic materials start to play a significant role at temperatures of >0.3 T_m leading to one of the most negative phenomena in the operation of structural materials in atomic power systems, i.e. swelling. This temperature range coincides with the range of preferential departure of defects to sinks (section 3.5), i.e. the preferential conditions for the growth of clusters, dislocation loops and voids are realised.

The theory of homogeneous nucleation of the voids in pure metals under irradiation was developed in [92,93]. Within the framework of the model [92], the main equations with special reference of the formation of nuclei of radiation voids have the following form [92, 131]:

$$I_k = z'\beta_k N_0 \exp\left(-\Delta G'_k / kT\right) \tag{3.90}$$

$$\Delta G'_k = kT \sum_{n=0}^{n'_k - 1} \ln \left\{ \frac{\beta_i}{\beta_v} + \exp\left[-\ln S_v^s + \left(36\pi\Omega^2\right)^{1/3} \frac{\gamma}{kT} \left(n+1\right)^{2/3} - n^{2/3} \right] \right\} \tag{3.99}$$

$$z' = \left[-\frac{1}{2\pi kT} \left(\frac{\partial^2 \Delta G'_n}{\partial n^2} \right) \right]^{1/2} \tag{3.92}$$

$$\tau = \frac{1}{2\beta_k z'^2} \qquad (3.93)$$

In equations (3.90)–(3.93) z' is the Zeldovich factor inversely proportional to the width of the size distribution of the free energy of formation of nuclei $\Delta G'_n$, corresponding to height $\Delta G'_n\, kT$ units lower than the maximum distribution of $\Delta G'_k$ for a nucleus of the critical size n'_k, β_k is the rate of absorption of the vacancies by the nucleus of the void of the critical size; β_i/β_v is the ratio of the rate of absorption of the the interstitials and vacancies by the nuclei of the void; γ is surface energy, $S_v^s = c_v/c_v^0$ is the vacancy supersaturation, where c_v^0 is the dynamically equilibrium concentration of vacancies during irradiation, and their thermodynamically equilibrium value c_v^0 is determined by the equations:

$$c_v^0 = \exp\left(S_f / k\right)\exp\left(-E_v^f / kT\right) \qquad (3.94)$$

S_f and E_v^f is the entropy and energy of formation of the vacancies.

The current theoretical considerations regarding the growth of voids and swelling have been proposed usually on the basis of quasiequilibrium conditions:

$$S_v D_v c_v = S_i D_i c_i \qquad (3.95)$$

which correspond to stationary solutions of the system of equations (3.3)–(3.4). In this approximation, the rate of growth of the void with radius r_v has the following form:

$$\frac{dr_v}{dt} = \frac{\Omega}{r_v}\left\{Z_v^v(r_v)D_v\left[c_v - c_v^\theta(r_v)\right] - Z_i^v(r_v)D_i c_i\right\} \qquad (3.96)$$

where $c_v^e(r_v)$ is the equilibrium concentration of vacancies on the surface of a void with radius r_v:

$$c_v^e(r_v) = c_v^o \exp\left[-\left(p - \frac{2\gamma}{r_v}\right)\Omega / kT\right] \qquad (3.97)$$

In equation (3.97) c_v^0 is the thermodynamically equilibrium concentration of the vacancies determined by equation (3.94), p is the pressure of gas in the void, and γ is surface energy.

The main relationships governing the formation of voids, their growth and swelling have been developed in considerable detail for pure metals and alloys with gas impurities. The results of these investigations have been published in many original and review studies (see, for example [43,44,77,80,92,93,127–137]). As one of the very important results for pure metals, it is important to mention the correlation between the susceptibility to void formation and the coefficient of self-diffusion of metals established in [138]. With an increase in the coefficient of self-diffusion, the susceptibility to void formation decreases as a result of a decrease in the level of vacancy supersaturation. The results of this theoretical analysis are in agreement with the existing experimental data [138].

The results of a large number of investigations show that gas impurities support the nucleation, growth and stabilisation of voids. The interpretation of this effect is possible in particular on the basis of analysis of equation (3.97) in which the members p and $2\gamma/r_v$ represent the actual and equilibrium pressure of gas in a void with radius r_v. It may be shown that to maintain equilibrium pressure in the void with increasing void radius, the ratio between the number of vacancies n and the atoms of the gas m must continuously increase in accordance with the dependence [139],

$$\frac{n}{m} = \frac{kTr_v}{2\Omega\gamma} \qquad (3.98)$$

Thus, the excess pressure of gas in the void results in vacancy preference stimulating the inflow of vacancies to its surface and, consequently, the system moves closer to thermodynamic equilibrium. This conclusion follows directly from equation (3.97). The excess pressure of gas in the void decreases the equilibrium concentration of vacancies on its surface c_v^e resulting in an increase in the vacancy gradient of the matrix–void system and, consequently, an increase in the intensity of the flux of the vacancies to the nucleating or growing void. These considerations are in complete agreement with the currently available experimental data.

The effect of non-gas impurities (metallic and non-metallic) in metals and alloys and of the phase and chemical composition of the components of alloys on the processes of nucleation and growth of voids is ambiguous, and their correct interpretation and prediction of defects are not always possible with the framework of the existing modelling considerations. To a large degree, this is associated with both the complicated nature of the given problem for alloys, in which it is necessary to take into account the superimposition of different process-controlling parameters, and also with the application of insufficiently correct or simplified assumptions in a number of models. In this case, it is important to note that there have been only a small number of relatively systematic experimental investigations of the effect of the type and concentration of impurity and alloying elements on the nucleation and growth of voids identical with the results on the formation and evolution of dislocation loops in alloys, investigated in the previous section. This complicates the correction and further development of modelling considerations in the given direction on the basis of comparison of the results of theory and experiments.

Prior to direct analysis of the processes of the nucleation and growth of voids in metals and alloys containing non-gas impurities, it is important to mention one universal feature of the effect of solutes: small concentrations of these elements affect void formation and swelling not only in relatively pure metals and diluted solid solutions but also in concentrated alloys. This has been shown convincingly on an example of special experiments (see, for example [140,141]) in which more extensive swelling was detected in concentrated alloys melted produced by melting pure components, in comparison with industrial alloys of the same composition.

3.8.1. Nucleation of voids in alloys

The effect of non-gas impurities in the nucleation of voids was investigated for the first time in [142]. Within the framework of the proposed model it was shown that in heterogeneous nucleation of a void on an impurity atom with binding energy $E_v^B(n)$, the rate of nucleation increases as a result of a decrease in the free energy of formation of a nucleus with the multiplicity n by the value:

$$\Delta G(n) = -\left(E_v^B(n) + kT \ln zc_t\right) \tag{3.99}$$

In equation (3.99), z is the coordination number characterising the number of equivalent areas on the surface of the void.

The mobile surface-active impurity also reduces the energy of formation of a nucleus as a result of a decrease of the surface energy

β (equation (3.91)) and, consequently, increases the rate of the void formation process.

In a number of subsequent investigations [143–147], the model proposed in [92] was used for numerical calculations of the effect of the variation of the diffusibility of point defects during the capture by the solute atoms on a number of parameters of void nucleation. In these calculations, the concentrations of the free point defects and of their complexes with the atoms of the solutes were estimated on the basis of stationary solutions of the rate equations of the type (3.3)–(3.4) in which, is necessary, the reactions of the defect–impurity interaction (formation and dissociation of complexes) were taken into account.

In the calculation of the nucleation parameters in [143], the investigators took into account only the concentration of free vacancies, and the vacancies captured by the impurity traps were not investigated. This approach resulted in a decrease of vacancy supersaturation S_v^s in equation (3.91) and, consequently, in an increase of the size of the critical nucleus n_k, an increase of the free energy of nucleation ΔG_k and, on the whole, in a decrease of the rate of nucleation I_k.

Identical results were also obtained in [144,145]. In these calculations, the traps were regarded as singly saturated, the corresponding defect–impurity complexes as immobile. In fact, these assumptions, like those in [143], excluded the defect–impurity complexes from the nucleation process. Within in the framework of the model [144], the calculated free energy of formation of the nuclei of vacancy voids increased with increasing binding energy of the interstitials with solutes as in the case of vacancy capture. Thus, the process of nucleation of the voids in [144,145], in accordance with the results of calculations are carried out in these investigations, was suppressed with a decrease of the mobility of both interstitials and vacancies.

The relationships obtained in [144,145] are interpreted on the basis of the intensification of mutual recombination of the point defects as result of a decrease of their diffusibility and an increase of the defect supersaturation. At the same time, an increase of defect supersaturation with a decrease of the diffusibility of point defects in addition to recombination also stimulates the process of nucleation of clusters. In a general case, the resultant effect may be determined by the selection of calculation parameters and, in particular, the power of dislocation sinks, irradiation temperature and intensity. If this problem of interpretation of the results is the subject of discussions, the exclusion of defect–impurity complexes from examination in the calculation of the process of nucleation, as carried out in [143,144], is according to the view of the authors of this book, completely unjustified.

Firstly, in the light of the available experimental and theoretical results, the impurity traps cannot be regarded as singly saturated. The formation of defect clusters, increase of their concentration with increasing concentration of solutes and subsequent growth appear to indicate that the atoms of the solutes are non-saturated traps. This is confirmed, for example, by the experimental data on the nucleation and growth of dislocation loops in aluminium-based alloys (section 3.7.2). It is also unjustified, in the evaluation of the dynamically equilibrium concentration of point defects, to exclude completely the point defects, trapped by the impurities, as carried out in [143]. In evaluation of this type, it is more accurate to calculate the diffusion properties of defects taking into account their interaction with solutes either with the framework of frequency models, or using considerations regarding the effective coefficients of diffusion (see Chapter 2).

Using the model proposed in [92], the authors of [146] carried out identical evaluations of the effect of the mobility of interstitials atoms and vacancies on the rate of void nucleation. In the calculations carried out in contrast [140, 144], diffusibility of the interstitials and vacancies was calculated directly from the formulae for the conventional coefficients of diffusion using the changing activation energy of the migration of point defects, simulating other same time, the capture of defects by the solute atoms. The selected initial parameters of migration of the point defects are characteristics of pure nickel. In addition to this, in accordance with [49], it was taken into account that the dislocation density ρ_d is an increasing function of the migration energy of interstitials E_i^m (as E_i^m increases, the value of ρ_d also becomes higher) which actually reflects the formation of dislocations structure in the period of homogeneous nucleation and growth of dislocation loop, preceding swelling.

Results of calculations are presented in Fig. 3.17 (a,b,c). Figure 3.17b shows directly that for the case of a decrease of the mobility of vacancies, the results are opposite of the calculated data from [143–145], i.e. in this case, the rate of nucleation, in contrast to [143-145], increases with a decrease of diffusibility. A decrease of diffusibility of the interstitials results in the suppression of void formation (Fig. 17a). Figure 3.17c also shows that the effect of a decrease of vacancy mobility on the process of void nucleation is stronger than a decrease of the diffusibility of the interstitials. The results of the calculations carried out in [146] also confirm the data calculated by Wolfer and Garner [147]. The study [147] was based on the assumptions

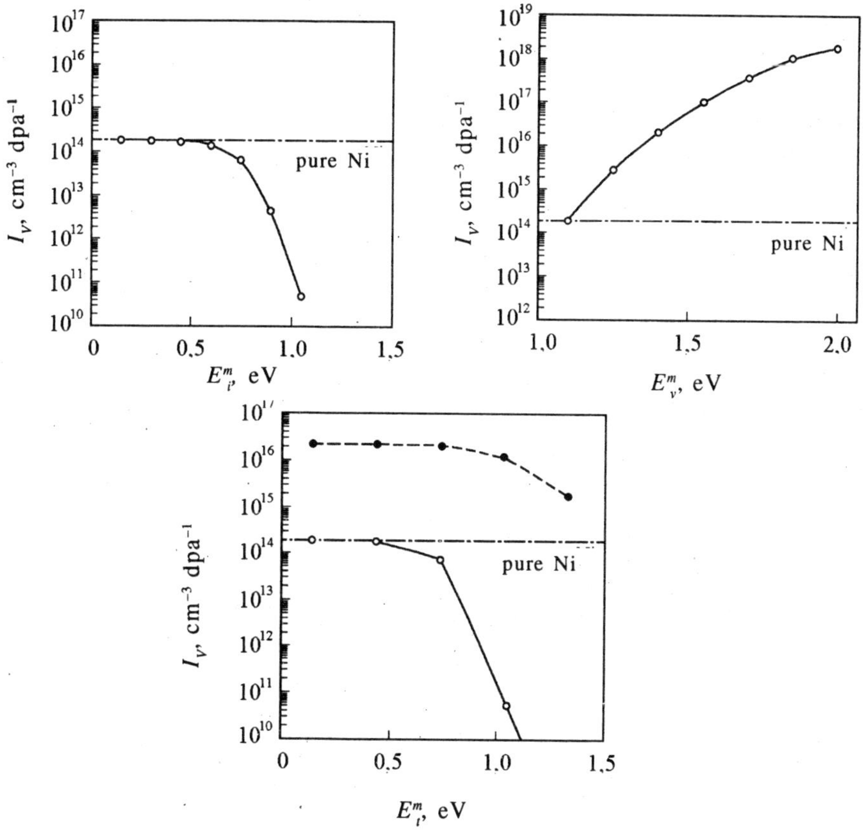

Fig. 3.17. Calculated dependence of the rate of nucleation of voids I_v on E_v^m [146]: a) E_v^m= 1.10 eV; b) E_i^m = 0.15 eV; c) E_v^m = 1.10 eV (o) and E_v^m = 1.40 eV (•).

on the possibility of acceleration of the diffusion of vacancies in an alloy during formation of their complexes with a rapidly diffusing impurity. In this case, the authors used the following expression for the effective coefficient of diffusion of vacancies:

$$D_v^{eff} = \frac{D_v + Kc_s D_s}{1 + Kc_s} \tag{3.100}$$

In formula (3.100), D_v is the diffusion coefficient of free vacancies, D_s is the diffusion coefficient of complexes, c_s is the concentration of complexes, and the value of K is determined by the expression:

$$K = 12\exp\left(E_v^B / kT\right) \tag{3.101}$$

where E_v^B is the binding energy in the vacancy–atom complex with the dissolved element. In numerical calculations, the value of D_v was represented by the diffusion coefficient of vacancies in nickel:

$$D_v = 0.015\exp\left(-1.4 \text{ eV} / kT\right) \tag{3.102}$$

The following dependence was used as the diffusion coefficient of complexes D_s:

$$D_s = 0.015\exp\left(-1.2 \text{ eV} / kT\right) \tag{3.103}$$

Figure 3.18 shows the relative rates of nucleation I/I_0, normalised in relation to the rate of nucleation of voids in pure nickel, in relation to the concentration of the fast-diffusing impurity at different temperatures for the migration energy of the impurity E_s^m = 1.2 eV and E_v^B = 0.05 eV.

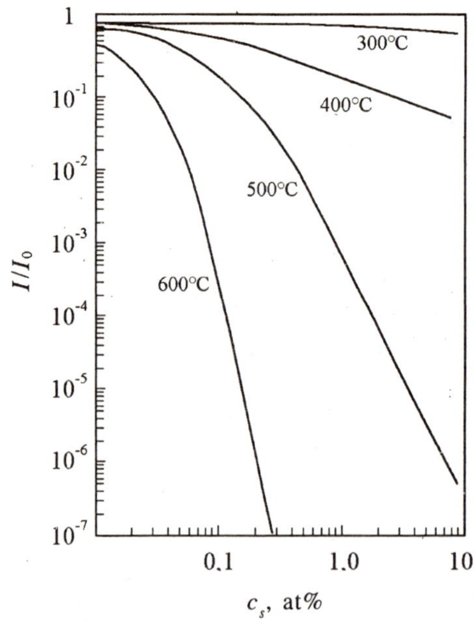

Fig. 3.18. Calculated relative rates of nucleation of voids in nickel at different temperatures in relation to the concentration of rapidly diffusing impurity c_s for E_s^m = 1.2 eV and E_v^b = 0.05 eV [147].

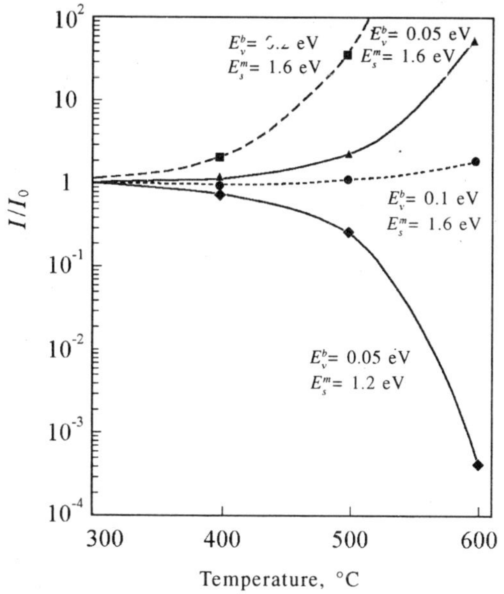

Fig. 3.19. Calculated temperature dependences of the relative rates of nucleation of voids in a nickel alloy with 1 at% of a dissolved element for different energies of migration of vacancy–impurity pairs E_s^m and their binding energy E_v^b [147].

In Fig. 3.19, the identical values of I/I_0 are presented in relation to temperature for different values of E_s^m and E_v^B.

The results of the calculations presented in [147] indicate that the acceleration of diffusion of vacancies in an alloy results in a significant increase of the free energy of formation of void nuclei, especially at high temperatures, and, consequently, in a decrease of the rate of nucleation. This conclusion, like the results of calculations carried out in [146], are in complete correlation with both theoretical and experimental data, presented for pure metals in [138]. In [138], as mentioned previously, it is shown that an increase of the coefficient of self-diffusion results in a decrease of the susceptibility to void formation as a result of a decrease of the value of vacancy supersaturation. Evidently, this explanation is also applicable to the results of [146,147].

Figure 3.19 also shows that in an alternative case, i.e. when the mobility of vacancies decreases, the rate of nucleation of void increases. This is also in agreement with the results published in [146] and contradicts the calculated data [143–145] examined previously. Analysis of possible reasons for the incorrect results of the calculations in [143–145] was already carried out previously in this book.

Because of the importance of the examined problem and the ambiguous nature of the results of [143–147], it is useful to carry out additional analysis of the effect of the variation of the diffusibility of point defects during their capture by the solute atoms on the nucleation of voids within the framework of a purely analytical approach in which the non-adequacy of the selection of the parameters in numerical calculations is prevented.

Equations (3.90) and (3.91) show that the main parameters, controlling the process of nucleation of voids in addition to surface energy γ are the values $\beta_k(\beta_{v,i})$, β_i/β_v and $S_v^s = c_v/c_v^0$. The rates of absorption of vacancies and interstitials by the nuclei of voids $\beta_k(\beta_{v,i})$ are proportional: $\beta_{v,i} \sim c_{v,i} D_{v,i}$, where $c_{v,i}$ is the same as in the equation for vacancy supersaturation, S_v^s are the dynamically equilibrium concentrations of point defects. Attention will be given to the functional dependence of the quantities $\beta_{v,i}$, β_i/β_v and S_v^s on $D_{v,i}$ for the three main mechanisms of buildup and annealing of radiation point defects.

1. The mechanism of mutual recombination

For the case in which the mutual recombination of point defects is dominant, i.e. the annealing of defects on fixed sinks is of no practical importance, the stationary solutions of the system of equations (3.3)–(3.4) give:

$$c_{v,i} = \left(\frac{G}{R}\right)^{1/2} \tag{3.104}$$

Consequently:

$$\beta_{v,i} \sim c_{v,i} D_{v,i} = D_{v,i} \left(\frac{G}{R}\right)^{1/2} \tag{3.105}$$

$$\frac{\beta_i}{\beta_v} = \frac{D_i}{D_v} \tag{3.106}$$

For the case of a decrease of D_v in the capture of vacancies by the solute atoms and on the condition that D_i is considerably higher than D_v (in this case, the constant of mutual recombination R in equation (3.104) is almost completely independent of D_v), β_v decreases, S_v^s =

c_v/c_v^0 does not change, and β_i/β_v increases. This results in an increase of ΔG and, on the whole, in a decrease of the nucleation rate. An increase of D_v results in an alternative result.

The identical analysis for the case of a decrease of the mobility of the interstitials in capture of these atoms by the atoms of the solutes indicates an increase of the nucleation rate.

The examined mechanism is more of theoretical interest because it dominates at temperatures below transition temperature T_t (equation (3.24), Table 3.1), and extensive swelling is detected at temperatures higher than this temperature at which the combined or linear mechanism is most realistic.

2. *The combined mechanism*

This mechanism operates when a significant role is played by the processes of mutual recombination and annealing of point defects on fixed sinks (equations (3.3) and (3.4)). In accordance with equations (3.16)–(3.21), for the quantities $\beta_c, \beta_{v,i}$ and β_i/β_v in this case:

$$c_v = \frac{1}{2}\frac{S_i D_i}{R}\left[(1+\eta)^{1/2} - 1\right] \tag{3.107}$$

$$\beta_v = \frac{1}{2}\frac{S_i D_i D_v}{R}\left[(1+\eta)^{1/2} - 1\right] \tag{3.108}$$

$$\beta_i = \frac{1}{2}\frac{S_v D_i D_v}{R}\left[(1+\eta)^{1/2} - 1\right] \tag{3.109}$$

$$\frac{\beta_i}{\beta_v} = \frac{S_v}{S_i} \tag{3.110}$$

For the case in which the capture of vacancies by the solute atoms results in a decrease of D_v in accordance with the equations (3.107)–(3.10), the value of β_v decreases, the value of β_i/β_v does not change, and S_v^s increases, decreasing ΔG. An increase of c_v and, correspondingly, of S_v^s is associated in this case with an increase of parameter η with a decrease of D_v (equation 3.22). Since the functional dependence of the nucleation rate on $\Delta G(D_v)$ in comparison with $\beta_v(D_v)$ in equation

(3.90) is dominant, in this case the intensity of the nucleation process increases. If D_v increases at capture, the value of S_v^s decreases, increasing ΔG and decreasing the rate of nucleation. This conclusion is in complete agreement with the results of theoretical analysis and experimental data [138] and numerical results of [146,147].

The capture of interstitials by solutes for the given mechanism should not affect the nucleation rate of voids until the value of D_i becomes equal to or lower than D_v. Because of equation (3.107), the realisation of this condition decreases the vacancy supersaturation S_v^s and, consequently, decreases ΔG and also the nucleation rate. This result is also in qualitative agreement with the data of numerical calculations [146].

3. The linear mechanism

This mechanism operates when the annealing of point defects on sinks is dominant and the process of mutual recombination plays no significant role. In practice, the linear mechanism may be realised in deformed metals with a high dislocation density at relatively high irradiation temperatures.

In this case, solving the system of equations (3.3)–(3.4) gives:

$$c_{i,v} = \frac{G}{S_{i,v} D_{i,v}} \tag{3.111}$$

Consequently:

$$\beta_{v,i} = \frac{G}{S_{v,i}} \tag{3.112}$$

$$\frac{\beta_i}{\beta_v} = \frac{S_v}{S_i} \tag{3.113}$$

For the case of decreasing D_v at capture of vacancies by the solute atoms, the values of β_v and β_i/β_v do not change, and S_v^s increases because of equation (3.111), decreasing the value of ΔG and, consequently, increasing the nucleation rate. If the value of D_v increases at capture, S_v^s decreases, ΔG increases, and the nucleation rate decreases. The identical effect is also observed in the combined mechanism and is also in agreement with the results of [138,146,147].

If the diffusibility of interstitials decreases at capture, then because of equation (3.113), this has no effect on the nucleation parameters.

On the whole, the analysis has confirmed the reliability of the numerical calculations carried out in [146,147]. Its results are in agreement with both the theoretical and experimental data, presented in [138]. It is important to note that the previously established relationships were obtained without using any calculation parameters and, consequently, there are no doubts regarding their accuracy and realistic nature.

Attention will be given to another important aspect, directly associated with the theory of nucleation of voids in alloys which has not been taken into account in any of the studies [143–147] nor in our previous analysis. The problem is that in alloys containing solutes, the thermodynamically equilibrium concentration of vacancies is higher than in the matrix pure metals and should be determined by means of the expression for binary solid solutions [148] and not in accordance with equation (3.94):

$$c_v^t = A_v\left[1-(z+1)c_t\right]+A_{tv}zc_t\exp\left(E_v^B/kT\right)\exp\left(-E_v^f/kT\right) \qquad (3.114)$$

In equations (3.114), A_v and A_{tv} are the pre-exponential factors for the formation of vacancies and impurity–vacancy complexes in the pure metal and alloy, respectively, z is the coordination number, E_v^f is the energy of formation of vacancies in a solvent, E_v^B is the impurity–vacancy binding energy.

In [149], in particular, this factor has been taken into account in the development of a model of the growth of voids taking the effect of segregation processes into account.

In the calculations carried out in [143–147] and in similar calculations, ignorance of this factor results in excessively high values of vacancy supersaturation S_v^s and, consequently, reduced values of ΔG in equation (3.91) and, consequently, in inaccurate estimates of the nucleation parameters. This will be shown on the example of estimates of S_v^s for the combined mechanism of buildup and annealing of point defects and for the case examined by the authors of the present book previously and in [143–147], where the capture of vacancies by the atoms of these elements resulted in a decrease of D_v. In accordance with the analysis and results of calculations [146,147], in this case the value of S_v^s increases, the value of ΔG decreases and the rate of void nucleation increases.

Figure 3.20 shows the characteristic examples of the numerical calculations of the values of ln S_{v1}^s = ln (c_v/c_v^0) and lnS_{v2}^s = (c_v/c_v^t) for irradiation temperatures of 400–550 °C of nickel with the concen-

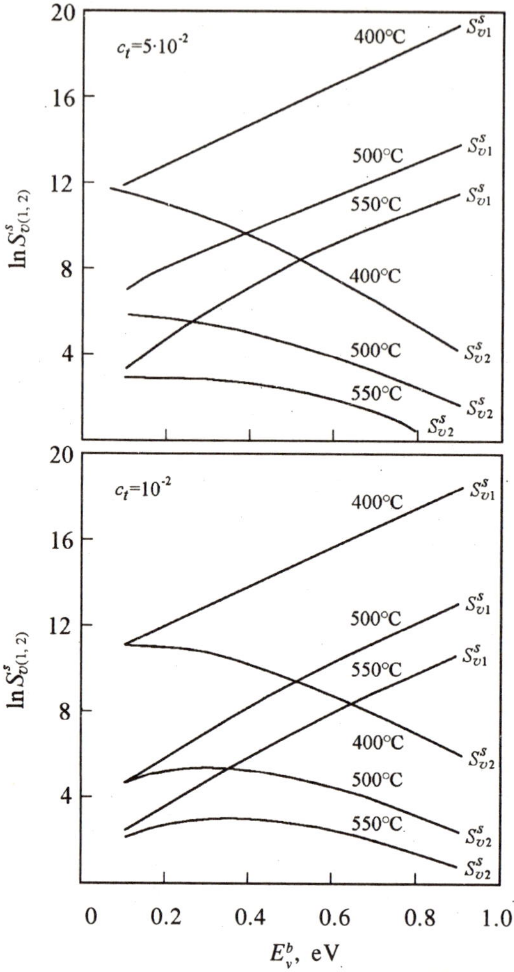

Fig. 3.20. Calculated values of vacancy supersaturation S^s_v in nickel for different concentrations of solutes c_t and their binding energy with vacancies E^b_v.

tration of solutes of 10^{-2} and 5×10^{-2}. The dynamically equilibrium concentrations of the vacancies c_v in the matrix were calculated using equations (2.6) and (3.20) and numerical parameters from Table 3.1. The thermodynamically equilibrium concentration of the vacancies c^0_v and c^t_v were evaluated using equations (3.94) and (3.114), respectively.

These results are not trivial. In contrast to the generally accepted considerations, corresponding to the nature of the variation of the vacancy supersaturation S^s_{v1} from Fig. 3.20, the values of S^s_{v2}, calculated taking into account equation (3.11 4) either do not change or decrease with an increase of the binding energy of the vacancies with the atoms of the solutes. For the case of nucleation of voids examined in our work, these results introduced a significant correction. Analysis using equations (3.107)–(3.110) shows that with a decrease of the diffusion coefficient D_v in capture by the solute atoms the rate of nucleation of voids may not increase (as shown previously us and obtained in the calculations in [146,147]) or remains constant or even decreases. The specific result, as indicated by Fig. 3.20, depends on the irradiation temperature and also on the concentration of the solute atoms and on their binding energy with the vacancies. This result is, in our view, of considerable practical importance because it shows that in alloying with elements, decreasing the mobility of vacancies, the void formation process may also be suppressed.

As indicated in [150], the nucleation of voids may be strongly affected by the nonequilibrium segregation of solutes on the surface of void nuclei. In this case, the effect of nonequilibrium segregation, stimulating the growth of the nucleus, is determined by the driving force which tries to decrease the degree of nonequilibrium segregation as a result of increasing the surface area of the void during vacancy absorption. This nucleus is resistant to healing because a decrease in the surface area may result in an increase in the degree of nonequilibrium segregation, i.e. in an increase of the surface energy of the void.

Figure 3.21 shows the results of numerical calculations of the variation of the effective barrier for the nucleation of voids at the saddle point Φ_s in relation to the segregation parameters $\lambda = \omega_{iB}\omega_{vA}/\omega_{iA}\omega_{vA}$, carried out for the A–B binary alloy with the framework of

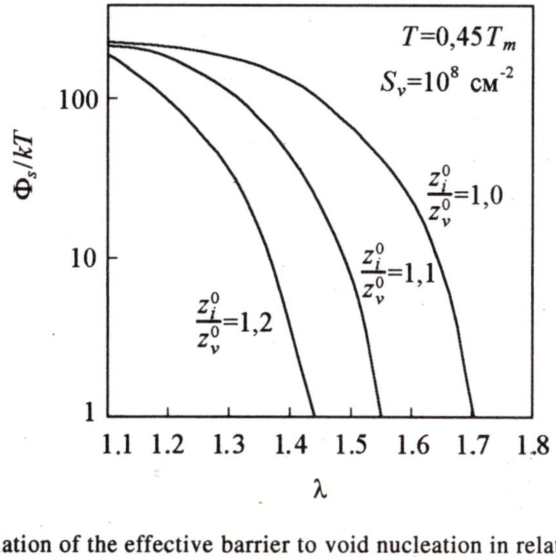

Fig. 3.21. Variation of the effective barrier to void nucleation in relation to segregation parameter λ and relative efficiency of capture of point defects by a void Z_i^o/Z_v^o [150].

the segregation model, proposed in [150]. The frequencies ω_{ij} in the segregation parameter λ are the frequencies of atomic jumps of the components A and B by the vacancy and interstitial mechanisms. At $\lambda > 1$ segregation of the solutes B on the sink takes place. The numbers at the curves in Fig. 3.21 characterise the efficiency of capture of the interstitials and vacancies by the void.

To a certain extent, the effect of nonequilibrium segregation, stabilising and stimulating the growth of the nucleus, is identical to the effect of helium examined in the introductory part of this section. At the same time, as claimed quite justifiably in [150], nonequilibrium segregation should increase the rate of nucleation of voids in early stages of irradiation in comparison with transmutation helium and, consequently, may prove to be a more important factor than helium, especially in early nucleation stages.

In single-phase systems, susceptible to radiation-stimulated ordering and breakdown, and in two-phase and more complicated alloys, there is an additional number of factors affecting the nucleation and growth of voids and clusters of defects as a whole.

The intensification of phase changes in irradiation of the type of ordering and breakdown of the solid solutions is possible, as shown in [151,152], not only as a result of the radiation intensification of diffusion but also as a result of the compensation, by the point defects, of coherent strains, accompanying the transformation processes. This mechanism decreases the dynamic concentration of the point defects and results in the suppression of radiation damage, including void formation. The growth of phases with the matrix–phase volume discrepancies of different sign is also accompanied by the annihilation of point defects of a specific type for the compensation of appropriate strains [153]. Evidently, in order to suppress void formation, the volume per atom in this hase should exceed the corresponding value for the matrix. A positive role in decreasing the dynamic concentration of point defects and in suppressing, in particular, void formation is played by the mechanism of mutual recombination at the phase boundaries of precipitates, acting as sinks for point defects [93,153,154]. At the same time, it should be mentioned that a specific type of precipitate may support the nucleation and growth of voids in the vicinity, concentrating the vacancies on the matrix-particle interfacial surface [153,155]. Analysis of the experimental data within the framework of the previously discussed mechanisms has been carried out, in particular, in [151–158].

3.8.2. Growth of voids in alloys

Expressing in equation (3.96) c_v and c_i by means of function $F(\eta)$ (equation (3.19)) for temperatures at which the emission of vacancies from voids is insignificant, the equation may be presented in the following form [149]:

$$\frac{dr_v}{dt}=\frac{\Omega F(\eta)G}{r_v S_v S_i}\left(Z_i^d Z_v^v - Z_v^d Z_i^v\right)\rho_d^e \tag{3.115}$$

In equation (3.115) $S_{v,i}$ is the power of the sinks for the vacancies and interstitials; $Z_{i,v}^d$ is the force efficiency of the dislocations with respect to the capture of point defects; $Z_{i,v}^v$ are the identical factors for the void; ρ_e^d is some effective density of the dislocations, which takes dislocations of all types into account.

The effect of impurities and solutes on the rate of void growth may be manifested through the following values in equation (3.115).

1. Function $F(\eta)$. It characterises the rate of mutual recombination or departure of defects to the sinks and changes in the range from 0 to 1. If $F(\eta)\to 0$, the growth of voids is the result of strong mutual recombination is greatly restricted. If $F(\eta)\to 1$, the recombination of defects has almost no effect on the rate of growth of the voids.

The effect of $F(\eta)$ on the rate of growth of the voids is manifested in accordance with equations (3.18) and (3.90) through the variable η (the recombination parameter):

$$\eta=\frac{4RG}{D_v D_i S_v S_i}=\frac{16\pi r\left(D_v+D_i\right)G}{D_v D_i S_v S_i} \tag{3.116}$$

As the value of η increases, the fraction of the point defects annihilated as a result of their mutual recombination increases, the value of $F(\eta)$ in equation (3.19) decreases and the swelling rate also decreases.

The solutes may increase the value of η also in the following manner:

a) decrease the diffusion coefficients D_v and D_i in (equation 3.116), capturing the point defects. Equation (3.116) shows that the increase of η and the increase of the rate of mutual recombination should be affected more efficiently, at least in the case of diluted alloys ($D_i \gg D_v$) by the capture of vacancies;

b) decrease the power of the sinks $S_{i,v} = Z_{i,v}\rho_{i,v}$, depositing on them, and decreasing at the same time their efficiency $Z_{i,v}$ ('poisoning' of the sinks).

2. The factor of preference of the sinks $Z_i^d Z_v^v - Z_v^d Z_i^v$. It characterises the preferential absorption of interstitials by dislocation sinks, and absorption of the vacancies by the voids in comparison with alternative processes. If sinks are 'clean', then $Z_i^d Z_v^v > Z_v^d Z_i^v$, because the interaction of dislocations with the interstitials is stronger than with the vacancies, and to a first approximation, the void is a neutral sink. The solutes may change these relationships in the following manner:

a) settling on the dislocations, they can decrease, suppress, and possibly can change the sign of preference.

b) settling on the voids, they may introduce preference either for the vacancies or for interstitials.

Thus, as a result of their effect on preference, the rate of growth of the voids may decrease, becoming equal to 0 and possibly change the sign.

We shall examine the role of each process in greater detail.

Equation (3.116) shows that one of the conditions for the maximum recombination (the value of η is maximum) is the condition of the equality of the diffusion coefficients of the vacancies and interstitials, or in the presence of traps – their effective diffusion coefficients $D^e_{v,i}$ (formula (2.6)).

For the case of strong capture, formula (2.6) gives:

$$D^e_{v,i} = \frac{D_{v,i}}{4\pi r_t^{v,i} c_t^{v,i} b^2 \exp\left[\left(E^m_{v,i} + E^B_{v,i}\right)/kT\right]} \tag{3.117}$$

For the energy condition of fulfilling the maximum of recombination, equating D^e_v to D_i^e, from equation (3.117) we obtain:

$$E_i^m + E_i^B = E_v^m + E_v^B - kT \ln \frac{D_v^0 r_t^i c_t^i}{D_i^0 r_t^v c_t^v} \tag{3.118}$$

In addition to the equality $D^e_v = D^e_i$, equation (3.118) is also the condition of equally probable participation of the vacancies and interstitials in the process of mutual recombination at capture by traps.

Equation (3.118) differs from the generally accepted equation (3.119) (see, for example, [145]):

$$E_i^m + E_i^B = E_v^m + E_v^B - kT \ln \frac{D_v^0 r_t^i}{D_i^0 r_t^v} \tag{3.119}$$

by the term:

$$kT \ln\left(\frac{c_t^i}{c_t^v}\right) \tag{3.120}$$

In this case, it is evident that relationship (3.119), in contrast to (3.118), is restricted by the following assumptions:

1) either the alloy contains only one type of trap for both interstitials and vacancies;

2) either the type of traps for the interstitials and vacancies differs but their concentrations identical.

Equation (3.118) is more accurate and as a result of the ratio of the concentrations it makes it possible to vary the binding energies of the traps in a wider range. At a high ratio of the concentrations and relatively high irradiation temperatures which may be reached in the case of refractory metals, the concentration correction (3.120) made reach several tens of eV, i.e. be comparable with the value of E_v^B.

A higher ratio c_t^i/c_t^v may, in particular, be realised if the impurity with a high value of E_v^B in high concentrations is either insoluble in the given alloy or is unacceptable because of technological or service requirements, and the impurity with the high value of the E_i^B is one of the alloying components of the alloys.

There is also a second condition for the realisation of equally probable participation of point defects in the process of mutual recombination. It is based on increasing the effective mobility of the vacancies during their interaction with rapidly diffusing components of the alloy [147,159], with a simultaneous decrease of the mobility of the interstitials at capture by the traps of solutes, i.e., as previously, the ratio $D_v^e = D_i^e$ should be satisfied. The effect of the increase of D_v with the introduction of a rapidly diffusing impurity on the nucleation of voids was examined by us in the previous section, taking also into account the results of [147].

This mechanism widens even more appreciably the possibilities of selecting alloying elements for the realisation of the conditions of equally probable participation of the point defects in the process of mutual recombination.

Theoretically, the possibility of reaching the conditions of equally probable participation of the point defects in the process of mutual recombination (the equivalence condition) evidently the most effective method of suppressing structure-phase instability and radiation damage of metallic materials during irradiation. The maximum effect may be obtained in this case either on the condition of complete absence of sinks for point defects or on the condition of their complete neutralisation.

Experimentally, the first condition may be fulfilled in the irradiation of efficiently annealed and relatively thick metallic foils in a high-voltage electron microscope. For the given case, the realisation of the condition, similar to the equivalence condition, was observed in particular in electron irradiation of an aluminium–zinc binary alloy with the zinc concentration close to 1 at.%. In this case, examination showed the nucleation and growth of vacancy dislocation loops, whereas at lower zinc concentrations dislocation loops of the interstitial type formed and grew [71, 160].

In more realistic cases, the irradiated materials almost always contain sinks of different type and the maximum suppression of radiation damage as a result of reaching the equivalence conditions may be realised only in combination with their neutralisation.

The fluxes of point defects to the dislocation sinks may lead to the enrichment of their surroundings by the impurity and alloying elements changing in this case their efficiency $Z^d_{i,v}$ for the capture of vacancies and interstitials and consequently, the growth rate of the voids.

As already mentioned, to a first approximation, a 'clean void' is usually regarded as a neutral sink and, consequently, $Z^v_v \cong Z^v_i$.

From equation (3.11 5) under this condition, we obtain:

$$\frac{dr_v}{dt} = \frac{\Omega F(\eta) G}{r_v S_v S_i} \left(Z^d_i - Z^d_v \right) Z^v_v \rho^e_d \qquad (3.121)$$

Equation (3.121) shows that the rate of growth of the voids decreases with a decrease of dislocation preference Z^d_i–Z^d_v, and when Z^d_i–$Z^d_v = 0$, it is completely interrupted. If the condition $Z^d_i < Z^d_v$ can reached, the voids may also 'heal'.

In accordance with the equations (3.11) and (3.14), the diffusion fluxes of the interstitials and vacancies to the dislocations are divided: the interstitials ($\Delta V > 0$) are preferentially absorbed in the field of expansion at the dislocation core, where ($\pi<\theta<2\pi$), and the vacancies ($\Delta V < 0$) in the compression region ($\pi < \theta < 0$). For the impurity atoms with $\Delta V > 0$ and $\Delta V < 0$, the favourable regions of segregations at the core of the dislocations are identical. Thus, it is possible to screen selectively the force potential of the dislocation in relation to the interstitials and vacancies.

If the impurities segregate in the vicinity of the dislocation core, they decrease the hydrostatic pressure of the dislocation at the point (r,θ) in accordance with the relationship [161]:

$$\Delta P = \frac{2\mu b \sin\theta}{3\pi r} - \frac{8}{3}\mu \Delta V n \qquad (3.122)$$

where n is the density of the impurity atoms in the atmosphere.

The equation for the concentration of the impurities, fully screening the elastic field of the dislocations, has the following form, as shown in [161]:

$$c_t = \frac{2\Omega}{\pi r_c^2} \int_0^{r_c} \int_0^{\pi} \frac{b \sin\theta}{4\pi \Delta V_t} \left(\frac{1}{r} - \frac{1}{r_c} \right) r dr d\theta = \frac{b\Omega}{2\pi^2 r_c \Delta V_t} \qquad (3.123)$$

in equation (3.123) b is the Burgers vector.

At $r_c = 3b$ and $\Delta V_t = 0.36\Omega$, for example, equation (3.123) gives the following value for c_t: $c_t = 4.7\times10^{-2}$.

Equation (3.123) shows that, in principle, varying the concentration and type of impurity atoms, it is possible to obtain both the complete suppression of the dislocation preference ($Z_i^d = Z_v^d$) and the growth of voids, and create suitable conditions for their dissolution ($Z_i^d < Z_v^d$). As already mentioned, the complete neutralisation of the sinks in combination with fulfilling the equivalence conditions (3.118) creates theoretically more suitable conditions for the suppression of structure-phase processes and radiation damage.

In practice, the situation may be considerably more complicated. It is very likely that in preliminary ageing the dislocations may be 'poisoned' partially or even completely by the impurity atmosphere. Evidently, this treatment also increases the incubation period of swelling in subsequent radiation. However, this may also be accompanied by the creation of favourable conditions for the nucleation and growth of dislocation loops and voids. New dislocations appear and preference is restored. The restoration of preference is also possible during release of the dislocations from the atmospheres, especially in holding under load.

We shall examine the processes associated with the formation of segregation zones directly in the vicinity of the voids.

At a constant dislocation preference, i.e. when $Z_i^d/Z_v^d = \alpha = \text{const}$, equation (3.115) may be written in the following form [149]:

$$\frac{dr_v}{dt} = \frac{\Omega F(\eta)G}{r_v S_v S_i}\left(\alpha Z_v^v - Z_i^v\right)Z_i^d \rho_d^e \tag{3.124}$$

If a void is 'clean, i.e. it does not contain a segregation band, then because of its neutrality as a sink ($Z_v^v \cong Z_i^v$) its growth rate will be determined by the value of the dislocation preference and other parameters of equation (3.124).

In a number of experimental investigations it was established [155-157, 162–167] that the growth of voids in both pure metals and in alloys is accompanied by the formation in their vicinity of segregation zones. Consequently, this results in changes of not only the efficiency $Z_{i,v}^v$ of the void as a sink but also in the characteristics of diffusion and absorption of the defects by the void. It is evident that this may also result in a significant change in the growth rate of the voids.

General considerations regarding the possible effect of segregation of the results elements on the growth of voids were published for the first time by Anthony [169]. Some models and mechanisms of the formation of segregations at the voids and their effect on the rate of growth of the voids in irradiation have been investigated in [149,163,170–176]. In the calculations carried out by Mansur and Wolfer [172,175], attention was given to the mechanical effect of a band of segregations around the voids on the efficiency of elastic interaction of the voids with the vacancies and interstitials $Z_{i,v}^v$. It is shown that, depending on the relative hardness of the segregation band and the surrounding matrix, the value of Z_i^v/Z_v^v and, consequently, the growth rate of the voids may change significantly. The strong effect on the growth rate of the voids with a decrease of the coefficient of diffusion of the vacancies in the segregation zones as a result of their capture by the solute atoms is shown in [170]. The identical results were also obtained in [149,155].

For the case of a homogeneous segregation band with thickness δ and the coefficient of diffusion of vacancies in the band and the matrix D_v^s and D_v^∞, respectively, the flow of vacancies to a void with radius r_v has the form [170]:

$$J_v = 4\pi r_v W D_v^\infty\left(c_v^\infty - c_v^\theta\right) \tag{3.125}$$

where c_v^e is expressed by equation (3.97) and the thermodynamically concentration of vacancies c_v^0 is determined by equation (3.94).

The efficiency of the void as a sink for vacancies in equation (3.125) is characterised by the quantity W:

$$W = \frac{1 + \dfrac{\delta}{r_v}}{1 + \dfrac{D_v^\infty \delta}{D_v^s r_v}} \qquad (3.126)$$

In equation (3.125), c_v^∞ is the dynamic concentration of the vacancies in the matrix, i.e. at a large distance from the void.

If the impurity traps in the segregation zone decrease the mobility of vacancies, i.e. $D_v^s < D_v^\infty$, and the mobility of the interstitials in the band is almost constant, i.e. $D_i^s \sim D_i^\infty$, the growth of the voids may be slowed down or interrupted completely.

One of the disadvantages of the current models of the nucleation and growth of the voids and also of swelling, including theoretical investigations into segregation at the voids, is, as stated in [149], the fact that the value of c_v^0 in equations (3.94) and (3.97) is assumed to be constant during irradiation and is usually calculated from equation (3.94). This is valid only in the absence of segregation.

In enrichment of the vicinity of the void with solutes, the value of c_v^0 does not remain constant and, depending on the concentration of the solutes and their energy of binding with vacancies, it may change. To calculate the value of c_v^0 in irradiated pure metals and diluted alloys, it is necessary to use equation (3.114) for the concentration of vacancies in diluted solid solutions, and not equation (3.94) as is usually the case. This problem has already been examined in the previous section in connection with analysis of the accuracy of the estimates of vacancy supersaturation in alloys.

The diffusion coefficient of vacancies in the segregation zone also changes depending on the concentration and binding energy of solutes with vacancies and may be evaluated from the equations for the effective coefficients of diffusion (2.6) or (3.117).

If $E_v^B > 0$, with an increase in the concentration of solutes at the void the vacancy supersaturation will increase and the diffusibility of the vacancies in the vicinity of the void will decrease. Consequently, the flow of vacancies to the void at the concentration of the voids in the matrix c_v^∞ $J \sim 4\pi r_v D_v (c_v^\infty - c_v^e)$ will decrease.

On the basis of considerations, presented above, the author of [149] derived an equation for calculating the growth rate of voids in the segregation conditions and carried out numerical calculations of equilibrium radii of the voids and the formation of segregation zones around the voids in relation to the concentration and binding energy of the solutes with vacancies. The following assumptions were made when deriving the equation for the rate of growth of voids.

1. During growth of a void with radius r_v in binary diluted alloys, a segregation band with thickness h forms around the void, and the concentration of the dissolved element in the band c_t is higher than its nominal concentration in the matrix c_t.

2. The concentration of the impurity in the segregation band with thickness h is uniform.

3. In a spherical band with thickness h, vacancies migrate to the void with diffusion coefficient D_v^s, and in a spherical layer with thickness $R-(r_v+h)$ with diffusion coefficient D_v^∞.

The outer radius of the sphere of the second band R corresponds to the distance from the void where the concentration of the vacancies is c_v and the concentration of impurities c_t.

With the framework of the accepted conditions, the following equation was obtained for the vacancy flow to the void with radius r_v, surrounded by a segregation band with thickness h:

$$J_{r_v} = 4\pi r_v D_v^\infty \left(c_v^\infty - c_v^s\right) \frac{R\left(1+\dfrac{h}{r_v}\right)}{R-\left(r_v+h\right)+\dfrac{RD_v^\infty h}{D_v^s r_v}} \tag{3.127}$$

In equation (3.127) c_v^s, c_v^h and c_v^∞ is the concentration of the vacancies on the spheres with the radii r_v, r_v+h and R.

For the case in which $R \gg r_v$, the last multiplier of the right-hand part of equation (3.127) is transformed to expression (3.126) which characterises the variation of the efficiency of the void as a sink for vacancies with the variation of their diffusibility in the segregation band.

The substitution of the density of the flow from expression (3.127) into equation (3.96), taking (3.126) into account, with a number of transformations, gives:

$$\frac{dr_v}{dt} = \frac{\Omega}{r_v}\left[Z_v^v D_v^\infty WB\left(c_v^\infty - c_v^\theta\right) - Z_i^v D_i c_i\right] \tag{3.128}$$

In the form, identical with equation (3.124), this equation has the form:

$$\frac{dr_v}{dt} = \frac{\Omega F(\eta) G}{r_v S_v S_i}\left(WB\alpha Z_v^v - Z_i^v\right) Z_v^d \rho_d^e \tag{3.129}$$

The value of B in equations (3.128) and (3.129):

$$B = \frac{c_v^\infty - c_v^s}{c_v^\infty - c_v^o} \tag{3.130}$$

like takes into account the variation of the efficiency of the void as a sink for vacancies as a result of vacancy supersaturation in capture of vacancies by the solute atoms of the segregation zone.

In equation (3.130), c_v^e is determined by equation (3.97) in which the quantity c_v^0 characterises the thermodynamically equilibrium concentration of the vacancies in the matrix or in the vicinity of the 'clean' avoid, and is calculated from equation (3.94). Quality the c_v^s is also determined by equation (3.97) in which, however, the value of c_v^0 is replaced by c_v^t, determined by equation (3.114). In the process of segregation of the solutes with $E_v^B > 0$ at the void, the values of c_v^t and c_v^s will increase, and the value of parameter B will decrease.

Figure 3.22 shows the results of numerical calculations of the variation of parameter B in nickel with the initial atomic concentration of the impurities of 10^{-5} in dependence on E_v^B and the concentration of admixtures in the segregation zone from 10^{-4} to 5×10^{-2}.

The graph indicates that the vacancy supersaturation of the vicinity of the void during segregation may greatly decrease the efficiency of the void as a sink for the vacancies, and as indicated by comparison with the results of calculations in [175], may have a similar strong effect on the kinetics of void growth as the variation of the force efficiency of the void as a sink for point defects.

Within the framework of the examined model, the author of [149] obtained an analytical expression for the concentration of solutes in a segregation zone with complete suppresion of the growth of voids c_t^s (the condition of saturation or equilibrium):

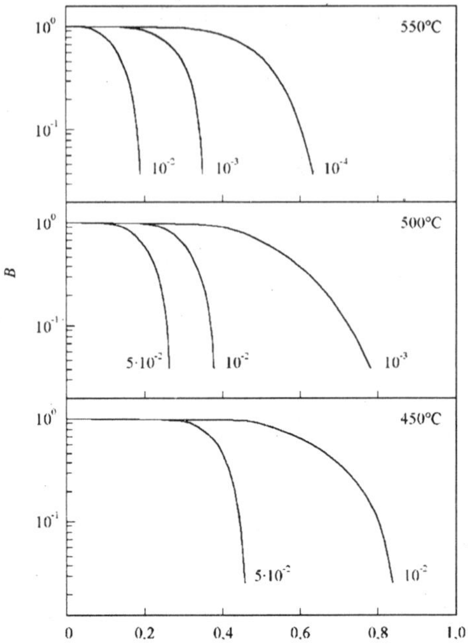

E_v^b, eV

Fig. 3.22. Variation of parameter *B* in nickel with the initial atomic concentration of impurities 10^{-5} in relation to their concentration in the segregation zone at voids and of their binding energy with vacancies for temperatures of 450, 500 and 550°C.

$$c_t^s = \frac{\left[c_v^{\infty} - B\left(c_v^{\infty} - c_v^{e}\right)\right]\exp\left(E_v^f / kT\right) - A_v}{A_{tv} z \exp\left(E_v^B / kT\right) - A_v\left(z+1\right)} \tag{3.131}$$

Identical expressions were also obtained for the ratio of the equilibrium radius of the voids to the width of the segregation zone r_v^s/h and the relative width of the segregations zone h/R. In this case, the expression h/R was obtained for the linear distribution of the concentration of solutes in the segregation zone. Numerical calculations were also carried out for these parameters for nickel and aluminium in relation to the initial concentration of solutes, temperature and binding energy of the vacancies with the solute atoms. The results make it possible to draw the following main conclusions [149]:

1. The vacancy supersaturation and the decrease of diffusion of the vacancies in the vicinity of the voids during segregation of solutes may greatly decrease both the rate of growth of the voids and and restricted the equilibrium size of the voids at saturation.

2. The effect of the given mechanism on the suppression of swelling is intensify with increasing the binding energy of the vacancy–dissolved element E_v^B and the initial concentration of the impurities c_v^∞.

3. The equilibrium size of the voids decreases with increasing values of E_v^B and c_t^∞.

4. As the values of E_v^B, c_t^∞ and T increase, the susceptibility of the alloy to segregation decreases.

5. As the energy of formation of the vacancies and the rate of buildup of defects increase, the intensity of segregation in the vicinity of the voids increases.

6. As a result of segregation during irradiation, the concentration of the dissolved element at the voids may be several orders of magnitude higher than its initial value in the matrix solid solution. At specific concentrations this may result in the precipitation of phases on the voids even in non-saturated solid solutions.

A similar effect was detected, in particular, in [69,177] in which after neutron irradiation of non-saturated aluminium–magnesium solid solutions at the voids examination showed phases Mg_5Al_8 and Mg_2Si (in these alloys, silicon is produced by neutron nuclear reactions).

The authors of [176] carried out theoretical analysis of the effect of segregation atmospheres in the vicinity of the dislocations and voids on the rate of swelling of interstitial solid solutions and the resultant analytical expressions were used for appropriate numerical calculations. The results show that the formation of segregates at the dislocations suppresses more efficiently the rate of growth of the voids than the formation of segregation bands in the vicinity of the voids as a result of concentrated and extended impurity atmospheres.

REFERENCES

1 Yung F.W., J. Nucl. Mater., 69/70, 310 (1978).

2 Corbett J.M., et al, Phys. Rev., 114, 1452 (1959).

3 Corbett J.M., In: Vacances and interstitials in metals, Amsterdam, North-Holland (1969), p.972.

4 Bystrov L.N., et al, Fiz. Khim. Obrab. Mater., No.6, 25 (1969).

5 Lucasson P., et al, Ann. Chim. Fr., 9. 15 (1984).

6 Piani C.S.B. and Aspeling J., Phys. Stat. Sol.(a), 96, 817 (1979).

7 Seeger A., Proc. Roy. Soc., A371, No.1744, 165 (1980).

8 Seeger A. and Frank W., Rad. Eff., 71, 157 (1983).

9 Seeger A. and Kronmuller H., In: Point defects and defects interactions in metals, Tokyo Univ. Press, Tokyo (1982), p.65.

10 Schule W., Mater. Sci. Forum, 97/99, 133 (1992).

11 Schilling W. and Sonnenberg K., J. Phys. F: Metal Phys., 3, 322 (1973).

12 Averback R.S. and Ehrhart P., J. Phys. F: 14, 1347 (1984).

13 Corbel C. and Moser P., Mater. Sci. Forum, 15/18, 641 (1987).

14 Dimitrov C., et al, J. Phys: Condens. Mater., 4, 10211 (1992).

15 Takamura S., Rad. Eff. Lett., 43,69 (1979).

16 Takamura S. and Kobiyma M., Rad. Eff., 49, 247 (1987).

17 Kijek M.M. and Palmer D.W., Mater. Sci. Forum, 15/18, 703 (1987).

18 Schultz H., In: Point defects and defect interactions in metals, Tokyo Univ. Press, Tokyo (1982), p.183.

19 Schultz H., Mater. Sci. Forum, 15/18, 733 (1987).

20 Petzold J. and Schultz H., Mater. Sci. Forum, 15/18, 733 (1987).

21 Neklyudov I.M., et al, Vopr. Atomn. Nauki i Tekhniki. Series: Fiz. Radiats. Povrezhd. Radiats. Materialoved., No.3 (1983), p.95.

22 Sen Gupta A., et al, Mater. Sci. Forum, 15/18, 931 (1987).

23 Sielemann R., et al, Phys. Lett., 117, No.2, 87 (1986).

24 Maury F., et al, J. Phys. B: Condens. Mater., 2, 9269 (1990).

25 Lucasson A., et al, J. Phys. F: Metal. Phys., 14, 1379 (1984).

26 Dworschak F., et al, Phil. Mag. A., 63, No.5, 849 (1991).

27 Habtetsion S., et al, J. Phys. B: Condens. Mater., 1, 9519 (1989).

28 Roitzheim R.D. and Vianden R.Z., J. Phys. B: Condens. Mater., 83, 71 (1991).

29 Maury F., et al, J. Phys. F: Metal. Phys., 18, 657 (1988).

30 Volobuev A.V., et al, in: Radiation Materials Science, Vol. 2, Khar'kov (1990), p.85.

31 Baily A.C., et al, Phys. Rev. B., 46, No.13, 8593 (1992).

32 Sadykov S.I.O. and Latov Yu.M., Fiz. Khim. Obrab. Mater., No.2, 98 (1974).

33 Platov Yu.M., et al, Fizika Metallov i Metallovedenie, 39, No.6, 1290 (1975).

34 Ivanov L.I., et al, J. Nucl. Mater., 69/70, 754 (1978).

35 Nakagawa M., Annual Repts. Res. Realt. Inst., Vol.13, Kyoto Univ. (1980), p.117.

36 Wolfer W.G. and Si-Ahmed A., J. Nucl. Mater., 99, 117 (1981).

37 Dettman K., et al, Phys. Stat. Sol.(a), 22, 423 (1967).

38 Kotomin E.A., et al, Izv. AN Latv. SSR, Fiz. i Tekhnich. Nauki, No.4, 114 (1984).

39 Iwase A., et al., J. Phys-Soc. Japan, 59, No.4, 1451 (1990).

40 Plani C.S.B. and Aspeling J., Rad. Eff., 45, 127 (1980).

41 Damask A. and Dienes J., Point Defects in Metals, Mir, Moscow (1966),

p.291.
42 Bystrov L.N., et al, Fiz. Khim. Obrab. Mater., No.1, 14 (1970).
43 Brailsford A.D. and Bullough R., J. Nucl. Mater., 44, 121 (1972).
44 Mansur L.K. Nucl. Technol., 40, No.1, 5 (1978).
45 Waite T.R, J. Chem. Phys., 28, No.1, 103 (1958).
46 Eshelby J., Continuous Theory of Dislocations, IL, Moscow (1963), p.247.
47 Yoo M.N., et al, In: Fundamental aspects of radiation damage in metals, USERDA, Vol.II (1976), p.804.
48 Held P.T., In: Radiation effects in Breeder reaction structural materials, AIME, New York (1977), p.781.
49 Held P.T. and Speight M.V., Acta Met., 23, 1389 (1975).
50 Margelashvili I.G. and Saralidze E.K., Fizika Tverdogo Tela, 15, 2665 (1973).
51 Wolfer W.G. and Ashkin M., J. Appl. Phys., 46, No.2, 547 (1975).
52 Wolfer W.G. and Ashkin M., J. Appl. Phys., 47, No.3, 791 (1976).
53 Wolfer W.G. and Si-Ahmed A., Preprint of Wisconsin-Madison University, UWFDM-327 (1979), p.7.
54 Ghoniem N.M. and Kulcinski G.L., Rad. Eff., 41 (1979), p.81.
55 Walker R.M., In: Proc. Int. School of Phys. "E. Fermi", Acad. Press, New York (1962), p.594.
56 Dworshak F., et al, Phys. Stat. Sol., 28, 75 (1968).
57 Kraut A., et al, Phys. Stat. Sol.(b), 44, 805 (1971).
58 Becker D.E., et al, Phys. Stat. Sol.(b), 54, 455 (1972).
59 Dworshak F., et al, J. Phys. F: Metal. Phys., 5, 400 (1975).
60 Dworshak F., et al, J. Phys. F: Metal. Phys., 6, No.12, 2207 (1976).
61 Dimitrov C., et al, J. Phys. F: Metal. Phys., 8, No.6, 1301 (1978).
62 Robrok K.G., In: Phase Transformations under Irradiation, Metallurgiya, Chelyabinsk (1989), p.94.
63 Ivanov L.I., et al, Rad. Eff., 32, 45 (1977).
64 Platov Yu.M., et al, Izvestiya AN SSSR, Moscow, No.2 (1972), p.101.
65 Wollenberger H., et al, In: School of Radiation and Other Defects in Metals, Vol. 1, Tbilisi (1974), p.73.
66 Wollenberger H., J. Nucl. Mater., 69/70, 362 (1978).
67 Platóv Yu.M. and Pletnev M.N. In: Physico-Chemical Investigations in Metallurgy and Metals Science using Computers, Nauka, Moscow (1974), p.124.
68 Dienes G.J., et al, Rad. Eff., 33, 59 (1977).
69 Platov Yu.M., et al, Mater. Sci. Forum, 97/99, 253 (1992).
70 Platov Yu.M. and Simakov S.V., Fiz. Khim. Obrab. Mater., No.6, 5 (1988).
71 Kiritani M. and Takata N., J. Nucl. Mater., 69/70, 277 (1978).
72 Halbachs M. and Yoshida N., Phil. Mag., 43, No.5, 1125 (1981).
73 Eyre B.I., et al, Metal Science (Jan. 1978), p.35.
74 Si-Ahmed A. and Wolfer W.G., In: Dislocation modelling of phys. systems, Pergamon Press (1980), p.142.
75 Ghoniem N.M., et al, In: Effects of radiation on materials. Proc. 11th Int. Symp., ASTM, Baltimore (1982), p.1054.
76 Katsnel'son A.A., et al, Fizika Metallov i Metallovedenie, 57, No.1, 138 (1984).
77 Effects of radiation on materials. Proc. 11th Int. Symp., ASTM, Baltimore (1982), p.1225.
78 Balandin Yu.F., et al, Structural Materials for Atomic Power Stations, Energoizdat, Moscow (1984), p.280.
79 Zelenskii V.F. and Neklyudov I.M. Vopr. Atomnoi Nauki Tekhniki. Ser.: Fizika

Radiats. Povrezhd. Radiats. Materialoved., No.1/2(30), 46 (1984).
80 Effects of radiation on materials. Proc. 12th Int. Symp., STM, Vol.1/2, Philadelphia (1985), p.1267.
81 Russell K.S. and Powell R.M., Acta Met., 21, 187 (1973).
82 Russell K.S. Advances in Colloid and Interface Science, No.3/4, 205 (1980).
83 Brager H.R., et al, In: Radiation damage in metals. Proc. Mater. Sci. Symp. (1976), p.194.
84 Brager H.R., et al, In: Radiation effects in breeder reactor structural materials, AIME, New York (1977), p.727.
85 Shimomura Y., Phil. Mag., 19, 773 (1969).
86 Chen J.H., et al, Rad. Eff., 18, No.10, 1891 (1979).
87 Shimomura Y. and Kuwabara S., J. Phys. Soc. Japan, 42, No.4, 1221 (1977).
88 Shimomura Y., J. Appl. Phys. (Japan), 18, No.10, 1891 (1979).
89 Lazorenko V.M., et al, Fizika Metallov i Metallovedenie, 50, No.1, 164 (1980).
90 Nakai K., et al, In: Point defects and defect interaction in metals, Tokyo Univ. Press, Tokyo (1982), p.887.
91 Nakai K., et al, Phil. Mag., 52, No.1, 115 (1985).
92 Russell K.C. Acta Met., 19, 753 (1971).
93 Katz J.L. and Wierdersich H., J. Chem. Phys., 55, No.3, 1414 (1971).
94 Pletnev M.N. and Platov Yu. M., Fizika Metallov i Metallovedenie, 40, No.2, 304 (1975).
95 Sprague J.A., et al, In: Fundamental aspects of radiation damage. Proc. Int. Conf., Gatlinburg (1975), p.1181.
96 Bondarenko A.I. and Konobeev Yu.V., Phys. Stat. Sol.(a), 34, 195 (1976).
97 Wolfer W.G., et al, In: Radiation effects in Breeder reactor structural materials, AIME, New York (1977), p.841.
98 Hall B.O., J. Nucl. Mater., 91, 63 (1980).
99 Wehner M.F. and Wolfer W.G., Phil. Mag., 52, No.2, 189 (1985).
100 Yacout A. and Stubbins J.F., J. Nucl. Mater., 141/143, 677 (1986).
101 Trinkhaus H. and Yoo M.N., Phil. Mag. A., 55, No.3, 269 (1987).
102 Ghoniem N.M., Phys. Rev. B., 39, No.16, 11810 (1989).
103 Kiritani M., J. Phys. Soc. (Japan), 35, 95 (1973).
104 Hayns M.R., J. Nucl. Mater., 56, 267 (1975).
105 Miller K.N. et al, J. Nucl. Mater., 110, 265 (1982).
106 Stathopoules A.Y., et al, J. Nucl. Mater., 110, 301 (1982).
107 Wiedersich H., Mater. Sci. Forum, 97/99, 59 (1992).
108 Brown L.M., et al, Phil. Mag., 19, 721 (1969).
109 Makin M.J., Phil. Mag., 20, 1133 (1969).
110 Norris D.I.R., Phil. Mag., 22, 1273 (1970).
111 Wolfeden A. and Yoo M.H., Rad. Eff., 22, 61 (1974).
112 Kinoshita C., et al, Mater. Sci. Forum, 15/18, 949 (1987).
113 Mayer R.M., J. Microscopy (Gr.Brit.), 110, 133 (1977).
114 Lazorenko V.M., et al, Fizika Metallov i Metallovedenie, 58, No.5, 943 (1984).
115 Platov Yu.M. and Simakov S.V., Fizika Metallov i Metallovedenie, 61, No.2, 213 (1986).
116 Platov Yu.M., et al, Computer Modelling of Defects in Metals, Nauka, Leningrad (1990).
117 Platov Yu.M., et al, Fiz. Khim. Obrab. Mater., No.3, 20 (1990).
118 Ivanov L.I., et al, J. Nucl. Mater., 191/194, 1075 (1992).
119 Bocquet J.I., Rapport CEA, R-5112, Saclay (1981), p.45.
120 Bocquet J.I., Acta Met., 34, No.4, 571 (1986).

121 Barbu A. and Ardell A.Z., Scripta Met., 9, 1233 (1975).
122 Kenik E.A. and Hojou K., J. Nucl. Mater., 191/194, 1331 (1992).
123 Fisher S.B., et al, Phys. Stat. Sol., 57, 281 (1980).
124 King H.W., J. Mater. Sci., 1, 79 (1966).
125 Hossain M.K. and Brown L.M., Acta Met., 85, 257 (1977).
126 Garner F.A. and Gelles D.S., J. Nucl. Mater., 159, 286 (1988).
127 Saralidze S.K. and Slezov V.V., Fizika Tverdogo Tela, 7 1605 (1965).
128 Katz J.H. and Widersich H., J. Nucl. Mater., 46, 41 (1973).
129 Mansur L.K. and Coghlum W.A., J. Nucl. Mater., 119, 1 (1983).
130 Parker C.A. and Russel K.C., Scripta Met., 119, No.1, 82 (1983).
131 Maydet S.I. and Russell K.C., Scripta Met., 14, 383 (1980).
132 Russell K.C., Acta Met., 26, 1615 (1978).
133 Bullough R. and Perrin R.C., In: Radiation-induced voids in metals, USAEC (1972), p.769.
134 Brailsford A.D. and Bullough R., J. Nucl. Mater., 69/70, 434 (1978).
135 Nayns M.R, J. Nucl. Mater., 79, 323 (1979).
136 Konobeev Yu.V., et al, Vopr. Atomnoi Nauki Tekhniki. Ser.: Fizika Radiats. Povrezhd. Radiats. Materialoved., Materialovedenie (1981).
137 Zelenskii V.F., et al, Radiation Defects and Swelling of Metals, Naukova dumka, Kiev 44 (1988).
138 Hadji M. and Russell K.C., J. Nucl. Mater., 92, 229 (1980).
139 Tompson M., Defects and Radiation Damage in Metals, Mir, Moscow (1971).
140 Leitnaker J.M., et al, J. Nucl. Mater., 49, 57 (1973/74).
141 Little E.A. and Stow D.A., J. Nucl. Mater., 87, 25 (1979).
142 Russell K.C., Scripta Met., 7, 755 (1973).
143 Smigt F.A. and Sprague J.A., Scripta Met., 7, 495 (1973).
144 Mansur L.K. and Yoo M.H., J. Nucl. Mater., 74, 228 (1978).
145 Mansur L.K., J. Nucl. Mater., 83, 109 (1979).
146 Nguy T., et al, Mater. Sci. Forum, 15/18, 675 (1987).
147 Garner F.A. and Wolfer W.G., J. Nucl. Mater., 102, 143 (1981).
148 Doyama M., J. Nucl. Mater., 69/70, 350 (1978).
149 Platov Yu.M. Fiz. Khim. Obrab. Mater., No.4, 48 (1983).
150 Chen I.W. and Taiwo A., In: Effects of radiation on materials. Proc. 12th Int. Symp., Vol.1, ASTM, Philadelphia (1985), p.507.
151 Kinoshita C., et al, Mater. Sci. Forum, 15/18, 1403 (1987).
152 Ivanov L.I., et al, Fiz. Khim. Obrab. Mater., No.1, 28 (1988).
153 Mansur L.K., et al, In: Phase stability during irradiation, AIME, Warrendale (1981), p.359.
154 Carpenter R.W. and Yoo M.H., Metallurgical Trans., 9A, 1739 (1978).
155 Bakai A.S., et al, J. Nucl. Mater., 185, 260 (1991).
156 Ivanov L.I., et al, Fiz. Khim. Obrab. Mater., No.3, 5 (1990).
157 Ivanov L.I., et al, J. Nucl. Mater., 191/194, 928 (1992).
158 Ivanov L.I., et al, Fiz. Khim. Obrab. Mater., No.4, 27 (1984).
159 Esmailzaden B. and Kumar A.S., In: Effects of radiation on materials. Proc. 12th Int. Symp., Vol.1, ASTM, Philadelphia (1985), p.468.
160 Kiritani M., et al, In: Proc. 8th Int. Congr. Electron Microscopy, Vol.1 (1974), p.622.
161 Norris D.I.R., In: The physics of the irradiation produced voids, AERE Report R-7934 (1975), p.134.
162 Farrell K., et al, Scripta Met., 11, 234 (1977).
163 Okamoto P.R., et al, Nucl. Technology, 22, 45 (1974).

164 Marwick A.D., et al, Scripta Met., 12, 1015 (1978).
165 Erck R.A., et al, J. Nucl. Mater., 80, 120 (1979).
166 Pokrovskii A.S., et al, Fizika Metallov i Metallovedenie, 48, No.4, 832 (1979).
167 Takeyama T., et al, Scripta Met., 14, 1105 (1980).
168 Ohnuki S., et al, In: Point defects and defects interactions in metals, Tokyo Univ. Press, Tokyo (1982), p.954.
169 Anthony T.R., In: Radiation-induced voids in metals, USAEC (1972), p.630.
170 Brailsford A.D., J. Nucl. Mater., 56, 7 (1975).
171 Lam N.Q., et al, J. Nucl. Mater., 74, 101 (1978).
172 Mansur L.K. and Wolfer W.G., J. Nucl. Mater., 69/70, No.1, 825 (1978).
173 Fikus J.H. and Johnson R.A., Rad. Eff., 40, 63 (1979).
174 Vershok B.A. and Gorbatov., Fizika Tverdogo Tela, 21, No.2, 508 (1979).
175 Wolfer W.G. and Mansur L.K., J. Nucl. Mater., 91, 256 (1980).
176 Pyatiletov Yu.S. and Edemskii H.M., Zh. Tekhn. Fiz., 62, No.11, 89 (1992).
177 Ivanov L.I., et al, Fiz. Khim. Obrab. Mater., No.4, 16 (1985).

Chapter 4

RADIATION-STIMULATED PHASE CHANGES IN ALLOYS

4.1. INTRODUCTION

In the previous chapter, special attention was given to the effect of impurity and alloying elements in pure metals and solid solutions on the processes of buildup of radiation defects in the formation of defect–impurity complexes, clusters of defects, dislocation loops and voids. The problems of phase changes and their effect on the structure of radiation damage have been studied only partially. In the present chapter, this problem will be given special attention, taking into account the fact that the phase changes, caused by the intensification of the processes of ordering, phase clustering and breakdown of solid solutions, coalescence and the dissolution of the phases may prevail over the purely imperfect effect of irradiation, controlling the degradation of the properties of irradiated metallic materials. In addition, the processes of formation of clusters of defects, dislocation loops and voids are usually inter-related with the variation of the chemical composition of the matrix solid solutions and even with the phase instability of single-phase and more complex alloys. The detailed analysis of these mechanisms makes it possible to interpret more efficiently and predict the radiation damage in alloys and develop methods of preventing it.

The current interest in the radiation-stimulated phase changes is associated mainly with the tendency for increasing the service parameters of materials in nuclear power engineering: temperature, intensity, fluence and service life of steels and alloys, and also the problem of development of fusion reactors in which the conditions of operation of structural metallic materials under the effect of irradiation in combination with other factors are close to extreme conditions.

A significant role in the intensification of structure–phase processes in the solid solutions and more complicated metallic systems is played by radiation-enhanced diffusion. Consequently, prior to transferring to the direct examination of the mechanisms of phase changes and phase instability of metallic materials under irradiation, we shall consider the current views regarding radiation-enhanced diffusion.

4.2. RADIATION-ENHANCED DIFFUSION

The volume radiation-enhanced diffusion is one of the main mechanisms of radiation-stimulated phase changes, accelerating the transfer of systems to the equilibrium condition.

For structural materials, this relates primarily to accelerating coalescence of phases under the effect of irradiation, controlled by the diffusion of solutes in matrix solid solutions. For resistive materials in electronic engineering, for example, where many of these materials are substitutional solid solutions, radiation-enhanced interdiffusion may result in a change of their physical parameters and, in particular, electrical resistivity as a result of acceleration of ordering, clustering or breakdown, stimulating the transition of these alloys to the equilibrium state. The characteristic time required to reach equilibrium rapidly decreases in this case. In Fig. 4.1 this is shown on the example of Ag–8.75 at.%Zn alloys, where the graph in the coordinates ρ–$1/T$ shows the time of establishment of the equilibrium values of electrical resistivity in the process of short range ordering during thermal annealing of the alloy (in brackets in Fig. 1) and during irradiation of the alloy with electrons with an energy of 2.3 MeV. Estimates were obtained on the basis of the results of experimental investigations [1,2] and diffusion parameters for vacancies from [3]. It may be seen that without irradiation below temperatures of 80–70 °C, the equilibrium values of electrical resistivity in this case for actual thermal conditions are in impossible to obtain, whereas in irradiation they are realised within reasonable experiment periods. A similar temperature dependence of the variation of electrical resistivity was presented in [4] for the Cu+30%Zn alloys. At the same time, it should be mentioned that the acceleration of a number of processes of approach of the systems to the thermodynamic equilibrium in irradiation, including ordering, may be additional, and associated with the compensation by point defects of coherent strains formed during transformations.

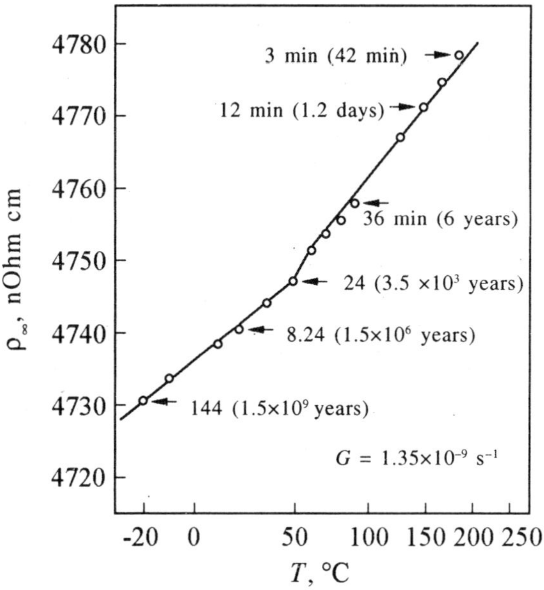

Fig. 4.1 Equilibrium values of electrical resistivity of Ag–8.75 at.%Zn ordering alloy.

4.2.1. The mechanisms of radiation-enhanced interdiffusion

The intensification of diffusion in irradiation is caused by the introduction of an excess concentration of interstitial atoms c_i and vacancies c_v. The coefficient of radiation-enhanced interdiffusion D^s is linked with c_t and c_v by the following obvious relationship:

$$D^s = D_v^s + D_i^s = f_v D_v c_v + f_i D_i c_i \tag{4.1}$$

where D_v^s and D_i^s are the coefficient of vacancy and interstitial interdiffusion, and the coefficients of diffusion of interstitials and vacancies $D_{i,v}$ have the following form:

$$D_{i,v} = D_{i,v}^0 \exp\left(-E_{i,v}^m / kT\right) \tag{4.2}$$

The corrèlation factors f_v and f_i for the FCC lattice are 0.78 and 0.44, respectively.

The main theoretical assumptions of the mechanisms of radiation-enhanced interdiffusion have been developed in [5–9]. On the basis of nonstationary, quasiequilibrium and stationary both general and partial solutions of the system of equations (3.3)–(3.4), attention was given to the following mechanisms of radiation-enhanced diffusion:

1. The combined mechanism. In this case, the establishment of the dynamically equilibrium concentration of point defects is controlled by both their mutual recombination and annihilation of defects on constantly acting sinks [5,6]. The variation of the concentration of point defects for this mechanism is described by a complete system of equations (3.3)–(3.4).

In a general (nonstationary) case, the system of equations (3.3)-(3.4) has no analytical solution. Its solution for a number of approximations has been examined in [7–9]. For a stationary case, the dynamically equilibrium concentrations of point defects in accordance with the equations (3.16), (3.17) and (3.21) are equal to:

$$c_{i,v} = \frac{F(\eta)G}{S_{i,v}D_{i,v}} \tag{4.3}$$

and the coefficients of radiation-enhanced interdiffusion have the form:

$$D^{s}_{i,v} = \frac{f_{i,v}F(\eta)G}{S_{i,v}} \tag{4.4}$$

Function $F(\eta)$ and the recombination parameter η in equations (4.3)-(4.4) are determined by equations (3.18) and (3.19).

If the diffusivity of the interstitials is considerably higher than the mobility of vacancies, which is always fulfilled for pure metals, then the time to establishment of the dynamically equilibrium concentrations of the point defects is:

$$\tau = \tau_v = \frac{1}{S_v D_v} \tag{4.5}$$

For the combined mechanism, the characteristic feature is that in relation to temperature and power of the constantly acting sinks, the activation energy of radiation-enhanced interdiffusion does not remain constant and with an increase of these two parameters increases from $1/2E^m_v$ to zero. This is clearly demonstrated by the resultant numerical calculations of the temperature dependence of the coefficient of radiation-enhanced interdiffusion D^s using equations (4.1)–(4.4) in aluminium, presented in Fig. 4.2. They are given for the rate of introduction of defects $G = 10^{-6}s^{-1}$ and the concentration of constantly acting sinks $S_{v,i}$ equal to 10^8, 10^{10} and 10^{12} cm^{-2}. These relationships reflect the variation of the ratio between the processes of mutual recombination of defects and their annihilation on the sinks with the variation of the temperature and power of the sinks. With increase of these parameters, the dependence of the coefficient of radiation-enhanced self-diffusion on the rate of introduction of point defects also changes from $G^{1/2}$ to G. The limiting values of the activation energy of interdiffusion $E = 0$ and the linear dependence of D^s on G are typical of the case in which the annealing of defects on constantly acting sinks prevails. The plateau in Fig. 4.2 also reflects the temperature special feature of this mechanism.

Figure 4.2 also shows that the increase of the power of the sinks greatly decreases the rate of radiation-stimulated diffusion and the temperature range in which it is dominant. Thus, preliminary deformation suppresses the effect of radiation under the diffusion processes in the same manner as in the case of swelling. In fact, for steels, operating

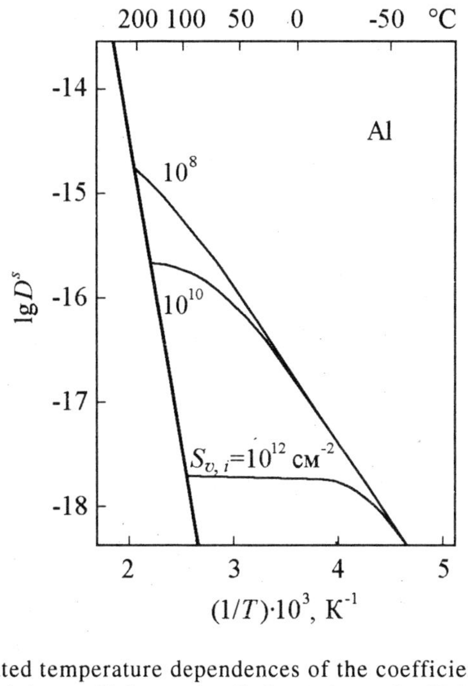

Fig. 4.2 Calculated temperature dependences of the coefficients of thermal diffusion (straight line) and the coefficients of radiation-enhanced self-diffusion of the combined mechanism in aluminium for different powers of constantly acting sinks.

in active zones of nuclear reactors, the method of preliminary plastic deformation is indeed used to reduce the degree of swelling.

2. The linear mechanism. For the case in which mutual recombination has no practical role, i.e., the number Rc_vc_i in equations (3.3)–(3.4) tends to zero and the annihilation of vacancies and interstitials on constantly-acting sinks is dominant, the corresponding solutions for nonstationary coefficients of radiation-enhanced interdiffusion have the following form:

$$D_{i,v}^{s} = \left[\frac{f_{i,v}G}{S_{i,v}}\right]\left\{1 - \exp\left(-S_{i,v}D_{i,v}t\right)\right\} \tag{4.6}$$

The time to establishment of dynamically equilibrium concentrations of the defects is:

$$\tau_{i,v} = \frac{1}{S_{i,v}D_{i,v}} \tag{4.7}$$

and the stationary coefficients of radiation-enhanced interdiffusion have the following form:

$$D_{i,v}^{s} = \frac{f_{i,v}G}{S_{i,v}} \tag{4.8}$$

As indicated by equations (4.7) and (4.8), in this case the rate of radiation-enhanced interdiffusion is independent of temperature and at a constant intensity of radiation it is inversely proportional to the power of the sinks.

3. The mutual recombination mechanism. When the mutual recombination of point defects is dominant, i.e. the terms $c_{i,v}D_{i,v}S_{i,v}$ in equations (3.3)–(3.4) tend to 0, the solutions for nonstationary coefficients of radiation-enhanced interdiffusion have the following form:

$$D_{i,v}^{s} = f_{i,v}D_{i,v}\left(\frac{G}{R}\right)^{1/2} th\left[(GR)^{1/2}t\right] \tag{4.9}$$

The time to establishment of dynamically equilibrium concentrations of the vacancies and interstitials is:

$$\tau = \frac{\pi}{(GR)^{1/2}} \tag{4.10}$$

and the stationary coefficients of radiation-enhanced interdiffusion have the form:

$$D^s_{i,v} = f_{i,v} D_{i,v} \left(\frac{G}{R} \right)^{1/2} \tag{4.11}$$

In the explicit form, taking into account the expression for R:

$$R = \frac{4\pi r_{vi} \left(D_i + D_v \right)}{\Omega} \tag{4.12}$$

the equation (4.11) for D^s_v and D^s_i and pure metals ($D_i \gg D_v$) may be represented in the following form:

$$D^s_v = f_v \left(\frac{G\Omega}{4\pi r_{vi} D^0_i} \right)^{1/2} D^0_v \exp \left[\frac{-\left(E^m_v - \frac{1}{2} E^m_i \right)}{kT} \right] \tag{4.13}$$

$$D^s_i = f_i \left(\frac{D^0_i G\Omega}{4\pi r_{vi}} \right)^{1/2} \exp \left(\frac{-\frac{1}{2} E^m_i}{kT} \right) \tag{4.14}$$

In the equations (4.13) and (4.14), the value of Ω is the atomic volume.

Comparison of this mechanism with the linear and combined mechanism shows that for the same parameters of irradiation and migration energies of the point defects, the interdiffusion coefficient has the highest value in this case.

4. The nonstationary combined mechanism. This mechanism has been proposed in [7,8] and is based on the quasistationary approximation when solving the system of equations (3.3)–(3.4) and the condition

that the point defects with higher mobility annihilate on both constantly acting sinks and as a result of mutual recombination, whereas annihilation of the defects with lower mobility on constantly-acting sinks plays almost no role.

Assuming that $D_i > D_v$, for the interstitial mechanism of radiation-enhanced interdiffusion the equation for D_i^s has the following form:

$$D_i^s = \left(\frac{f_i G}{S_i}\right)\left(1 + \frac{4\pi r_{vi} G t}{S_i \Omega}\right)^{-1/2} \tag{4.15}$$

Equation (4.15) is valid for the time range $\tau_i < t < \tau_v$ in which the characteristic time $\tau_{i,v}$ is determined by the expression:

$$\tau_{i,v} = \frac{1}{S_{i,v} D_{i,v}} \tag{4.16}$$

The upper boundary of this time range τ_v characterises the condition in which the flow of vacancies on the constantly-acting sinks becomes significant.

In the examined case, the coefficient of radiation-enhanced interdiffusion, as in the case of the linear mechanism, is independent of temperature and at relatively long times is proportional to $t^{-1/2}$ and $G^{1/2}$.

4.2.2. Radiation-enhanced diffusion of solutes and interdiffusion

In the section, attention will be given to the possibility of evaluation of the coefficients of radiation-enhanced diffusion of solutes.

As previously, the coefficient of radiation-stimulated interdiffusion will be denoted by $D^s_{v,i}$, and the coefficient of thermal interdiffusion (or the coefficient of thermal diffusion of the solvent in a diluted alloy) and the coefficient of thermal diffusion of solutes by the vacancy mechanism, as in Chapter 2, will be denoted by the values D_0 and D_2. The notation of the coefficient of radiation-enhanced diffusion of the solutes by the vacancy mechanism will be D_2^v.

It is assumed that in isothermal irradiation some mean dynamically equilibrium concentration of vacancy c_v is established in the diluted solid solution. Consequently, taking into account equations

(2.15) and (2.16):

$$\frac{D_2^v}{D_v^s} = \frac{\left(D_2 / c_v^0\right) / c_v}{\left(D_0 / c_v^0\right) / c_v} = \frac{D_2}{D_0} \tag{4.17}$$

In equation (4.17), the thermodynamically equilibrium concentration of vacancies c_v^0 is determined by expression (3.94).

Finally, for the coefficient of volume radiation-enhanced diffusion of the dissolved element in the diluted alloy:

$$D_2^v = \frac{D_2 D_v^s}{D_0} = \frac{D_2 c_v}{c_v^0} \tag{4.18}$$

A similar equation for $D^v{}_2$ was published in [10].

For the alloy in which the concentration of the dissolved element c_s is such that the probability of formation of pairs of impurity atoms cannot be ignored (the equation (2.21)), and also taking into account the contribution of thermal diffusion, in this case, for the total coefficient D_2^v:

$$D_2^v = \left(\frac{c_v^0}{c_v} + 1\right)\left(1 + B_1 c_s\right) \tag{4.19}$$

Equations (4.18) and (4.19) also show that if the coefficients of thermal diffusion are available either from the experiments or are calculated using frequency models [11] (see section 2.3), the estimates of D_2^v require either the experimental data for the coefficient of radiation-enhanced interdiffusion of the solvent D_v^s or complex calculations using corresponding equations for D_v^s, $c_{v,i}$, c_v^0 and $D_{v,i}$.

The coefficients of radiation-enhanced diffusion of solutes in the diluted alloys by the interstitial mechanism can be determined using two models: the Barbu model [12] and the Bocquet model [13,14]. The main assumptions of these models were examined in section 2.2. The Bocquet model can also be useful for approximate evaluation of the coefficients of radiation-enhanced diffusion by the interstitial mechanism in concentrated alloys.

As in the case of diluted alloys, it may be shown that the expressions for partial coefficients of radiation-enhanced diffusion of components A and B by the vacancy mechanism, and also the coefficient of radiation-enhanced interdiffusion have the following form:

$$D^{v}_{A,B} = \left(\frac{D^{T}_{A,B}}{c^{0}_{v}} \right) c_{v} + D^{T}_{A,B} = D^{T}_{A,B} \left(\frac{c_{v}}{c^{0}_{v}} + 1 \right) \quad (4.20)$$

$$\tilde{D}^{v} = \tilde{D}^{T} \left(\frac{c_{v}}{c^{0}_{v}} + 1 \right) \quad (4.21)$$

The partial coefficients of thermal diffusion $D^{T}_{A,B}$ and the coefficients of interdiffusion $\tilde{D}$ are linked with the coefficients of interdiffusion of components $D^{*T}_{A,B}$, their activity coefficients $\gamma_{A,B}$ and the concentration relationships [15]:

$$D^{T}_{A,B} = D^{*T}_{A,B} \left(1 + \frac{\partial \ln \gamma_{A,B}}{\partial \ln c_{A,B}} \right) \quad (4.22)$$

$$\tilde{D}^{T} = c_{A} D^{T}_{B} + c_{B} D^{T}_{A} = \left(c_{A} D^{*T}_{B} + c_{B} D^{*T}_{A} \right) \left(1 + \frac{\partial \ln \gamma_{A}}{\partial \ln c_{A}} \right) \quad (4.23)$$

The temperature dependences of the coefficients of thermal diffusion D^{T}_{A}, D^{T}_{B}, and $\tilde{D}^{T}$ may be obtained using the results of measurements of D^{*}_{A} and D^{*}_{B} by the method of radioactive isotopes, and the thermodynamically equilibrium concentration of vacancies c^{o}_{v} is determined on the basis of the appropriate corresponding quenching or equilibrium experiments.

In the estimation of the concentrated alloys, greater difficulties are encountered in the case of evaluation of c_{v}. If in the examined alloys the coefficient of diffusion of interstitial atoms D_{i} is considerably higher than the coefficient of diffusion of the vacancies D_{v}, the dynamically equilibrium concentration of the vacancies c_{v} of the combined or linear mechanism is independent of D_{i} (equations (3.16) and (3.111)) and is only the function of D_{v} in this case. Its value may also be evaluated from the experiments with annealing of quenched alloys.

4.2.3. The experimental data for radiation-enhanced diffusion, their analysis and interpretation

The main bulk of the experimental studies into radiation-enhanced diffusion was carried out using indirect investigation methods in analysis of the results for the investigation of the processes of short-range ordering and short-range clustering in alloys under irradiation. Basically, these investigations were carried out on the silver–zinc [7,16–26], copper–zinc [27–32], copper–aluminium [13,33], copper–nickel [34-38], and gold–silver [30,39,14] system using the methods of measurement of electrical resistivity and Zener relaxation. Far less investigations into radiation-enhanced diffusion were carried out using direct methods, such as the method of radioactive isotopes, Auger spectroscopy and mass spectroscopy of secondary ions. The results of these experiments for self-diffusion in silver, diffusion of lead in silver, interdiffusion in nickel, nickel in copper, copper in nickel and aluminium in nickel have been published in the studies [41,42], [43], [32] and [44], [45-47], [48] and [49]. In these studies, the mechanism of radiation-enhanced diffusion is interpreted mainly on the basis of analysis of the temperature dependences of relaxation time τ_{50} or the coefficients of diffusion and the dependences of these quantities on irradiation intensity. Some of the results of the investigations discussed here have been reviewed in [9,10,50].

The main problems in the analysis of the experimental data are associated with the correct interpretation of the mechanisms of radiation-enhanced diffusion. This process is especially important for investigations carried out using the results of indirect methods, based on the measurement of the variation of the properties of alloys in radiation-stimulated clustering and ordering. Usually, analysis of the mechanisms of radiation-enhanced diffusion in the system is carried out on the basis of the temperature dependences of the relaxation time of measured properties τ_{50}. Taking into account the fact that these alloys are usually concentrated solid solutions in which the diffusion mobility of interstitials and vacancies in contrast to pure metals may be very similar (sections 2.2, 3.2 and 3.6), the interpretation of the mechanisms of radiation-enhanced diffusion only on the basis of the temperature dependences of the relaxation times (or diffusion coefficients) may lead to erroneous results. In fact, this problem is also important for the interpretation of experimental data obtained using direct experimental methods.

In order to illustrate clearly this situation, Fig. 4.3 shows the results of numerical calculations for a hypothetical aluminium-based alloy

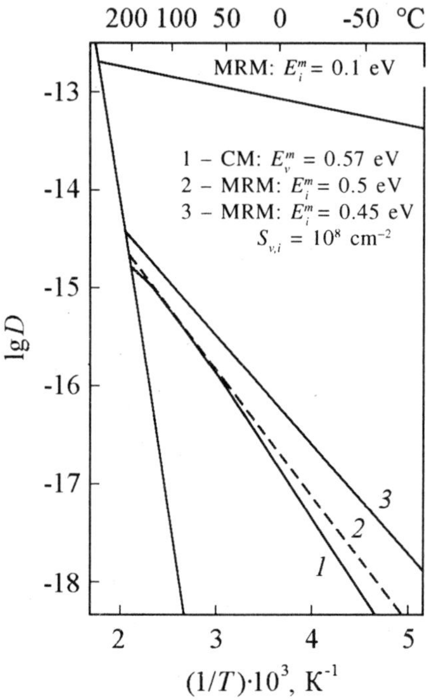

Fig. 4.3 Calculated temperature dependences of the coefficients of radiation-enhanced diffusion in an aluminium-based alloy for the combined mechanism (CM) and the mutual recombination mechanism (MRM).

in which the value of E_v^m is 0.57 eV, and the energy of migration of interstitials is assumed to be 0.1; 0.45 and 0.5 eV. The values of E_v^m = 0.57 eV and E_i^m = 0.1 eV correspond in the present case to the energy parameters of the migration of defects for pure aluminium. The calculations were carried out for the combined mechanisms (CM) and the mechanism of mutual recombination of defects (MRM) at the rate of introduction of free defects of $G = 10^{-6}\text{s}^{-1}$.

Figure 4.3 shows that for the case in which E_i^m = 0.5 eV it is almost impossible to select the mechanism of radiation-enhanced diffusion, especially for temperatures of >~0.3 T_m. In this case, on the basis of analysis of the process kinetics, it is necessary to evaluate the time to establishment of dynamic equilibrium whose value for the mechanism of mutual recombination is considerably lower than for the combined mechanism. This graph also shows that the convergence of the diffusion mobilities of interstitials and vacancies decreases the value of the coefficient of radiation-enhanced diffusion as a result of the intensification of the mutual recombination of defects.

The detailed analysis of the kinetic relationships makes it possible not only to interpret correctly the type of mechanism of radiation-enhanced diffusion but also obtain information on its stationary nature. When using averaged relaxation times τ_{50}, this analysis is not possible.

Characteristic examples are represented by the results of analysis of kinetic dependences of electrical resistivity in Ag–8.75 at.% alloy with the variation of the degree of the short range order in the process of electron irradiation [26]. The experimental data for the alloy subjected to efficient annealing (780 °C) and the irradiation temperature range from –20 to +190 °C are presented in Fig. 4.4 and 4.5 [2,18].

The results were processed on the basis of the well-known expression for the variation of the electrical resistivity in the short range ordering or clustering:

$$\frac{d\rho}{dt}=-\frac{(\rho-\rho_\infty)^\gamma}{(\rho_0-\rho_\infty)^{\gamma-1}\tau} \tag{4.24}$$

where γ is the order of the reaction, and ρ_0, ρ_∞ and ρ are the initial, equilibrium and actual values of electrical resistivity.

Relaxation time τ in equation (4.24) is associated with the temperature and time dependence of the coefficient of radiation-enhanced diffusion by the relationship [6]:

$$D=a\tau^{-1} \tag{4.25}$$

where a is a constant.

Analysis of the kinetic relationships in Fig. 4.4 and 4.5, carried out using equations (4.24) and (4.25) shows that in the temperature range 10–130 °C, the coefficient of radiation-enhanced diffusion in the Ag–8.75 at.% Zn alloy increases with radiation time, and at temperatures of –20, –10, 150 and 170 °C it is decreases or does not change. On the basis of the previously examined theoretical considerations, and the increase of the coefficient of radiation-enhanced diffusion with the radiation time is characteristic of the nonstationary mechanism of mutual recombination in metals with a low concentration of constantly acting sinks, i.e. efficiently annealed, in the exit of the concentration of the point defects to dynamic equilibrium. As shown by analysis, the almost stationary kinetics of the variation of electrical resistivity in this alloy at temperatures of –20 and –10 °C is evidently determined by the very small time variation

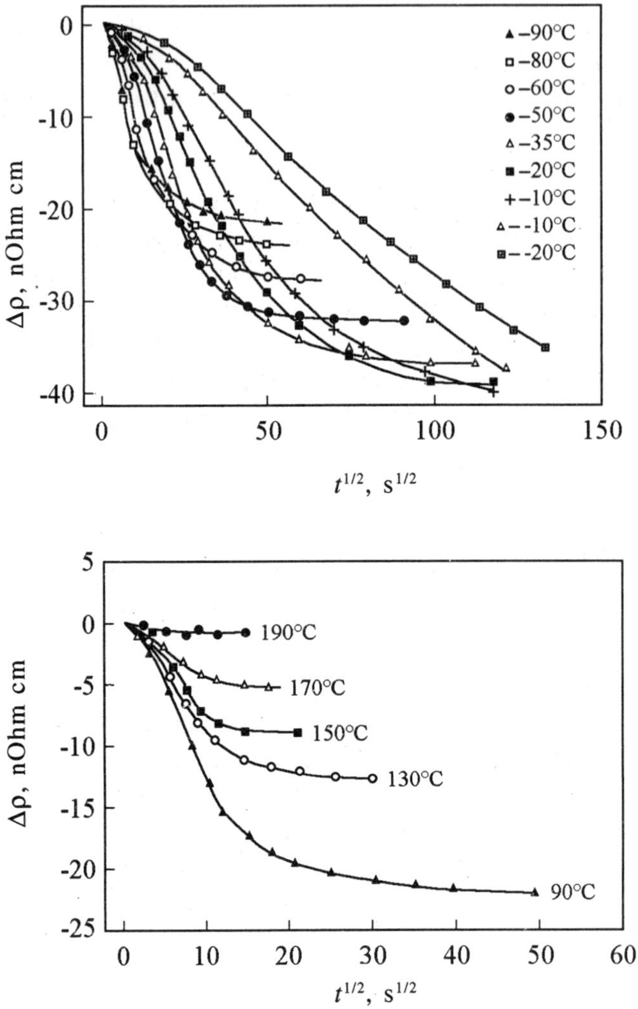

Fig. 4.4 Kinetic dependences of the variation of the electrical resistivity in annealed Ag–8.5 at.%Zn alloy during electron irradiation in the temperature range from -20 to +190°C [2,18]. Electron energy 2.3 MeV, intensity of irradiation 1.5×10^{17} $m^{-2}s^{-1}$.

of the coefficient of radiation-enhanced diffusion as a result of a low mobility of point defects.

The nonstationary solution of the system of equations (3.3) and (3.4) for the mechanism of mutual recombination (the last terms in the right-hand part of these equations are equal to zero) has the following form:

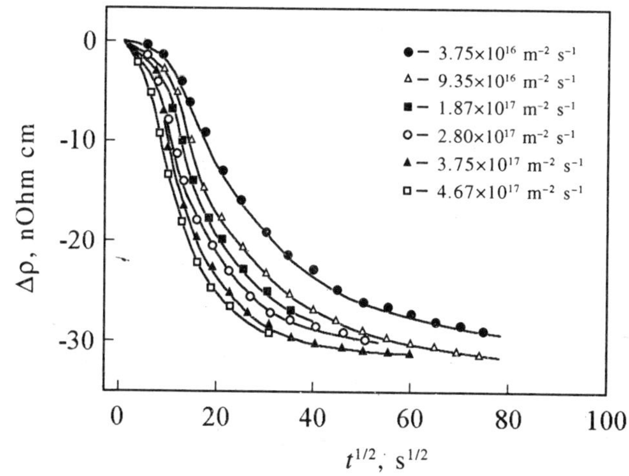

Fig. 4.5 Kinetic dependences of the variation of the electrical resistivity of the annealed Ag–8.75 at.% Zn alloy during electron irradiation at 50°C with different intensities [2,18].

$$c_v = c_i = \left(\frac{G}{R}\right)^{1/2} th\left[(GR)^{1/2} t\right] \tag{4.26}$$

where quantity $(G/R)^{1/2}$ is the quasiequilibrium or dynamically equilibrium concentration of point defects.

Taking the expressions (4.24) and (4.26) into account, it may be shown that relaxation time τ in equation (4.24) for the nonstationary mechanism of mutual recombination of defects is:

$$\tau(t) = \frac{\tau_d}{th\left[(GR)^{1/2} t\right]} \tag{4.27}$$

where τ_d is the relaxation time of the process, corresponding to the dynamically equilibrium mechanism of the mutual recombination of the defects.

The solution of equation (4.24) taking (4.27) into account, for the reaction with the order $\tau > 1$, may be represented in the following form:

$$\Delta\rho_t = \Delta\rho_0 \left\{ 1 - \frac{1}{\left\{1+(\gamma-1) b \ln ch\left[(GR)^{1/2} t\right]\right\}^{1/(\gamma-1)}} \right\} \tag{4.28}$$

where

$$\Delta\rho_t = \rho_0 - \rho_t \tag{4.29}$$

$$\Delta\rho_0 = \rho_0 - \rho_\infty \tag{4.30}$$

$$b = \frac{1}{(GR)^{1/2} \tau_d} \tag{4.31}$$

where ρ_t is the actual value of electrical resistivity.

For the stationary process (τ = const), the solution of equation (4.24) has the following form:

$$\Delta\rho_t = \Delta\rho_0 \left\{ 1 - \frac{1}{\left(1+\frac{\gamma-1}{\tau} t\right)^{1/(\gamma-1)}} \right\} \tag{4.32}$$

The processing of the kinetic dependences in Fig. 4.4 and 4.5 shows that they are efficiently approximated by the equations (4.28) (for the temperature range 10–130 °C) and (4.32) for the temperatures of –20, –10, 150 and 170 °C. These results show quite convincingly that the radiation-enhanced diffusion in the annealed alloy Ag–8.75 at.% Zn in the temperature range 10–130 °C and at temperatures of 150 and 170 °C is controlled by respectively nonstationary and stationary mechanisms of the mutual recombination of point defects.

The dependences of the quantities ρ_∞, $(GR)^{1/2}$, τ_d and b on temperature and the intensity of irradiation, obtained in the processing of experimental data using equations (4.28) and (4.32), are presented in Fig. 4.1, 4.6 and 4.7. The most reliable order of the reaction γ was obtained in processing and was equal to γ = 1.6. The activa-

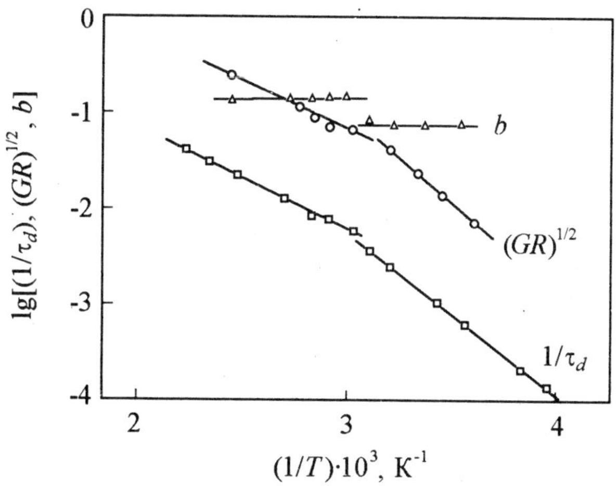

Fig. 4.6 Temperature dependences of parameters τ_d, $(GR)^{1/2}$ and b in the electron-irradiated annealed Ag–8.75 at.% Zn alloy [26].

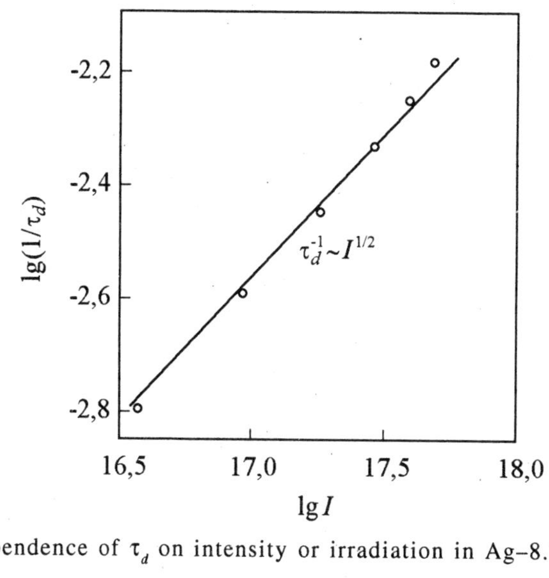

Fig. 4.7 Dependence of τ_d on intensity or irradiation in Ag–8.75 at.%Zn alloy.

tion energies of radiation-enhanced diffusion according to the data, presented in Fig. 4.6 for low- and high-temperature ranges, are 0.3 and 0.23 eV, respectively.

In the analysis of the activation energies of radiation-enhanced diffusion in the Ag–8.75 at.% Zn alloy, the authors of [26] used traditionally the experimental data for the annealing of the electrical resistivity in the quenched Ag–8.75 at.% Zn alloy and results of electron microscope examination of Ag–8.75 at.% Zn alloy, irradiated in the temperature range of 20–300 °C with the electrons with an energy of 1 MeV in a high-voltage microscope. According to results of quenching experiments, the activation energy is 0.62 eV, which is in good agreement with the activation energies of migration of single vacancies in the Ag–8.14 at.% Zn (0.64 eV) and Ag–9 at.% Zn (0.6–0.65 eV) alloys, obtained in the experimental investigations in [23] and [51], respectively.

The experiments carried out in an electron microscope show that the irradiation of the alloys in the temperature range 80–250 °C leads the formation of dislocation loops of the interstitial type, and at temperatures of <80 °C to the formation and growth of tetrahedral stacking faults of the vacancy type. Identical results were also obtained in [51] for Ag–9 at.% Zn alloy.

Thus, the results of quenching and electron microscope experiments make it possible to interpret, on the basis of equation (4.14), the activation energies of radiation-enhanced diffusion of 0.3 eV and 0.23 eV in this alloy for low- and high-temperature ranges as half of the activation energy of migration of monovacancies and interstitials, respectively. These results show that the mechanism of radiation-enhanced diffusion at temperatures of ≤70–80 °C is controlled by the diffusivity of vacancies, whereas at higher temperatures it is controlled by the diffusivity of interstitials. The temperature of transition from the interstitial to the vacancy mechanism of radiation-enhanced diffusion is also clearly recorded on the temperature dependence of the equilibrium values of the electrical resistivity in this alloy (Fig. 4.1).

The identical kinetic analysis of the results for the variation of electrical resistivity in irradiation of Ag–Zn alloys with an increased density of dislocations at temperatures of 10–35 °C shows that the acceleration of diffusion during irradiation in this case is controlled by a nonstationary combined mechanism [7,19,20]. The solution of equation (4.24) for the given mechanism and a relatively long period of time may be described by the equation:

$$\Delta\rho_t = \Delta\rho_0 \left(\frac{\Omega G t}{2\pi r_{vi} S a^2} \right)^{\frac{1}{2}} \qquad (4.33)$$

It is characteristic that, in the given approximation, the kinetics of variation of the electrical resistivity is independent of the order of the reaction.

A typical experimental dependence of the variation of the elec-

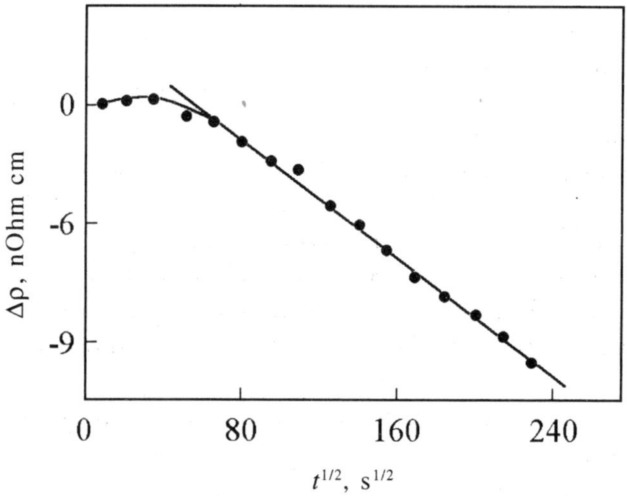

Fig. 4.8 Variation of the electrical resistivity of cold-worked Ag–8.75 at.% Zn alloy during electron irradiation at 10°C [7,19,20]. Electron energy 2.3 MeV, intensity of irradiation 2.8×10^{17} $m^{-2}s^{-1}$.

trical resistivity of a cold-worked Ag–8.75 at.% Zn alloy in the process of electron irradiation at 10 °C, corresponding to equation (4.33), is presented in Fig. 4.8. Special analysis of the experimental data [7, 19, 20] shows that in this case the diffusion mobility of the vacancies is a controlling factor.

The transition from the interstitial mechanism of radiation-enhanced diffusion to the vacancy mechanism, observed in [26], is also in agreement with the corresponding theoretical considerations. In accordance with [13,14,52], the effective mobility of interstitials in binary alloys is the superimposition of the mobilities of the dumbbell configuration of the interstitials of different type, whose concentrations are the functions of composition and temperature. The comparison of modelling ideas [13,14,52] with the experimental data on a high-voltage electron microscopy makes it possible to conclude that at temperatures >~70–80 °C the processes of structural-phase changes and the mechanism of radiation-enhanced diffusion in the Ag–Zn alloys is controlled by the migration of more mobile interstitial dumbbell configurations of the Ag–Ag type with the activation energy of 0.46 eV. A decrease in temperature in accordance with the equations (2.12) and (2.13) is accompanied by the preferential formation of less mobile dumbbell configurations of the Ag–Zn and Zn–Zn type. Consequently, the effective mobility of the interstitials in the alloy greatly decreases and at specific temperatures (in the present case

<70–80 °C) it becomes lower than the mobility of vacancies. In the experiments, this is confirmed by the preferential formation and growth of vacancy clusters – the tetrahedrons of the stacking faults and temperatures below ~80 °C [26,51].

The theoretical calculations and also the results of direct and indirect measurements show that, depending on the intensity of radiation and a number of other factors, the radiation-enhanced diffusion in metals and alloys may prevail at temperatures of up to 0.5–0.6 T_m, and the coefficients of radiation-enhanced diffusion may change in this case in the range from 10^{-20} to 10^{-14} $cm^2\ s^{-1}$ (see, for example [6,10,41,43-50].

Adding up the results of analysis, it is important to note the following main assumptions:

1. The radiation-enhanced diffusion in the metallic materials may intensify structural-phase changes up to temperatures of ~0.5–0.6 T_m, and the coefficient of radiation-enhanced diffusion radiation values of ~10^{-15}–10^{-14} $cm^2\ s^{-1}$, depending on intensity of irradiation.
2. The self-diffusion coefficients and the coefficients of diffusion of solutes, the partial diffusion coefficients and the interdiffusion coefficients by the vacancy mechanism increase with irradiation in proportion to the value of vacancy supersaturation.
3. Intensification of the diffusion of solutes by the interstitial dumbbell mechanism is efficient mainly for the elements with a negative dilation volume. This conclusion follows directly from the results presented in the sections 1.3 and 2.2.
4. A decrease in the coefficient of radiation-enhanced diffusion may be achieved by introducing a higher concentration of diffusion sinks and by alloying with elements reducing the difference between the diffusion coefficients of the interstitials and the vacancies.
5. The approximate estimates of the coefficients of radiation-enhanced diffusion may be made of the basis of a number of the currently available models. In the case of the vacancy mechanism it is also possible to use the experimental data obtained in the investigations of thermal diffusion, phase transformations, and the results of measurements of the parameters of the vacancies in equilibrium and nonequilibrium experiments.

4.3. INTENSIFICATION OF THE PROCESSES OF ORDERING, SHORT-RANGE CLUSTERING AND BREAKDOWN OF SOLID SOLUTIONS IN IRRADIATION

When the process of bulk radiation-enhanced diffusion is dominant, the irradiated metallic systems tend to a thermodynamic equilibrium. However, in this case, the relaxation mechanisms, associated with the excess concentration of the point defects, the dynamic processes of the type of chains of substituting collisions, cascades of displacement, etc., may greatly modify the equilibrium state of the irradiated alloys in comparison with the state of their thermal equilibrium.

4.3.1. Ordering

In a binary disordered *A–B* alloy, the short-range order in each of *i* coordination spheres adjacent to the selected central atom *A* may be characterised in a general case by the Warren–Cowley parameters [53]:

$$\alpha_i = 1 - \frac{N_i^{AB}}{Nc_A c_B c_i} = 1 - \frac{P_i^{AB}}{c_B} \qquad (4.34)$$

In equation (4.34) N_i^{AB} is the number of pairs of *A–B* atoms at a distance r_i, $Nc_A c_B c_i$ is the number of pairs of *A–B* atoms at a distance r_i for the completely disordered distribution of the atoms in the lattice nodes, *N* is the total number of the atoms, c_i is the number of nodes on the *i*-coordination spheres, P_i^{AB} is the probability of atoms *B* being situated in the *i*-coordination sphere in relation to the atoms *A*, situated at the origin of the coordinates.

Direct measurements of parameters α_i by the method of diffusion x-ray scattering were carried out, in particular, in [54] for an ordering Al–14.5 at.% Zn alloy, irradiated with neutrons at 40 °C (fluence 2×10^{29}m^{-2}). The results show that the degree of short order according to the estimates of α_i in the five nearest coordination *C* increases during irradiation. Identical results of radiation stimulation of the process of short-range ordering have been obtained in the investigations of the variation of the properties of ordering alloys at irradiation temperatures in the vicinity and above room temperature [2,4,6,7,10,16–33,39,40,55–58). Typical kinetic dependences of the variation of electrical resistivity in short-range ordering in the radiation process are presented in Fig. 4.4 and 4.5.

In similar investigations, special attention is given to the investigations of the mechanisms of radiation-enhanced diffusion and analysis of the kinetics of the variation of electrical resistivity in order to estimate the type and energies of migration of defects, controlling the short range ordering process under irradiation. At the same time, the temperature dependences of the residual electrical resistivity of irradiated ordering alloys ρ_∞ may also be used for evaluating the ordering energy and the heat of mixing of components of alloys ΔH. This will be examined on an example of analysis of the temperature dependence ρ_∞ of the Ag–8.75 at.% Zn irradiated alloy (Fig. 4.1).

In accordance with theory [59], the residual electrical resistivity of the binary disordered *A–B* alloy for the case if it is determined mainly by the parameters of correlation of the first coordination sphere, is:

$$\rho_\infty = A(c_A c_B - z\varepsilon_{AB}) = A\left[c_A c_B - z c_A^2 c_B^2\left(\frac{\omega}{kT}\right)\right] \tag{4.35}$$

where A is a constant, z is the coordination number.

In a general case, the correlation parameters $\varepsilon(r_i)$ in equation (4.35) are linked with the Warren–Cowley parameters α_i by the relationship [60]:

$$\varepsilon(r_i) = c_A c_B \alpha_i \tag{4.36}$$

For the first coordination sphere [59]:

$$\varepsilon_{AB} = c_A^2 c_B^2\left(\frac{\omega}{kT}\right) \tag{4.37}$$

where ordering energy ω is:

$$\omega = 2\omega_{AB} - (\omega_{AA} - \omega_{BB}) \tag{4.38}$$

In equation (4.38) ω_{AB}, ω_{AA}, ω_{BB} is the binding energy of the pairs of the corresponding atoms.

Ordering energy ω is directly linked with the heat of mixing of the alloy ΔH by the relationship [61]:

$$\Delta H = -\frac{N_{AB}\omega}{2} \qquad (4.39)$$

The number of the nearest A–B pairs is:

$$N_{AB} = zNc_A c_B \left(1-\alpha_1\right) \qquad (4.40)$$

If an alloy is characterised by a tendency for the formation of the short-range order (preferably of the A–B bonds), the order parameter in the first coordination sphere α_1 and the heat of mixing of the alloy ΔH are negative, and the ordering energy ω is positive. For the examined system (preferential bonds A–A and B–B) ΔH and $\alpha_1 > 0$, and $\omega < 0$.

In the subsequent coordination spheres, the sign of α_i and ω_i may change ambiguously in relation to the system [60].

The experimental results for the variation of the residual electrical resistivity of the Ag–8.75 at.% Zn alloy in relation to temperature are correctly described by equation (4.35), as indicated by Fig. 4.1. Processing of the results using equation (4.35) provides the following values for ω: $\omega_1 = 5.8 \times 10^{-4}$ eV (for the temperature range from 60 to 170 °C), $\omega_2 = 3.5 \times 10^{-4}$ eV (for the temperature range from –20 to 50 °C).

The estimates of the heat of mixing ΔH using equation (4.39) and the values of ω_1 and ω_2 give: $\Delta H_1 = -6.8$ cal mol^{-1}, $\Delta H_2 = -4.0$ cal mol^{-1}.

The value of the heat of mixing for the high-temperature range ΔH_1 calculated from equation (4.39), is in agreement with the value of $\Delta H = -(7–10)$ cal mol^{-1} for the Ag–8.75 at.%Zn, according to the results of experiments carried out to determine the experimental values of the heat of mixing [62].

With a decrease in irradiation temperature, the role of radiation-enhanced diffusion decreases, and that of the dynamic processes, leading to disordering, increases. The strongest effect is exerted by the dynamic or ballistic effects [63–65] on the structures with the long-range order.

Since the rate of ordering is proportional to the coefficient of radiation-enhanced diffusion, and the rate of disordering is a linear function of the degree of the long-range order S (see, for example, [66,67]), the resulting effect in irradiation below critical temperature T_c will be determined mainly by the initial condition of the alloy, irradiation temperature and irradiation parameters. On the whole, this is in agreement with the currently available experimental data [66–79].

Detailed analysis of the experimental data on the thermal and radiation ordering of the disordered Ni_3Mo system with the application of computer modelling of the processes of ordering within the framework of the nucleation domain and spinodal model has made it possible to determine a number of characteristic special features of the process of variation of the long-range order under irradiation [67]. One of the main conclusions is that, depending on the ratio of the rate of ordering and disordering, the irradiated system may be in different stages of equilibrium. The consecutive evolution of the stages depending on the radiation conditions may be determined by an increase in the amplitude of the concentration spinodal waves, the interaction of these waves with the formation of nuclei of domains and the break-down of domains during disordering.

On the whole, the results of experimental and theoretical investigations show that due to dynamic or ballistic effects, the relative thermodynamic stability of the structures with the long-range order in irradiation below critical temperature T_c decreases, characterised by the stabilisation of the disordered state or the state with the short-range order. In addition to [63–65], mentioned previously, the theoretical considerations regarding the influence of dynamic effects on the processes of ordering may also be found in a number of reviews and original articles [66,80–85].

The problems of disordering in irradiation are interesting also because of the problems of stability of fine-dispersion precipitates of hardening ordered phases in industrial steels and alloys. Special attention has been given, in particular, to the instability mechanism, based on the fact that the free energy of the disordered state is higher than that of the ordered particle and, consequently, the disordered phase in irradiation shows a tendency for dissolution or transition to another stoichiometric composition (for example, [66]). An experimental example of this phase transition has been presented in [66]. In ion irradiation of the films of Ni_3Si at a temperature of ~200 °C, examination showed disordering of the initial phase and its transformation to the Ni_5Si_2 phase.

4.3.2. Short-range clustering and break-down of supersaturated solid solutions

The main bulk of the experimental investigations into the effect of radiation on short-range clustering using the methods of measurement of electrical resistivity, diffusion and small-angle neutron scattering, transmission electron microscopy and electron diffraction has been

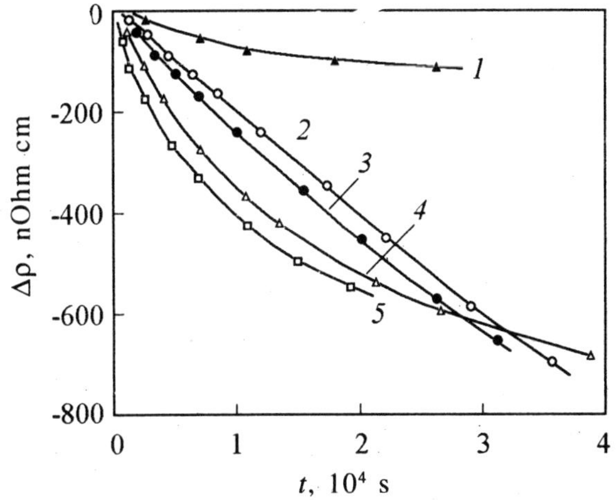

Fig. 4.9 Kinetic dependences of the variation of the electrical resistivity of cold-worked (1) and annealed (2,3,4,5) Cu–44 at.%Ni alloy during electron irradiation [38]. Electron energy 2.1 MeV; irradiation intensity $4\times10^{17}\ m^{-2}s^{-1}$. Irradiation temperature, °C: 134 (1), 128 (2), 153 (3), 215 (4), 295 (5).

carried out on the copper–nickel system [37,38,86–89]. In these investigations, it was established that irradiation greatly accelerates the short-range clustering process and the process of breakdown of the solid solution in the given system.

Figure 4.9 shows the kinetic dependences of the variation of the electrical resistivity of an efficiently annealed (800 °C) and slowly cooled and also cold-worked Cu–44 at.%Ni alloy in the process of electron irradiation. Analysis of a number of experimental data shows that the copper–nickel system undergoes short-range clustering and breakdown by the spinodal mechanism under irradiation [38,87–89], with subsequent formation of a modulated structure [89].

According to the theory of spinodal breakdown [90,91], which, as shown in [92–94], is also suitable for describing the processes of ordering, short-range clustering and breakdown in the metastable region, the distribution of the concentration of the component $c(\vec{r},t)$ in the solid solution of the mean concentration c_0 during time t is described by the equation:

$$c(\vec{r},t)-c_0=\sum_{\vec{k}}A(\vec{k},t)\exp(i,\vec{k},\vec{r}) \tag{4.41}$$

where $\vec{k}$ is the wave vector of the concentration heterogeneity with the wavelength λ.

The variation of the amplitude of the concentration wave with the vector $\vec{k}$ with time has the following form:

$$A\left(\vec{k},t\right)=A\left(\vec{k},0\right)\exp\left[a\left(\vec{k}\right),t\right] \tag{4.42}$$

The amplification factor $a\left(\vec{k}\right)$ in equation (4.42) within the framework of the discrete model [92–94] is:

$$a\left(\vec{k}\right)=-\frac{\tilde{D}}{f''}\left[f''+2\eta^2Y+2KB^2\left(\vec{k}\right)\right]B^2\left(\vec{k}\right) \tag{4.43}$$

In equation (4.43), $\widetilde{D}$ is the interdiffusion coefficient, f'' is the second derivative of the free Helmholtz energy with respect to concentration; η is the deformation of the lattice with respect to the unit difference as regards composition; Y is the effective modulus of elasticity; K is the coefficient of the energy gradient. For the systems with short-range clustering and ordering, the sign of K is usually positive or negative [94]. Consequently, in breakdown inside the coherent spinodal ($f''+2h^2Y \leq 0$), the following inequality is realised:

$$\frac{f''+2\eta^2Y}{K}\leq 0 \tag{4.44}$$

The lattice sum $B^2\left(\vec{k}\right)$ is:

$$B^2\left(\vec{k}\right)=\left(\frac{1}{a_0^2}\right)\sum_{\vec{r}}\left[1-\cos\left(\vec{k},\vec{r}\right)\right] \tag{4.45}$$

For the low values of $\vec{k}$ (long wavelength), the value of $B^2\left(\vec{k}\right)$ can be replaced by k^2, leading to a continual model [90,91].

Direct analysis of the equations (4.41)–(4.43) shows that the intensification of the phase changes in irradiation is possible not only as a result of radiation enhancement of diffusion (self-diffusion coefficient of $\widetilde{D}$) but also as a result of the compensation, by point defects, of the coherent strains, accompanying the transformation

Table 4.1 Values of $(f'+2\eta^2Y)/K$ in Cu–44 at.%Ni alloy [38]

T, °C	215	295	400
$(f'+2\eta^2Y)K$, nm^{-2}	–38	25	110 [96]

processes ($2\eta^2Y$). Evidently, the effect of irradiation on the value of K may operate only at very high concentrations of the radiation defects.

Within the framework of the previously described considerations and the theory of diffusion scattering [95], the authors of [38] analysed the experimental data for the copper–nickel system, presented in Fig. 4.9. The values of $(f''+2\eta^2Y)$/K, obtained in processing together with the value of $(f''+2\eta^2Y)$/K for the Cu–44 at.%Ni alloy from [96], are presented in Table 4.1 and Fig 4.10.

The critical temperature of breakdown on the coherent spinodal T_c in the Cu–44 at.%Ni alloy, determined on the basis of equation (4.44), is ~262 °C. For thermodynamic equilibrium conditions in the absence of radiation, T_c in the alloy of this composition, according to the thermodynamic calculation results [97], is ~200 °C. Evidently, the detected temperature shift is associated primarily with the compensation, by the radiation point defects, of the coherent strains, accompanying the processes of short-range clustering and breakdown (the member $2\eta^2Y$ in equation (4.44)). An identical conclusion was

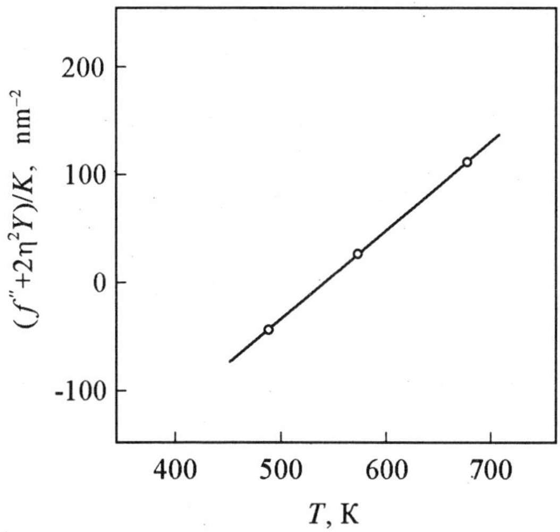

Fig. 4.10 Temperature dependence of $(f'+2\eta^2Y)/K$ in alloy Cu–44 at.%Ni [38].

made in [89], where in addition to the copper–nickel alloy, the increase of the temperature of spinodal breakdown in electron irradiation was also detected in copper–titanium and iron–molybdenum solid solutions.

According to the theory [90, 91], the difference of the temperatures of coherent and non-coherent spinodals ΔT is:

$$\Delta T = \frac{2\eta^2 Y c_A c_B \Omega}{k_B} \tag{4.46}$$

According to theoretical and experimental estimates for Cu–Ni alloys, including the studies [88,89,97], the value of the ΔT is in the range 30–70 K, which is in good agreement with the temperature shift $\Delta T \approx 62$ K, obtained in [38].

The experimental results show that the compensation of coherent strains by point defects plays a significant role in both intensification and modification of the processes of short-range clustering and breakdown of the solid solutions in irradiation, displacing the transformations from the metastable to the unstable region. This may be accompanied by a change in the breakdown mechanism, transforming from the classic mechanism of nucleation, typical of the metastable region [91], to the spinodal mechanism.

In addition to the previously mentioned Cu–Ni, Cu–Ti and Fe-Mo solid solutions, the spinodal-type breakdown with the formation of spatially-oriented or spatially-modulated structures in irradiation was also detected in matrix solid solutions of a number of steels and alloys of the systems Fe–Ni [98,99], Fe–Cr–Ni [98–102], Fe–Cr–Mn [103] and V–Ti–Nb–N [104]. Interpretation of these results is also possible on the basis of theoretical considerations, examined previously, and a number of other theoretical concepts (see, for example [99]). Analysis of the radiation-stimulated breakdown in the iron-nickel system has resulted in the conclusion according to which the zone of the miscibility gap in this alloy in irradiation greatly expands and is displaced into the range of higher temperatures [99].

In a series of investigations of the Cu–Ni–Fe system, characterised by the breakdown by the spinodal mechanism, it was shown [105-107] that, depending on the conditions and parameters of the experiments (temperature, type of particles, particle energy and intensity), irradiation may both stimulate the process of breakdown and also cause dissolution of the precipitate as a result of cascade mixing. As in the case of the previously examined effects of the instability of the ordered structures, this transformation is characteristic of high-intensity ion irradiation and can hardly play an appreciable role for structural reactor materials. In particular, the previously mentioned experimental data for the neutron irradiation of iron–nickel and iron–chromium–nickel alloys [98,101,102] confirmed this conclusion.

In the initial stages of the development of radiation solid-state physics and materials science, the main direction of investigations into the phase transformations in irradiation was associated mainly with the examination of the kinetics and mechanism of breakdown of supersaturated solid solutions based on copper, aluminium and nickel (for example, [108–111]) in which the breakdown process takes place by the classic mechanism of formation of nuclei. Special attention to radiation-stimulated phase transformations in similar systems has been paid in recent years and investigations are continuing (for example, [82, 112–121]). These investigations are of both purely fundamental [112–117] and fundamental–applied nature [118–121]. In fundamental investigations, special attention is given to the special features of radiation-stimulated breakdown, the relationships governing the transformation of the Guinier–Preston zones into more stable precipitates, depending on the conditions and parameters of irradiation. In fundamental-applied investigations, the breakdown mechanisms are analysed in connection with the problems of radiation damage of the materials of the given grade. In the previously mentioned studies [118-121], the mechanisms of phase transformations under irradiation have been examined and discussed in connection with the problems of swelling of vanadium alloys [118,119] and embrittlement of vessel steels [120,121].

The main theoretical representations regarding the radiation-stimulated breakdown of supersaturated solid solutions within the framework of the mechanism of formation and growth of nuclei of the precipitates have been developed in [122–125]. In a general case, they are also suitable for the analysis of the phase instability of metallic systems under irradiation, including under-saturated solid solutions.

In [122,123], the theoretical analysis of the processes of the nucleation and growth of non-coherent particles in irradiation has been carried out taking into account the factors of supersaturation of solid solutions by vacancies and solutes, and also the parameters of dimensional mismatch of the matrix solution and the precipitate. The evolution of precipitation is described by the trajectory of its movement in the phase space with 'the concentration of the vacancies in a cluster – concentration of atoms of solutes' coordinates. The rate of the process is determined by the rate of capture and emission of the vacancies and solutes by the precipitates, and also by the rate of absorption of the interstitials. A significant role in the acceleration of the nucleation and growth of the precipitates within the framework of the given model is played by the mechanism of compensation of elastic strains at the matrix-particle boundary and by the vacancies. With an increase of vacancy supersaturation, the rate of breakdown of the solid solution rapidly increases.

Using the basic assumptions and kinetic formalism, identical with the models in [122,126], the authors of [124,125] proposed a recombination mechanism of breakdown with the precipitation of coherent particles. This model examines the clusters of atoms of solutes present at the periphery and containing either excess interstitials or vacancies. As a result of the arrival of the solutes by the vacancy (in complexes with vacancies) and interstitial mechanisms to clusters with defects with the opposite sign and subsequent mutual recombination of defects, the number of atoms of solutes in clusters increases. The process of departure of the atoms of solutes by the interstitial and vacancy mechanisms is also taken into account in this case. The given model was used for deriving an equation for the shift of the limit of solubility in the irradiation conditions depending on the ratio of the coefficient of diffusion of the dissolved element by the interstitial and vacancy mechanisms and on the parameter, determining the relative probability of the localisation of vacancies and interstitial atoms on the clusters.

The results of analysis of the experimental data for supersaturated solid solutions within the framework of the previously examined mechanisms [122–125] can be found, in particular, in [81,112,113, 116].

The process of intensification of ordering, short-range clustering and breakdown with the tendency for the approach of the irradiated metallic systems to a more equilibrium condition may be used as a method for preliminary irradiation treatment of alloys, stabilising the properties of materials in service. The authors of the present book used in practice this method for stabilising the electrical resistivity of resistive alloys, such as Nichrome (Fe–Ni–Cu), Constantan (Cu–Ni), Ag–Pd and Pd–Ag. The first two systems relate to systems with short-range clustering, and the two other to ordering metallic materials. Preliminary electron or γ-irradiation of the resistors based on these alloys resulted in the stabilisation of their properties during subsequent operation of these elements in electronic systems, including space stations.

During the service of metallic materials in the irradiation conditions, the process of intensification of phase transformations, being one of the factors of destabilisation of their properties, may also lead to a decrease in the radiation damage in the irradiated alloys. Primarily, this is associated with the annihilation of point effects during the compensation of elastic strains, accompanying the processes of ordering, short-range clustering and breakdown, and also mutual recombination of the defects at the interphase boundaries. These mechanisms decrease the dynamically equilibrium concentration of point defects, thus suppressing the processes of nucleation and growth of clusters of defects, dislocation loops and voids. In particular, in [118, 119], the radiation-stimulated breakdown of vanadium–titanium β-solid solutions with the formation of the α-Ti phase is regarded as one of the important factors for the high swelling resistance of the alloys of the vanadium–titanium system. According to the results obtained in [127], at the specific values of the wavelength of the modulated structure it is possible to obtain the minimum value of swelling during radiation-stimulated breakdown in a nickel–titanium alloy. A very large number of experimental examples of similar types may be found in the currently available scientific literature on the radiation damage of metallic materials.

4.4. PHASE INSTABILITY OF METALLIC MATERIALS UNDER IRRADIATION

In the previous section, attention was given to the intensification, during irradiation, of phase transformations with a tendency for reaching a thermodynamic equilibrium. This is caused by the fact that, in this case, the controlling mechanism is radiation-enhanced diffusion, the mechanism is associated with the excess concentration of point defects and dynamic processes may, as will be shown, greatly modify and even change qualitatively the condition of thermodynamic equilibrium of irradiated alloys. Some similar mechanisms, such as ordering, cascade mixing and compensation of elastic strains by point defects during phase transformations have already been discussed to some extent in the previous chapter. In this section, special attention will be given to the processes controlling the phase instability and the displacement of the phase equilibrium of metallic systems under irradiation. In a general case, phase instability is the deviation of the phase composition of alloys from their real or practically attainable thermodynamically equilibrium condition in the absence of irradiation, and also the variation of the chemical composition of solid solutions under irradiation.

The theoretical assumptions regarding the mechanisms of phase instability during irradiation are being developed at the present time on the basis of thermodynamic and kinetic approaches.

The calculations of the shift of phase equilibrium under irradiation within the framework of the thermodynamics of equilibrium processes are based mainly on the evaluation of the difference of free energies of two analysed phases as a result of the difference of the concentration of point defects in the phases and of the energy of interaction with the components of the phases [128–131]. This value is compared with the energy required for the phase transition which is $\sim 10^{-4}$– 5×10^{-2} eV atom^{-1} [128,130,131]. According to similar estimates, the large displacement of the phase equilibrium is possible only on the condition if the difference of the concentration of point defects in analysed phases is $>10^{-4}$–10^{-3} [130,131]. The cascade mechanisms, thermal peaks and other processes may additionally change the free energy of the irradiated material and support a shift of the phase equilibrium [64,132].

The main successes in the interpretation of phase instability under irradiation have been achieved within the framework of the kinetic approach. The kinetic mechanisms will also be examined in further considerations.

4.4.1. Instability of under-saturated solid solutions

4.4.1.1. The mechanisms of instability of under-saturated solid solutions

In a relatively general case, the variation of the concentration of interstitials, vacancies and atoms of solutes in the solid solution may be represented in the following form:

$$\frac{dc_i}{dt} = G - Rc_v c_i - c_i D_i S_i - divJ_i \qquad (4.47)$$

$$\frac{dc_v}{dt} = G - Rc_v c_i - c_v D_v S_v - divJ_v \qquad (4.48)$$

$$\frac{dc_s}{dt} = -divJ_s \qquad (4.49)$$

Depending on the specific models, calculations of the concentration of point defects and analysis of the instability of solid solutions with respect to the chemical and phase composition were carried out using different modifications of the system (4.47)–(4.49). At present, equations identical to (4.47)–(4.49) are being used to develop the following main mechanisms of instability of under-saturated solid solutions.

1. The segregation mechanism

The kinetic mechanism of the variation of the concentration of solutes at sinks was proposed for the first time in the studies by Antony [133,134]. The main assumptions of the model were examined in section 2.3. Equation (2.24) shows that when the vacancies are strongly bonded with the atoms of solutes ($\omega_3 \ll \omega_1$), the flux of dissolved atoms J_B has the same direction as the flux of vacancies J_v. In fact, this condition indicates the transfer of solutes to the sink in the form of complexes with vacancies. In the opposite case, when there is almost no interaction of the impurities and vacancies ($\omega_1 \cong \omega_2$), the flux of the dissolved atoms is opposite to the flux of the vacancies. In contrast to the direct Kirkendall effect (see, for example [135]), when the appearance of a concentrational heterogeneity results in the formation of a vacancy flux, this mechanism is referred to as the reversed Kirkendall effect [136]. In contrast to the diffusion of complexes, this effect is typical of solutes, migrating by the volume vacancy mechanism.

In the model proposed by Johnson and Lam [137,138], developed for the surface segregation in diluted alloys, attention is given to the possibility of transfer of solutes to a sink and by the interstitial mechanism. The interstitial mechanism of segregation of solutes of sinks may also be regarded as the reversed Kirkendall effect [138] because in this case the flux of the defects to a sink brings into existence the concentrational heterogeneity.

The further development of modelling assumptions regarding the segregation mechanism during irradiation in diluted and concentrated binary and also multicomponent alloys, and formation of phases during segregation in under-saturated solid solutions, was carried out in [52, 139–149]. The main relationships of the segregation mechanism, characteristic of both diluted and concentrated alloys – the controlling role of the interstitial mechanism, enrichment and depletion of the sinks with the solutes with negative and deposited dilation volumes, respectively, have been confirmed in many experimental investigations of radiation-stimulated segregation. The appropriate experimental data have been published in many original and review studies (for example, [26,64,82,150–166]).

The preferential enrichment of sinks with elements with $\Delta V < 0$ also indicates the significant role of the interstitial segregation mechanism where the transfer of solutes is realised as a result of fluxes of mixed dumbbells. Typical experimental examples of the breakdown of under-saturated solid solutions on the basis of nickel, aluminium, palladium, silver, etc and other systems with the formation of phase precipitates at sinks (surfaces, grain boundaries, radiation defects) as a result of the operation of this mechanism can be found in, for example, [26,64,82,150,151,157–159,164,167,168]. At the same time, for solutes with $\Delta V > 0$, associated with vacancies, transfer of these elements to sinks in the form of vacancy complexes, with subsequent formation of segregation precipitates, is also possible. The characteristic experimental examples of a similar type have been published in, in particular, [167,168], where phases Mg_5Al_8 and Mg_2Si formed at voids in neutron-irradiated under-saturated solid solutions of the aluminium–magnesium system (ΔV Mg in Al > 0 [169]).

It should also be mentioned that the radiation-stimulated segregation processes with the participation of impurity or transmutation elements may modify the chemical composition of primary precipitates. In the previously cited studies [167,168], for example, the initial segregation phase Mg_5Al_8 at voids transform partially to the Mg_2Si phase as a result of the segregation of the atoms of transmutation silicon. In irradiated vanadium–titanium alloys, according to [118, 119], the formation of compounds of titanium with carbon, oxygen and nitrogen is preceded by the radiation-stimulated breakdown of the vanadium–titanium β-solid solution with the formation of precipitates of the α-Ti phase. Subsequent segregation of the impurities on the precipitates of the α-Ti phase leads to its transformation to oxides, carbonitrides or oxycarbonitrides of titanium, typical of the irradiated vanadium–titanium alloys. It is characteristic that the reflections from the precipitates of the α-Ti phase are fixed only for relatively low radiation doses and are not detected in further irradiation [118,119].

The mechanism of radiation-stimulated segregation (like all other mechanisms of phase instability) may have a significant effect on the radiation resistance of structural metallic materials, changing the power of the sinks, and the chemical composition of the matrix solid solutions thus causing changes in the dynamic concentration of point defects and affecting, at the same time, the processes of the nucleation and growth of clusters, dislocation loops and voids. Segregating at sinks of different type, the atoms of dissolved or transmutation elements may change the strength of obstacles to the movement of dislocations, concentrated in the form of unfavourable impurities at the grain boundarics, initiating at the same time the processes of radiation hardening and embrittlement. The prediction and control of radiation-stimulated segregation phenomena in real metallic materials are important factors of decreasing their radiation damage.

The segregation mechanism is also important in the ion treatment of metallic materials where the result of modification of the irradiated surface is determined, in the final analysis, by the competing processes of segregation (and formation of phase precipitates) and sputtering (see, for example, [170,171]). This complex mechanism of the changes of the surface properties may also be controlling for materials of fusion reactors and, in particular, for divertor systems subjected to the high-intensity effect of ions.

2. The recombination mechanism

This mechanism has been proposed in [124,125] and relates to homogeneous (in contrast to the heterogeneous segregation mechanism) mechanisms of phase instability of under-saturated solid solutions in irradiation. The main modelling assumptions of this mechanism have been examined briefly in section 4.3.2 The interpretation of the experimental data on the breakdown of aluminium–zinc, copper–beryllium, nickel–beryllium and tungsten–rhenium under-saturated solid solutions, in irradiation within the framework of the given mechanism, has been published in [64,124,125,172–174].

3. The spinodal mechanism

This mechanism was examined for the first time in studies by Martin [175,176]. It is based on the application of assumptions on the spinodal breakdown of solid solutions [90,91] for describing the formation of an instability of the homogeneous concentration of point defects with respect to its spatial fluctuation. Taking into account the relationship of the fluxes of point defects and components of the alloys (fluxes $J_{i,v}$ and J_s in the equations (4.47)–(4.49)), the appearance of concentrational heterogeneities in the modulation of point defects should result in the instability of solid solutions with respect to the chemical and phase composition. This instability increases with a decrease in the irradiation temperature and with an increase in the speed of introduction of defects and the concentration of the second component. As mentioned in [177], this homogeneous mechanism may be used for interpreting the formation of radiation-stimulated precipitates in under-saturated aluminium–zinc, aluminium–silver, aluminium–silicon, copper–beryllium and tungsten–rhenium solid solutions. The experimental results on the breakdown of solid solutions in the systems are summarised in [178].

4. The cascade mechanism

The cascade mechanism was proposed in [179,180] and is based on the diffusion redistribution of components of the alloys during annealing of interstitials and vacancies in the process of relaxation of displacement cascades. As a result of the effect of this mechanism, the phase instability of under-saturated solid solutions is also possible.

4.4.1.2. Analyses of the experimental data

To conclude this section, attention will be given to a number of characteristic experimental data on the breakdown of under-saturated diluted and concentrated solid solutions based on aluminium and silver under irradiation with electrons and neutrons, taken from studies [26,167, 168,181–185] by the authors of the present book, and they will be analysed on the basis of the previously examined theoretical considerations. These results are summarised in Table 4.2, with a brief description of the experimental conditions [168].

1. Silver–zinc alloys

The irradiation of alloys of this system in a high-voltage electron microscope at temperatures of >80 °C results in the formation of dislocation loops of the interstitial type. The irradiation temperatures lower than 80 °C are characterised by the formation and growth of tetrahedrons of the vacancy-type stacking faults (Fig. 3.16) with the second appearance of fine-dispersion pre-precipitates in the zone of the tetrahedrons. Further irradiation leads to the formation of a spatially oriented structure in the regions of the order of several hundreds of nm. The formation of precipitates in the zone of the tetrahedrons is accompanied by the appearance of diffraction reflections, corresponding to the hexagonal ζ-phase. This phase borders with the silver-zinc α-solid solutions on the equilibrium diagram [186].

The mechanisms of breakdown in the systems are correctly explained on the basis of the segregation model [168,184], examined in Chapter 3 (section 3.7.2.). The kinetics of breakdown is controlled by the flux of silver–zinc mixed dumbbells to the vacancy tetrahedrons

Table 4.2 Phases in irradiated under-saturated solid solutions based on Ag and Al [168]

Phase	Alloy, at.%	Irradiation, annealing	Type of precipitate	Literature
ζ	Ag–(1–25)Zn	el., Φ >4×10^{25}m^{-2}, <80°C	Precipitates in the zone of vacancy tetrahedrons	26, 184
Mg_5Al_8	Al–0.06Mg Al–0.54Mg	el., Φ >10^{25} m^{-2} ~20°C	Plate-shaped precipiates with tetrahedrons in the matrix	181, 182
Mg_5Al_8	Al–0.54Mg Al–2.1Mg	n, Φ =2.6×10^{24} m^{-2} ~100°C, annealing>150°C	Needle-shaped precipitates in the matrix precipitates at voids	167, 181
Mg_2Si	Al–2.1Mg	n, Φ = 2.6×10^{24} m^{-2} ~100°C, annealing >400°C	Precipitates at voids and grain boundaries	167, 181
$Mg_{32}(Al,Zn)_{49}$	Al–0.02Zn–0.01Mg	n, Φ = 2.6×10^{24} m^{-2}, ~100°C	Plate-shaped precipitates in the matrix	167, 181
$Mg_{32}(Al.Zn)_{49}$	Al–0.1Zn–0.06Mg	el., Φ >10^{26} m^{-2} 100–130°C	Plate-shaped precipitates with tetrahedrons in the matrix	185
Al–Mg–Zn	Al–0.1Zn–0.06Mg	el., Φ >10^{26} m^{-2} 100–130°C	Plate-shaped precipitates with tetrahedrons in the matrix	185

of the stacking faults (equation (3.87)). The estimates of the incubation period of the start of precipitation of zinc atoms in the zone of the tetrahedrons according to equation (3.87) give the value $\tau_i \cong 15$ min which is in good agreement with the experimental values (12–20 min).

The nucleation and growth of the ζ-phase during irradiation are also supported by the crystallographic structure of the resultant tetrahedrons. In fact, the vacancy-type tetrahedral stacking faults, observed in [26], have crystal faceting in the {111} close-packed planes of the FCC lattice of the matrix. In contrast to the normal volume of the matrix of the FCC lattice with the alternation of the close-packed planes in the [111] *ABCABC*... directions, the crystal structure in the regions of the tetrahedrons, containing the stacking faults, represents the alternation of the close-packed {111}*ABAB*... planes in the <111> direction. Identical packing is also found in the hexagonal close-packed structure in the [0001] directions. Thus, the formation of tetrahedrons creates suitable crystallographic conditions for the structural transition from the FCC to the HCP lattice. The inflow of the zinc atoms in the form of mixed dumbbells to the tetrahedrons causes stabilisation and growth of the nuclei of the hexagonal ζ-phase, because the ratio the concentrations c_{Zn}/c_{Ag} for this phase is higher than for the silver–zinc α-solid solution.

It should be mentioned that this transformation takes place at temperatures at which the mobility of interstitials is lower than the mobility of vacancies (see sections 2.2; 3.2; 3.6 and 3.7). It is of the purely radiation nature, because the process of the breakdown is controlled by the flux of the Ag–Zn mixed dumbbells to the sinks.

2. Aluminium-based alloys

Table 4.2 shows the generalised data on the precipitation of phases during irradiation in aluminium–magnesium and aluminium–magnesium-zinc under-saturated solid solutions. The type of phases, presented in the Table 4.2, has been identified in decoding electron diffraction patterns on the basis of comparison of the measured planar spacings and the orientation relationships between the phase and the matrix with the corresponding standard values.

It should be stressed that the formation of Mg_5Al_8, $Mg_{32}(Al,Zn)_{49}$ and Al–Mg–Zn phases under electron irradiation was observed at high rates of introduction of point defects (~0 $3\times10^{-4}s^{-1}$), when irradiation was applied in the channelling regime. In Al–0.06 at.% Mg and Al–0.5 at.%Mg alloys, the formation of pre-precipitates of Mg_5Al phase in the form of the zones with uniform contrast (Guinier–Preston type zones) was preceded by the formation of dislocation loops of the interstitial type. At doses of ~3×10^{25} m^{-2}, tetrahedral stacking faults appeared inside the zones of uniform contrast. The loops and tetrahedrons were characterised by spatial correlation: the mean distance between them was three times smaller than in the case of orderless distribution.

It is believed that the formation of the Mg_5Al_8 phase can be efficiently interpreted on the basis of the spinodal mechanism [175, 176]. In this case, the formation of regions with a high local concentration of point defects is accompanied by an increase of the content in the zones of the solute atoms, associated with the point defects. Since the vacancies in aluminium–magnesium alloys are preferentially bonded with magnesium atoms [133,134], an increase in the concentration of these atoms in specific regions of the crystal will be accompanied by a simultaneous increase in the magnesium concentration. The formation of precipitates in the zones is beneficial: the vacancies compensate the volume dilations, formed in the solid solution with an increase in the magnesium content. When the magnesium concentration in the zones reaches the limit of homogeneity for the Mg_5Al_8 phase, its formation will take place with a large decrease in the volume per atom (1.9×10^{-2} nm^3 and 1.77×10^{-2} nm^3 for the solid solution and phase, respectively). As a result of this transformation, the zone of formation of the phase becomes supersaturated with vacancies which may coalesce into loops or tetrahedrons. With an increase in the magnesium content, the stacking fault energy decreases [187]. This is beneficial for the formation of tetrahedrons. The general schema of this transformation is as follows: GP $\rightarrow$ phase $Mg_5Al_8 \rightarrow$ tetrahedron.

The most probable mechanisms of the formation of Mg_5Al_8 phase during neutron irradiation are the segregation mechanisms of different types. The formation of the precipitates of the Mg_5Al_8 phase of the needle shape in the matrix of Al–0.54 at%Mg and Al–2.1 at% Mg is observed together with the vacancy loops and high density of the dislocations. This dislocation structure corresponds to the formation of a dislocation network in the stage in which coalescence of dislocation loops of the interstitial type has already taken place. The formation of needle-shaped precipitates of the Mg_5Al_8 phase is the result of the formation and evolution of this dislocation structure. In the initial stage of irradiation, the nucleation of this phase takes place on the dislocation loops of the interstitial type, as a result of the fluxes of the magnesium atoms to the loops in the form of complexes with vacancies [52]. The growth of the segregation zones and the precipitation of the Mg_5Al_8 phase take place in the process of coalenscence of loops and the formation of the dislocation network.

Formation of the Mg_5Al_8 phase at voids was detected only after annealing of neutron-irradiated alloys at temperatures greater than 150 °C. The characteristic features is that immediately after irradiation, the voids in these alloys, in contrast to pure aluminium, were not detected. In subsequent annealing, the increase of the concentration of the magnesium atoms, accompanying the process of coalenscence of the voids, results in the formation of the Mg_5Al_8 phase. Possibly, the nucleation of this phase at the voids also stimulates the cascade mechanism of instability of under-saturated solid solutions [179,180].

The formation of the Mg_2Si phase on the voids was detected together with the Mg_5Al_8 phase but at higher annealing temperatures ($\geq$ 400 °C) and only in the Al–2.1 at.% Mg alloy with the highest concentration. The principal possibility of the formation of this phase is associated with a large cross-section of production of transmutation silicon in accordance with the nuclear reaction $^{27}Al\,(n,\gamma)^{28}Si$. Previously, this phase was detected in aluminium–magnesium alloys

(> 1% Mg) after irradiation with a considerably stronger neutron fluence (greater than $10^{26}\ m^{-2}$) [188]. In the experiments carried out by the authors of this book, irradiation with smaller fluences resulted in the formation of the Mg_2Si phase only after annealing in the conditions of intensive coalescence of the voids – the segregation centres for the atoms of magnesium and silicon in aluminium.

Special attention must be given to the formation of $Mg_{32}(Al,Zn)_{49}$ and Al–Mg–Zn phases under electron and neutron irradiation. The formation of plate-shaped $Mg_{32}(Al,Zn)_{49}$ phase in the matrix was detected for the first time after neutron irradiation of the Al–0.02 at.%Zn–0.01 at.% Mg alloys. The authors of the present book believe that, in this case, the mechanism of the formation of this phase is identical with the previously examined mechanism of formation of the needle-shaped Mg_5Al_8 phase in the matrix of neutron-irradiated aluminium–magnesium solid solutions. Taking into account the fact that in the under-saturated solid solutions of the aluminium–magnesium system, the formation of the Mg_5Al_8 phase was detected in both neutron and electron irradiation [167,182], Table 4.2, the authors of [185] carried out experiments with the formation of $Mg_{32}(Al,Zn)_{49}$ phase in under-saturated aluminium–magnesium–zinc solid solutions under electron irradiation. The second reason for conducting these experiments was the evaluation of the possibility of simulation of the transmutation phase transformations during neutron irradiation by the method of preliminary alloying and subsequent electron irradiation. In aluminium-based alloys, the production of, in particular, transmutation magnesium takes place by the (n,α) reaction.

The results of these experiments are presented in Table 4.2. In contrast to neutron irradiation, the electron irradiation of the Al–0.1 at.% Zn–0.06 at.% Mg under-saturated solid solution is characterised by the formation of two phases: $Mg_{32}(Al,Zn)_{49}$ and Al–Mg–Zn. The formation of an additional phase in comparison with neutron irradiation is associated in this case with a higher magnesium content of the initial alloy (the magnesium concentration of the Al–Mg–Zn phase is higher than in the $Mg_{32}(Al,Zn)_{49}$ phase. Since the magnesium and zinc atoms in the solid solutions of aluminium are bonded with vacancies [133,134], it may be assumed that the mechanism of breakdown of the under-saturated aluminium–magnesium–zinc solid solutions during electron irradiation is identical with the previously examined mechanisms for identical conditions of irradiation of the aluminium–magnesium alloys.

On the whole, the results of these experiments indicate the possibility of simulation of phase transformations during electron irradiation (including transmutation effects), using electron irradiation.

4.4.2. Variation of the phase composition in compensation by point defects of deformation effects of phase transformations

The fundamentals of the mechanism have already been examined in the section 4.3.2 in the analysis of the processes of intensification of short-range clustering and breakdown of the solid solutions within the framework of the spinodal and nucleation models. During the formation of non-coherent nuclei, as shown in [122,123], the flux of vacancies stabilises oversized precipitates of the phases with $\delta = [(\Omega_p - \Omega_m) / \Omega_m] > 0$ and destabilises the undersized particles, for which $\delta < 0$. In this expression, Ω_p and Ω_m are the corresponding volumes per atom in the precipitates and the matrix. For thermally equilibrium conditions, the particles with $\delta < 0$ may be stable. Consequently, not only the great acceleration of the process of breakdown of the supersaturated solid solution is possible, but there may also be changes of the phase composition of the irradiated material in comparison with the thermodynamically equilibrium condition as a result of changes in the stoichiometry of the precipitates.

4.4.3. Phase instability, determined by dynamic radiation defects

The problems of phase instability, associated with this problem, have already been examined in the section 4.3.1 when discussing the process of radiation-stimulated ordering. They include the changes of the phase equilibrium, mainly as a result of the formation of chains of substituting collisions and cascades of atomic displacements. In contrast to radiation-enhanced diffusion, stimulating the transition of the irradiated systems to the equilibrium condition, the dynamic or ballistic effects prevent the establishment of thermodynamic equilibrium. When these effects are dominant, the phase equilibrium may be greatly shifted. The characteristic manifestation of the dynamic effects are processes of disordering, the dissolution of equilibrium phases and the formation of metastable compounds, amorphisation, cascade mixing. Typical experimental examples, confirming the previously mentioned effects, are included also in [64,66,67,80–83, 106,107,132,150,189–195].

The initial simulating considerations regarding the influence of dynamic effects on the structure of a two-phase alloy were published in [189]. The instability of the particles of the second phase was analysed within the framework of two mechanisms: the disordering of the crystal lattice of precipitates, and dissolution of particles during the ballistic transition of the solutes from the precipitate back into the solid solution. Analysis of the process, taking into account the mechanism of radiation-enhanced diffusion, results in the modification of the structure of the irradiated alloy: the small particles show a tendency for growth, large ones for dissolution. This dimensional tendency of the particles is directly opposite to the dimensional effects, characteristic of the thermal coalescence mechanism, proposed by Lifshits, Slezov and Wagner [196,197], in which the large particles grow as a result of dissolution of smaller ones. At specific temperatures, below the critical temperatures, the dynamic effects are dominant and the rate of dissolution of the particles exceeds the rate of growth of the particles as a result of radiation-enhanced diffusion. The critical temperature increases with increase of the rate of atomic displacements.

The simulating considerations of the influence of the dynamic effects on phase instability under irradiation were developed further in [63-65,83–85,198–206]. The theoretical analysis of the phase instability of the binary alloy within the framework of the theory of spinodal breakdown [90,91] shows [63] that the phase condition of the alloy, irradiated at a temperature of T, corresponds to the phase condition of the same system at temperature T' in the absence of irradiation. The ratio between the temperatures T and the T' is [63–65]:

$$T' = (1+\Delta)T \tag{4.50}$$

$$\Delta = (\phi\sigma_d)^{1/2}\frac{\sigma_r}{\sigma_d}\exp\left(\frac{E_m^v}{kT}\right) \tag{4.51}$$

In formula (4.51), ϕ is the intensity of irradiation, σ_d and σ_r are the cross sections of displacement and substitution, respectively; E_m^v is the migration energy of vacancies. This equation corresponds to the case in which the mechanism of mutual recombination of point defects is dominant in irradiation.

The physical interpretation of the results is very convincing: at the radiation temperatures at which the ballistic effects play a certain role, the more nonequilibrium state of the system is realised. This state is the one which in the absence of irradiation the system shows at a higher temperature.

On the whole, the influence of dynamic effects on the phase stability of irradiated alloys may be summarised as follows:

1. In a general case, the more nonequilibrium state of the irra-

diated system is realised;

2. The limit of solubility of the particles of the second phase increases.

3. The more disordered state is stabilised.

4. The two-phase region of coexistence of the disordered and ordered phases is expanded.

5. Nonequilibrium and metastable phases, including amorphous phases, form and are stabilised [132,191–195].

4.4.4. Phase instability caused by transmutation effects in nuclear reactions

The authors of [207] examined for the first time the mechanism of instability of the *A–B* two-phase alloy with the precipitates of A_mB_n particles, caused by the burning-out of the dissolved element *B* as a result of transmutation nuclear reactions during irradiation. The burning out of the element *B* in the alloy during its transmutation transformation $B \rightarrow C$ introduces the diffusion flow of the atoms *B* from the particle of the second phase to the solid solution for maintaining the concentration of component *B* in the solid solution in the vicinity of its solubility limit. As a result, this process together with a decrease in the size of the precipitates during direct burning out of atoms *B* in the particles of the second phase, stimulates their dissolution during irradiation.

Figure 4.11 shows the results of numerical calculations of the evolution of distribution of the $ScAl_3$ particles with respect to the size in an aluminium–scandium alloy during burning out of scandium by the $(n,2n)$ and (n,α) reactions for different neutron fluxes, carried out on the basis of the theory proposed in [207]. The kinetics of variation of the relative mean size of the $ScAl_3$ particles and their density during irradiation is presented in Fig. 4.12. The calculation results show directly (Fig. 4.12) that the particles of $ScAl_3$ in the aluminium–scandium alloys may, as a result of burning out of scandium, fully dissolve after irradiation with neutron fluxes of $\sim 5\times10^{27}$ m^{-2}.

In a general case, the results of [207] show that the time and the corresponding fluences of the neutrons, required for the complete dissolution of the particles of the second phase, especially at relatively large cross sections of burning out and small deviations of the concentration of component *B* in the alloy from its solubility limit, may be comparable with the actual service fluences. The phase changes in the process of burning out of components of the alloys may result in a significant degradation of the mechanical and other properties

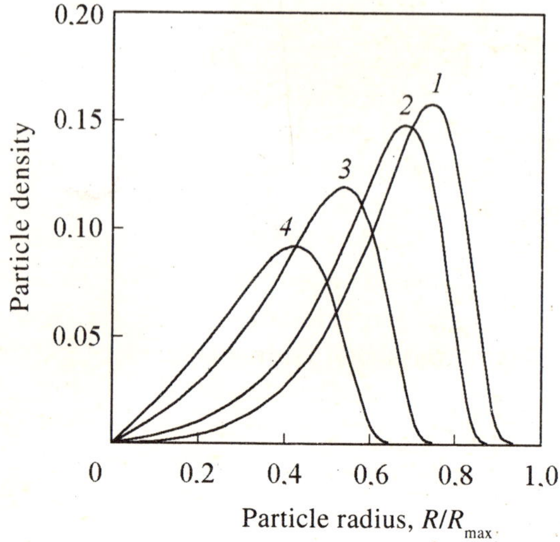

Fig. 4.11 Size distribution of $ScAl_3$ particles in Al–0.058 at.% Sc alloy for different neutron fluences Φ: 1) Φ = 0; 2) Φ = 1.25×10^{27} m^{-2}; 3) Φ = 3.5×10^{27} m^{-2}; 4) Φ = 4.5×10^{27} m^{-2} [207].

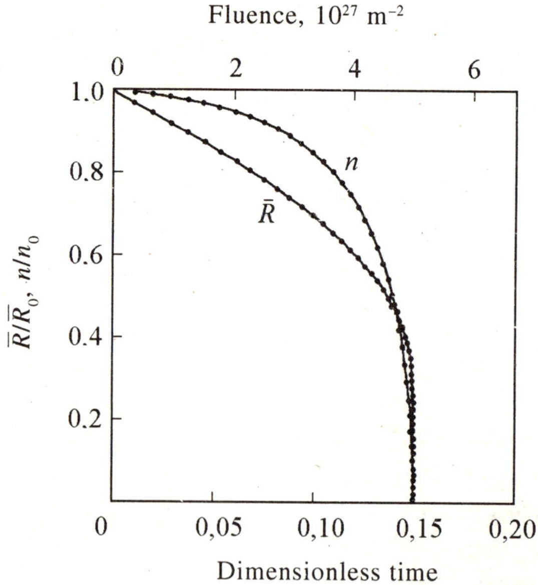

Fig. 4.12 Kinetics of relative changes in the mean radius $\bar{R}/\bar{R}_0$ and density n/n_0 of second phase particles in Al–0.085 at.% Sc alloy under neutron irradiation [207].

of the irradiated materials. This may be a limiting factor in their application at high neutron fluences, in particular, in fusion reactors of different type.

4.5. COALESCENCE

At relatively low intensities and irradiation doses, the process of growth of the particles of the second phase may be described on the basis of the Lifshits–Slezov–Wagner theory of thermal coalescence [196,197], modified taking into account radiation-enhanced diffusion (see, for example [66,208,209]). The main assumptions of the theory of thermal coalescence are reduced to the following [196]:

1. The growth of the particles of the second phase is controlled by the coefficient of bulk diffusion of the dissolved element D_B.

2. The mean radius of the particles of the second phase r_t changes with time in accordance with the expression:

$$r_t^3 = r_0^3 + \frac{8D_B \sigma V_m c_\alpha(\infty)}{9RT} \qquad (4.52)$$

3. The volume fraction of the particles of the second phase in coalescence remains unchanged.

In equation (4.52), σ is the interfacial surface energy, V_m is the molar volume, $c_a(\infty)$ is the limit of solubility of element B in the α-solid solution, r_0 is the initial radius of the particles, t is time, R is the gas constant.

Equation (4.52) shows directly that the process of coalescence under irradiation may be accelerated as a result of an increase in the diffusion coefficient of the dissolved element.

The general analysis of the susceptibility of alloys to radiation-stimulated coalescence on the basis of the model proposed in [196, 197] may be carried out on the basis of the diffusion considerations presented in Chapter 2 and section 4.2.

With an increase in the radiation dose, the special features of radiation-stimulated coalescence start to show themselves: the large particles dissolve with the simultaneous formation and growth of new precipitates [66,190,208–210]. The largest size of the particles in relation to the radiation dose may pass through a maximum [66,209,210], and in further irradiation there may be a tendency for reaching the equilibrium size distribution of the particles [190].

One of the directions of the theoretical interpretation of the special features of radiation-stimulated coalescence is associated with taking into account, in addition to the mechanism of radiation-enhanced diffusion, also dynamic (or ballistic) effects, examined in the sections 4.3.1, 4.3.2 and 4.4.3. The original mechanisms of radiation-stimulated coalescence have been presented in [211–213]. They include the recombination mechanism proposed by Urban and Martin [211], and the mechanisms based on the analysis of the fluxes of point defects and solutes to the matrix–particle interface [212,213]. It is evident that the mechanism of transmutation burning out of the solutes [207] may also lead to the modification of the coalescence process during irradiation.

4.6. PHASE CHANGES IN INDUSTRIAL AND ADVANCED CONSTRUCTIONAL MATERIALS FOR NUCLEAR AND THERMONUCLEAR ENGINEERING

To conclude the present chapter, it is convenient to pay special attention to radiation-stimulated phase changes in a number of steels and vanadium-based alloys used or examined at the present time as promising structural materials for application in fission and fusion reactors, including the possibility of development, on the basis of materials of this type, steels and alloys with accelerated decrease of induced radioactivity (reduced-activation materials).

This analysis is essential because phase changes in the industrial steels and alloys for nuclear engineering are closely linked with the mechanism of their radiation damage and, to a large degree, determine the relative stability or accelerated degradation of the properties of these materials under irradiation.

A very detailed analysis of the experimental data for radiation-stimulated phase changes in prototypes of reduced-activation martensitic, ferritic–martensitic and austenitic steels (conventional Cr steels, alloys with molybdenum, niobium and nickel) was published in [82].

4.6.1. Low-alloy ferritic steels

Low-alloy ferritic steels (< 1% Cr and other elements) are used in atomic engineering for pressure vessels and other structural elements of reactors. The phase instability of pressure vessel ferritic steels in irradiation at temperatures of <300 °C is reflected in radiation-stimulated processes of breakdown of matrix solid solutions and the segregation of solutes at sinks of different type (see, for example [121,122,214–217]).

The stabilisation of clusters of radiation defects and dislocation loops by solutes in segregation, and also the formation of radiation-stimulated pre-precipitates and phases are important factors controlling the radiation hardening and embrittlement of pressure vessel ferritic steels. According to the results of a large number of investigations, the especially negative role in radiation hardening and embrittlement is played by impurities of copper: pre-precipitates and Cu-based phases are effective obstacles to the movement of dislocations.

In Fe–Cu binary alloys, radiation-stimulated precipitates represent Cu-enriched zones, with vacancies in their composition [120,216]. In addition to copper, nickel and manganese can also be introduced into the composition of radiation-stimulated precipitates of irradiated multicomponent modelling alloys and industrial ferritic steels [120,121,216]. The presence in the precipitates of nickel, in addition to copper, is possibly one of the main factors of the mutually related effect of these elements on the processes of radiation embrittlement and hardening. It is well-known [214,215,217] that an increase in the content of one of these elements in the ferritic steels in the presence of another element increases the degree of radiation hardening and embrittlement of the materials. It should also be mentioned that both copper and nickel belong to unfavourable elements from the viewpoint of activation of metallic materials for atomic and thermonuclear power engineering [218] (see Chapter 5). Therefore, refining of pressure vessel steels to remove impurities of copper and replacement of nickel as the alloying element in the steels is an important task not only for suppressing the process of embrittlement but also for solving a general problem – the development of reduced-activation radiation-resistant materials.

4.6.2. Bainitic, martensitic and ferritic–martensitic steels

Cr steels of the bainitic grade Fe–CrMoWNb with 2–3%Cr, martensitic and ferritic–martensitic with 9–12%Cr of the grades FV448 (DIN 4914), EM12, IT-9, CRM-12, and others are used at present as materials of the pressure vessels and structural elements of the active zones of nuclear reactors. The steels of this grade and their reduced-activation compositions, in which molybdenum is replaced by tungsten and vanadium, niobium is replaced by tantalum, and nickel and nitrogen by manganese, are regarded at present as promising structural materials for fusion reactors [218].

As reported in a number of investigations (see, for example [219, 220]), the initial phase composition, the nature of phase changes and the phase stability of the reduced-activation ferritic Cr steels and industrial prototypes of these steels in both heat treatment and under irradiation is almost identical. For example, the initial structure of a reduced-activation steel with 2.0% Cr and with 2% W and 0.5% Mn is identical with the structure of its prototype with 1–2% Mo and after quenching from 930 °C and tempering and 730 °C it consists of bainite and the phases $M_{23}C_6$, M_6C and W_2C [220]. Comparison of the properties of the steels after simulation ion irradiation at temperatures of 350–500 °C up to 200 displacements per atom (dpa) shows [221] that the reduced-activation composition is characterised, in particular, by higher phase instability.

Irradiation, especially at temperatures of >400 °C, stimulates the processes of breakdown in the bainitic Cr steels. The precipitates are localised mostly at the grain boundaries, sub-boundaries and the boundaries of the prior austenite grains. In the reduced-activation bainitic steel 2.25Cr–2W–0.5Mn–0.2V–0.07 Ta, irradiated with neutrons up to 36 dpa at 698 K [222], the type of radiation-stimulated phases is basically identical to the type of phases in the previously examined steel 2Cr–2W–0.5Mn after quenching and tempering. In the structure of irradiated steels examination showed precipitates of the $M_{23}C_6$ phase, enriched with Cr, small spherical precipitates of M_6C enriched with Ta (W) and small carbides of M_2C, enriched with W(Ta). The bainitic Cr steel with an increased vanadium content (0.5–1.5%) after irradiation with a high neutron flux (> 200 dpa) at 400 °C showed the precipitates of V_4C_3 phase inside the grains and subgrains [223]. As shown in a number of investigations into the neutron irradiation of these materials, including [224], an increase of the Cr content of the steels of the examined grade increases the density of radiation-stimulated precipitates at the grain boundaries, sub-boundaries and lath boundaries of martensite, and the size of the precipitates shows a tendency for saturation.

The matrix structure of high-chromium (10–13%C) ferritic steels in the non-irradiated condition after normalising (~1000–1050 °C) and tempering (~700–750 °C), in relation to the type and concentration of other main alloying elements (molybdenum, niobium, nickel, manganese, tungsten, vanadium, tantalum) consists either completely of lath martensite with a high dislocation density or contains additional grains of δ-ferrite. This type of ferrite is almost always present in the steels at a Cr content higher than 12% [225].

A detailed analysis of radiation-stimulated phase changes in typical martensitic (FV448 and CRM-12) and ferritic–martensitic (FI) steels was carried out in [226]. The composition of the steels is presented in Table 4.3.

In the non-irradiated condition after normalising and tempering, the structure of martensite and δ-ferrite in all steels contained the

Table 4.3 Composition of FV448, CRM-12 and FI steels

Steel	C	Si	Mn	S	P	Cr	Ni	Mo	V	Nb
FV448	0.1	0.46	0.86	0.009	0.02	10.7	0.65	0.60	0.14	0.26
CRM-12	0.19	0.45	0.54	0.017	0.016	11.7	0.62	0.96	0.30	–
FI	0.46	0.22	0.46	0.011	0.014	13.3	0.38	–	–	–

precipitates $M_{23}X$ (X = C, Si), M_2X (FI and FV-448), MX(CRM-12) and NbC(FV-448).

In the process of subsequent ageing at 460–600 °C, simulating (as regards temperature and time) the identical parameters of reaction irradiation of these steels in the present work, examination showed the following structure-phase changes: coalescence of the $M_{23}C$ phase in all steels with a tendency for dissolution at 600 °C, the dissolution of M_2X in FV448 steel and in the grains of δ-ferrite of FI steel. At a temperature of 600 °C, the M_2X phase in FV448 steel dissolved with simultaneous formation of the MX phase. Ageing at 600 °C also resulted in the formation of an intermetallic compound, i.e. the Laves phase (Fe_2Mo base) in the steels FV448 and CMR-12. At the same time, the dislocation structure recovered.

After irradiation of these steels with fission neutrons ($E > 0.1$ MeV) in the temperature range 380–615 °C with a fluence equivalent to 30 dpa, examination showed the accelerated coalescence of the $M_{23}C_6$ phase, the dissolution of the M_2X phase at temperatures ≥420 °C, dissolution of the MX phase in CRM-12 and its formation in the FV448 steel. The CRM-12 steel was characterised by the formation of the Laves phase, whereas in FV448 steel, the formation of this phase was suppressed, in contrast to thermal ageing.

In addition to the irradiation-stimulated changes, the following new phases were found in the structure of the steels: M_6X (in all steels), the χ-phase with the composition similar to M_6X but with a difference in the content of iron and nickel (FV448 and CRM-12), the σ-phase with the composition similar to the composition of the χ-phase but with different concentration of molybdenum and iron (CRM-12 and FV448), Cr-enriched ferrite (α'-phase) (in all steels) and phosphides M_3P (FI) and MP (CRM-12 and FV448). Identical results were obtained previously in [227]. In the steels Fe–(2.25–17)Cr–(0.04–0.2)Mo–(0.04–0.4)C, irradiated at a temperature of 400–650 °C with a neutron fluence up to 1.76×10^{23} cm^{-2} (84 dpa), examination showed the phases M_6C, MoC_6, χ, Laves phases, $M_{23}C_6$, α' and a number of unidentified phases. As in [226], the specific phase composition depended on the composition of steels and irradiation temperature.

The results of further investigations into the phase composition of martensitic and ferritic–martensitic Fe–CrMoV and Fe–CrMoVNb steels in the initial condition and its changes under irradiation (see, for example [225,228–230]), which are on the whole in agreement with the experimental data obtained in [226,227], may be summarised as follows: in the non-irradiated condition after normalising or quenching and subsequent tempering, the main precipitates in the structure of the steels are the phases $M_{23}C_6$, M_2X (X = C,Si), MC. After long-term ageing in the temperature range ~300–650 °C, the Laves phase (Fe_2Mo base) may form in the material. Irradiation may be accompanied by the transformation of the initial structure with the formation, depending on the composition of steels, irradiation temperature and fluence, of the precipitates of new phases M_6X (X = C,Si), χ, *G*, Laves and α'. The structure of tempered martensite in Fe–CrMoV and Fe–CrMoVNb steels is stable in irradiation at temperatures of ~<500 °C [226,231].

The irradiation-stimulated phases in the martensitic and ferritic-martensitic Cr steels with Nb and Mo form preferentially at the boundaries of different types (grains, subgrains, prior austenite grains, lath boundaries of martensite, phase boundaries). This indicates the significant role of the segregation processes in the variation of the phase composition of these steels. The phase instability of Fe–CrMoV and Fe–CrMoVNb steels in irradiation is determined to a large degree by diffusion transfer, mainly of elements such as Si, Ni, Mn, S, P, Mo and by the formation of intermetallic and carbide embrittling phases at the boundaries (primarily M_6C, $M_{23}C_6$, χ, Laves) [226,228, 232–235]. Intensive radiation embrittlement results in the formation of the α'-phase in the ferritic steels (in both ferrite and martensite [222,226]) which without irradiation is observed only in Cr steels with $\geq$14 %Cr [226].

In a general case, for operation in the irradiation conditions, the martensitic steels with 7–9 %Cr are preferred to steels with $\leq$12% Cr. Firstly, in the steels with $\geq$12%Cr without additional alloying with austenite stabilising elements (carbon, nickel, manganese) it is difficult to avoid the formation of residual δ-ferrite which increases the ductile–brittle transition temperature [230,236]. The additional introduction of nickel and manganese into ferritic steels is undesirable because, as already mentioned, they rapidly segregate during irradiation, resulting in the formation and growth of the embrittling phases at different boundaries.

Intensive investigations have been carried out in recent years into reduced-activation compositions of martensitic and ferritic–martensitic steels (Fe–(7–12)CrWVTa and Fe–(7–12)CrWMnV) show that as regards the technological and service properties, they are not inferior and are often superior to the identical characteristics of their prototypes with Mo and Nb (Fe–CrMoV and Fe–CrMoVNb). The reduced-activation ferritic steels in comparison with their Fe–CrMoV and Fe–CrMoVNb prototypes are also characterised by high phase stability under irradiation which also controls their relatively high radiation resistance. Primarily, this relates to the reduced-activation compositions in which the tungsten concentration does not exceed 2%, and the Mn content is limited at ~≤0.5%.

The restriction of the tungsten concentration is associated mainly with the need to prevent the formation of residual δ-ferrite and decreasing the probability of precipitation of the Laves phase (Fe_2W), increasing the temperature of the ductile–brittle transition [236]. Mn, like in ferritic steels with Mo and Nb, together with Si, P and S rapidly segregate during irradiation at boundaries of different type, causing embrittlement of reduced-activation steels mainly as a result of the formation of the χ-phase [223,229,234,237]. Since manganese is characterised by intensive stimulation of the formation of precipitates at the boundaries during irradiation and, consequently, results in the degradation of the mechanical properties, this element is generally not recommended (as stressed in [223]) for alloying in the development of reduced-activation ferritic steels. The vanadium content should also be restricted because, like manganese, vanadium increases the phase instability of reduced-activation steels as a result of the formation of radiation-stimulated precipitates of the V_4C_3 phase inside the grains [223].

The currently available promising reduced-activation martensitic steels with the Cr concentration in the range 7–9 %, W ≤ 2%, and a restricted content of V (≤0.2%) and Mn (≤0.5%), are characterised by very high phase stability in the reactor irradiation conditions. As reported in [224], the neutron irradiation of the martensitic steels of the previously mentioned composition with a fluence of up to 8×10^{22}cm^{-2} (35 dpa) in the temperature range 390–520 °C does not change their initial phase composition with the precipitates of carbides $M_{23}C_6$, MC and M_6C. At all temperatures, the structure of the irradiated steels was characterised by the dominant effect of the precipitates of the $M_{23}C_6$ phase, enriched with Cr. The composition of the identified particles was almost independent of irradiation temperature. In this investigation, in contrast to [222,238], the formation of the Laves phase, enriched with tungsten, in the steels of the same composition after irradiation at temperatures of 420 °C [222] and

750 °C [238] was not detected. At the same time, in the ferritic-martensitic steel with 12%Cr, the precipitation of the Laves phase was observed after irradiation at a temperature of 520 °C and the α'-phase precipitated after irradiation at 390 °C [224]. The precipitates of the α'-phase in the ferritic–martensitic steel the same composition after identical irradiation at 420 °C were also reported in [222].

In the reduced-activation martensitic and ferritic–martensitic Fe–Cr2WVTa steels, the structure of the martensite, like in Fe–CrMoVNb steels, is stable in irradiation at a temperature of 500 °C [224,239].

It is important to stress that in the reduced-activation martensitic Fe–(7–9)Cr2WVTa steels with the restricted Mn content in contrast to their Fe–CrMoV and Fe–CrMoVNb prototypes, the embrittling χ, G and α'-phases did not form during irradiation. The high phase stability of these steels is one of the main reasons for the minimum degradation of their mechanical properties in irradiation in comparison with both the martensitic Fe–CrMoV and Fe–CrMoVNb steels, and reduced-activation bainitic and ferritic–martensitic steels. In comparison with these materials, the reduced-activation Fe–9Cr2WTa steels are subjected to the smallest radiation hardening and embrittlement [222,224]. After reactor irradiation at a temperature of ~365 °C, with a fluence equivalent to 10 dpa, the shift of the temperature of the ductile–brittle transition in the steel was only 4 °C [240], and after irradiation at 390 °C (35 dpa) reached 45 °C [239].

In comparison with the bainitic and ferritic–martensitic reduced-activation steels of the same composition, the irradiation hardening of martensitic Fe–(7–9)Cr2WVTa steels is minimum and is associated mainly with the precipitation of the M_6C carbides [222]. The degree of radiation hardening of the reduced-activation bainitic steels is higher and is determined mainly by radiation-stimulated precipitates of the M_2C carbide [222]. The strongest radiation hardening takes place in the reduced-activation ferritic–martensitic steel Fe–12Cr2WVTa as a result of the formation of the α'-phase with a strong embrittling effect [222].

4.6.3. Austenitic steels

The results of investigations of the phase composition of different austenitic stainless steels for nuclear applications after heat treatment, deformation and irradiation have been published in many original and review articles (see, in particular, [82,241–252]). In this section, special attention will be given to the phase stability of 316 steel and its different analogues and modifications, the stability of steels with a high nickel content and reduced-activation compositions based on the Fe–Cr–Mn system. These steels in different national and combined materials science programmes are regarded as promising materials for the active zones of nuclear reactors and the first wall of the fusion reactor. Table 4.4 shows the typical compositions of austenitic steels of the examined grade of materials: steels AISI-316, 316-LN, USPCA, JPCA and AMCR-0033.

In the programmes of the USA, Japan and European Community, the Cr–Ni steels of the type PCA and 316 (USPCA, JPCA, EC316L and 316LN) are regarded as promising materials for the first wall of the first demonstration nuclear reactor ITER. The reduced-activation Cr–Mn steel AMCR-003 has been proposed for this purpose by the European Centre of Joint Investigations [218]. The programmes for the reduced-activation materials based on the Fe–Mn–Cr system in Russia and the USA are oriented to the selection of reduced-activation compositions of the type Fe–20Mn–12Cr–C, additionally alloyed with W, Ti, V, P or B [218].

4.6.3.1. Austenitic Cr–Ni steels

The phase composition of the 316 type steels (Table 4.4) and their modification with titanium, especially in the normalised or quenched condition, is highly unstable not only under the effect of irradiation but also during thermal ageing or in operation under load at elevated temperatures. After normalising at ~1150 °C or after normalising and cold deformation, subsequent ageing or holding under load results in the breakdown of austenite of the steels with the formation, depending on temperature, of phases of different type (carbides, intermetallics, etc). These are the Laves phases, M_2P, MC, η, σ, $M_{23}C_6$, χ and TiO.

In irradiation, the temperature of formation of a number of these phases, their composition, stability, morphology and location change. In comparison with the thermal conditions, the concentration and size of radiation-stimulated precipitates may also greatly change.

In addition to the previously mentioned phases, which also form

Table 4.4 Composition of several austenitic steels

Steel	C	Ni	Cr	Ti	Mo	Mn	Si	P	S	Nb	Al	Others
AISI 316	0.05	13.5	17.5	–	2.4	1.6	0.56	<0.01	0.003	0.1	–	N
316 LN	0.021	12.3	17.5	–	2.41	1.79	0.53	0.025	0.009	<0.005	–	Cu,Co,N,B
USPCA	0.05	16.5	14.3	0.31	1.95	1.8	0.52	<0.01	0.003	0.06	0.05	N
JPCA	0.06	15.6	14.8	0.24	2.28	1.80	0.5	0.027	0.0045	–	–	N,B
AMCR–0033	0.1	<0.1	10.0	–	–	17.5	0.6	0.016	0.008	–	–	N,Cu

during thermal ageing and under irradiation, the irradiated steels of the 316 type also contain the $\gamma'(Ni_3Si)$ and G-phases, with the latter having the usual formula $T_6Ni_{16}Si_7$, where T is the transition element [241–246, 253]. The γ'-phase is a typical component of steels and alloys with Ni≥25%, but in materials with a low nickel content the phase forms only under irradiation [243]. With increase in the fluence the upper boundary of formation of this phase increases, and is restricted by a temperature of 540°C. The composition of this phase in the 316 type steels does not correspond to its normal stoichiometry, but tends to it with increasing irradiation temperatures [243]. During annealing at temperatures of the formation of this phase in the course of irradiation, the γ'-phase shows a tendency for dissolution [242]. It is well known that this special feature is typical of irradiation-induced phases.

In the conditions of thermal ageing, the G-phase like the γ'-phase forms only in alloys with a high (≥25%) nickel content. In irradiated AISI-316 steel, this phase was detected for the first time in [253]. The composition of the G-phase changes with temperature, and like the γ'-phase, the G-phase is unstable and dissolves during annealing at the temperatures of its formation [242,243].

In [242,243] different types of phases in irradiated steels of the 316-type were classified as follows:

– phases introduced by irradiation. Evidently, they include the phases γ' and G;

– the phases modified in composition during irradiation. Thcy include the Laves phase and the M_2P phase;

– phases which form at lower temperatures during irradiation.

Using this classification, Table 4.5 shows a number of parameters and characteristics of the previously examined phases in irradiated type-316 steels. This table was taken from [252]. It generalises the experimental results, including those from [241–244] and a number of other investigations. The notations (IR), (IM), (LT) and (TP) in Table 4.5 relate to the phases introduced during irradiation, the phases with modified composition during irradiation, the phases during irradiation at lower temperatures, and thermal phases. It is evident that the index TP can be applied to all previously examined phases, with the exception of γ' and G. The index (PP) in Table 4.5 indicates the phases formed in the 316 type steels during their production.

As indicated by the experimental investigations [241–244,246] one of the main relationships of phase instability of the 316-type steels under irradiation is based on the irradiation-stimulated formation of phases enriched with nickel and silicon. These also include the phases

Table 4.5 Precipitates of phases found in austenitic stainless steels [252]

Phases	Crystal structure	Lattice parameter, nm	Atoms/unit cell	Typical morphology
γ	Cubic Al, FM3m	$a_0 = 0.36$	4	Matrix
γ' (PB)	Cubic Li_2, Fm3m	$a_0 = 0.35$	4	Small sphere
G (PB)	Cubic Al, Fm3m	$a_0 = 1.12$	116	Small sphere
Laves (PM)	Hexagonal, C_{14}, $P6_3$/mmc	$a_0 = 0.47$ $c_0 = 0.77$	12	Faulted lath
M_2P (PM)	Hexagonal, C_{22}, P_{321}	$a_0 = 0.604$ $c_0 = 0.36$	6	Thin lath Rod–shaped
MC (PY)	Cubic, B1, Fm3m	$a_0 = 0.433$	8	Small sphere
η (PY)	Cubic $E9_3$, Fd3m	$a_0 = 1.08$	96	Rhombohedral
σ (PY)	Tetragonal, $D8_6$,P_4/mnm Cubic, $D8_4$, Fm3m	$a_0 = 0.88$ $c_0 = 0.46$	30	Various
$M_{23}C_6$ (TF)	Cubic A12, $\bar{1}$43m	$a_0 = 1.06$	92	Rhombohedral platelet
χ (TF)	Cubic, B1, Fm3m	$a_0 = 0.89$	58	Various
TiO (TF)	Hexagonal, D51, R$\bar{3}$ s	$a_0 = 0.425$	8	Small rod
Cr_2O_3 (PP)		$a_0 = 1.358$	30	Globular
SiO (PP)	Cubic, B1, Fm$\bar{3}$m	$a_0 = 0.46$	–	Globular
ZrO (PP)	Cubic, B1, Fm3m	$a = 0.46$	8	Globular
NbO (PP)	Cubic, B1, Fm3m	$a = 0.46$	8	Globular
TiN (PP)	M_xS_y (PP)		8	Large cuboid
M_xS_y (PP)		Many stoichiometric relationships		

introduced by irradiation, i.e. the γ'-phase, *G*-phase, and the η-phase (M_6C or M_5SiX). It is a characteristic that the η-phase forms during irradiation also like an isolated phase and a phase on precipitates of $M_{23}C_6$ carbides [242,246]. The Ni and Si content of the $M_{23}C_6$ phase is minimum which in irradiation causes its destabilisation in relation to the η-phase enriched with Ni and Si.

Formation of the γ'(Ni_3Si) phase is usually detected in the vicinity of dislocation loops and voids [241,242,252] and may be interpreted with sufficient accuracy on the basis of special features of the mechanisms of migration and radiation-stimulated segregation of nickel and silicon.

The dilation volume of Ni in Fe is $\Delta V = 0.046$ [169]. Because of considerations, presented in the Chapters 1 and 2, it does not form stable mixed dumbells and may migrate, basically by the vacancy mechanism. In Fe–Cr–Ni alloys, nickel has the lowest coefficient of thermal diffusion ($D_{Cr} > D_{Fe} > D_{Ni}$) [254]. Consequently, the vacancy flows to seams should, because of the reversed Kirkendahl effect, lead in Fe–Cr–Ni phases to preferential enrichment with nickel. This was also detected in experiments (see, for example, [255]). In contrast to nickel, the dilation volume of Si in Fe<0 ($\Delta V = -0.078$ [169]), and under irradiation it may migrate by both the vacancy and interstitial mechanisms. If the interstitial mechanism of diffusion transfer of silicon is dominant, then silicon in Fe–Cr–Ni alloys should enrich sinks during irradiation. This has been confirmed by experiments [246,256]. The microscopic mechanism of formation of the γ'-phase on loops may be interpreted on the basis of the model proposed in [52]. This mechanism has been examined quite extensively in section 3.7.

In addition to the segregation mechanism, a significant role in the formation of radiation-stimulated phases in the 316-type steels (including those enriched with nickel and silicon) is played, in our view, by the modified mechanism of spinodal breakdown [38]. The modification of the classic mechanism of spinodal breakdown [90-94] is based (in irradiation) on the application of considerations regarding radiation-enhanced diffusion and the compensation of coherent strains by radiation point defects during the formation of concentration waves [38]. This mechanism results in the widening of the concentration zone of the transformation, the shift of spinodal breakdown to higher temperatures and, on the whole, acceleration of the breakdown process. This mechanism has been examined in great detail in section 4.3.2.

In experiments, the breakdown of a similar type in irradiation with the formation of spatially-oriented and modulated structures in the Fe–Cr–Ni system, including directly in 316-type steels, was observed in [98,102]. It should be mentioned that, in contrast to [98,99,101,102], in which the concentration fluctuations of iron, chromium and nickel in the Fe–7.5Cr–35.5Ni alloy were observed after irradiation with relatively high neutron or ion fluences ($\geq$38 displacements per atom or 7.6×10^{22}n cm^{-2}) at temperatures of ~600–630°C, the formation of the modulated structure in type 316 steels [100] was observed after a considerably lower dose (10×10^{19} cm^{-2}) of irradiation with electrons with an energy of 21 MeV at 450°C. The modulation period in neutron-irradiated Fe–7.5Cr–35.5Ni alloy was ~200 nm [101], whereas in electron-irradiated 316 steel it was only ~20 nm [100].

As reported in [256], after ion irradiation of FV548 steel (similar in composition to 316 steel) at 600°C the formation of fluctuations of chromium and nickel steel is stimulated by the introduction of He. After irradiation with a dose corresponding to 5 dpa, the period of composition fluctuations with respect to chromium and nickel in the steel was ~50 nm.

The experimental results show that the formation of radiation-stimulated phases in 316-type steels can be interpreted with high accuracy on the basis of the modified mechanism of spinodal breakdown.

Combined with the segragation mechanism and/or the mechanism of radiation-enhanced diffusion of solutes, this mechanism explains not only the heterogeneous (on sinks: loops, voids, grain boundaries, etc.) but also homogeneous formation of the phases.

In fact, the formation of concentration fluctuations of iron, chromium and nickel during irradiation creates suitable conditions for the nucleation of phases enriched with the appropriate component. The diffusion flows to these pre-precipitates of these elements stimulates the transformation of these concentration domains in the precipitates of the appropriate phases. Evidently, the fluctuations of nickel with subsequent segragation of silicon atoms on them create thermal conditions for the formation of radiation-stimulated precipitates of the phases γ', G and η. On the other hand, the formation of concentration zones, enriched with iron and chromium, and subsequent diffusion of appropriate elements to them, support the formation of Laves, σ and χ phases (enriched with iron and chromium). Qualitatively, this model is in quite a good agreement with the experimental data for the phase stability of 316 steel under irradiation.

The intensive formation of nickel-enriched phases in 316 steel under irradiation greatly reduces the nickel content in the matrix solid solution (see, for example, [253]). The destabilisation of austenite with a decrease in the nickel concentration in the solid solution stimulates the $\gamma \rightarrow \alpha$ transformations leading to the formation of embrittling ferrite and martensite phases. In particular, this is characteristic of austenitic steels with a relatively low nickel content. The strong radiation embrittlement of industrial steels of active zones of nuclear reactors of the type 304(Fe–9Ni–18Cr) as a result of the $\gamma \rightarrow \alpha$ transformations is one of the main reasons for evaluating the possibilities of replacing this steel by 316-type steels and a further modification with a higher nickel content.

As already mentioned, these promising steels for both the active zones of nuclear reactors and the first wall of thermonuclear synthesis reactors studied in programmes in the USA and Japan include steels of the type PCA (Table 4.4): USPCA and JPCA. In comparison with 316-type steels, the nickel content of these steels has been increased to ~16% and that of chromium decreased to ~14–15%. The content of the alloying elements of the USPCA and JPCA steels is almost identical.

A number of experimental investigations have been carried out to investigate the effect of irradiation on the structure and properties of type PCA steels (see, for example, [257–263]).

The radiation-stimulated precipitation of fine-dispersion MC carbides, enriched with titanium, was observed in both annealed (1100 °C) and cold-worked JPCA steel already at a temperature of 300°C, after irradiation with a neutron fluence corresponding to 10 dpa [261]. With increasing fluence, their density initially increased (to 34 dpa) and then decreased. As indicated in [261], this shows the tendency for the dissolution of MC carbides with increasing fluence which, especially in annealed Fe–Cr–Ni steels, was also observed in other investigations (for example, [257,264].

The precipitates MC, $M_{23}C_6$, η, (M_6C) and γ'-phases dominate in JPCA steel in relation to temperature in almost the entire examined irradiation temperature range (370–650°C). Precipitates of the *G*-phase were found in a number of investigations [257,262]. It should be mentioned in the above-cited studies [257–263] there is no direct data on the formation of Laves, σ and χ phases in irradiated steel of the JPCA type, although in [257] the results also show the formation of a phase enriched with chromium in cold-worked JPCA steel irradiated at 500°C (57 dpa). A tendency for surpressing the formation of the radiated-stimulated Laves σ and χ phases enriched with chromium in JPCA type steels in comparison with type 316 steels is highly likely, since the chromium content in these steels has been reduced to 14–15%.

It is also important to mention that the nature of phase changes and phase stability of the PCA-type steel depend greatly on their initial structure. For example, in the annealed JPCA steel, after irradiation at 500 °C, the controlling phase is the M_6C phase in the presence of MC and *G* phases in small concentrations [257]. In this case, an increase of the fluence is accompanied by a tendency for the growth and dissolution of the MC phase and formation of phases enriched with silicon and nickel (M_6C and G). The irradiated cold-worked JPCA steel is characterised by the preferential formation of the MC phase (basically TiC) and also by a small concentration of the M_6C phase and the phase enriched with Cr. This tendency was also reported in [262]. After irradiation at temperatures of 390 and 410 °C, the density of the γ'(Ni_3Si) phase in the annealed JPCA steel was higher in comparison with the cold-worked steel, and the density of the MC carbides at all radiation temperatures (390–600 °C) was higher in the cold-worked steel.

This result shows directly that the introduction of the dislocations suppresses the processes of radiation-stimulated segregation of nickel and Cr and, consequently, complicates the formation of the γ', *G* and η phases, enriched with these elements. As reported, in particular, in [257, 263], of the introduction of dislocations in cold deformation stabilises the phase structure of the PCA-type steel and, on the whole, increases their radiation resistance.

In this connection, it is necessary to pay special attention to the role of phosphorus; of the extent of application of phosphorus as the alloying elements of austenitic steels has greatly increase in recent years.

With special reference to the formation and stabilisation of the dislocations structure in irradiation, the role of phosphorus may be reduced to the following. It is well-known that the introduction of dislocations sinks decreases the dynamic the concentration of radiation point defects. This result in the suppression of the processes of radiation damage and the phases stability of metallic materials. The formation of a dislocation network, playing the role of dislocations sinks, is the result of interaction of the initial dislocations structure with a dislocation loops formed during irradiation. As shown in a number investigations (see, for example [265, 266]), the introduction of phosphorus into austenitic steels greatly increases the density of dislocation loops of the interstitial type is as supporting the formation of a dislocation network with a high density of sinks. It is characteristic that in the Fe–16Cr–17Ni steel alloyed with phosphorus (0.1%), the dislocation density in the cold-worked (20%) and annealed specimens was almost the same at any radiation fluence (up to 12.4 dpa) at a temperature of 410 °C [267]. With an increase in the concentration of phosphorus, the density of the dislocations in the irradiated specimens increase. At higher temperatures (≥450–500 °C), the radiation-stimulated precipitates of phosphides in the authentic steels suppresses the processes of recovery of the dislocations structure [268].

The positive effect of phosphorus on the phases stability and the radiation resistance of austenitic Cr–Ni steels is also determined by the effect of a number of other mechanisms. As an undersized atom in the iron lattice ($\Delta V = -0.136$ [169]), phosphorus forms stable mixed dumbbells, decreasing the effective diffusion mobility of the interstitial atoms in steels. Consequently, the degree of mutual recombination of the vacancies and interstitial atoms increases and their dynamic concentration decreases.

The introduction of phosphorus also suppresses the segregation of nickel and the irradiation [265, 269]. The mechanism of suppression of these segregations associated with the formation of the Ni–P or Ni–P–Ti complexes [269].

A similar interpretation is highly likely, especially for the mechanism of formation of the Ni–P complexes. In fact, nickel, like titanium ($\Delta V = 0.144$ [169]) may capture intrinsic or mixed dumbbells. In capture of a mixed Fe-P dumbbell by a nickel atom the formation of a stable Ni–P complexes is possible, because the localisation of the undersized atom of phosphorus in the oversized atom of nickel is advantageous from the energy viewpoint. The formation of these complexes may also stimulate the nucleation of phosphides (usually enriched in nickel [243, 264]) and suppress in this case the formation of various high-nickel phases, such as *G*, and γ'. A similar effect was observed in experiments in [262].

The undesirable segregation of phosphorus on the grain boundaries [257], leading to embrittlement, can be suppressed as a result of adding titanium [257]. These can also be interpreted within the framework of the previously examined mechanism of formation of the Ni–P complexes. The localisation of phosphorus at the titanium atoms is even more advantageous from the energy viewpoint than at nickel, because ΔV_{Ti} is considerably higher than ΔV_{Ni}. The formation of these complexes may stimulate the process of nucleation of dislocation loops in the austenitic steels, alloyed with titanium and phosphorus [270].

Further prospects of the development of Cr–Ni austenitic steels are associated with the evaluation of the radiation resistance of steels with a higher nickel content (≥25%), modified with titanium, silicon and phosphorus. As shown in [263], the cold-worked (20%) steels of the type Fe–15Cr–(20–25)Ni, alloyed with Si (~0.8%), Ti (~0.2%) and P (0.02–0.026%) are characterised by high radiation resistance in the temperature range 405–670 °C in irradiation up to a fluence corresponding to ~150 dpa. The swelling of these alloys after irradiation with a fluence of ~120 dpa in the given temperature range is <1%. The phase composition of the irradiated high-nickel steels and of the irradiated steel 316 in the cold-worked condition, modified with Si (0.8%), Ti (0.1%), and P (0.025%), is identical.

After irradiation with a a fluence corresponding to 92.5 displacements per atom in the temperature range 470–620 °C, the structure of the steels contained radiation-stimulated MC phases (mostly titanium carbide), M_6C and the needle-shaped phosphides with a very high concentration. The phosphide precipitates are stable up to the maximum fluence of 210 dpa.

These results show that in the authentic steels with higher nickel content, modified with silicon and titanium, cold plastic deformation combined with alloying with phosphorus increases the phase stability of these materials and greatly increases their radiation resistance as a whole.

4.6.3.2. Austenitic chromium–manganese steels

The development of austenitic steels with a faster decrease of induced radioactivity is based on the partial or complete substitution of the main alloying elements, i.e. nickel by manganese. In addition, as in the development of reduced-activation compositions of ferritic steels, tungsten and vanadium are introduced instead of molybdenum with complete exclusion of niobium and its replacement by tantalum.

As the alloying element of austenitic steels, manganese is a close analogue of nickel as regards a number of properties. Like nickel, Mn is also an austenite stabilising element, although its stabilising efficiency is considerably lower (by a factor of 2 according to the criterion of the nickel equivalent). Manganese and nickel have almost the same dilation volumes in iron (0.049 and 0.047, respectively [169]), which determines the identical mechanisms of the interaction with vacancies and interstitial atoms, and the mechanisms of their radiation-enhanced diffusion and segregation. Manganese, like nickel, does not form stable mixed dumbbells and its diffusion transfer under irradiation takes place, as in the conventional temperature conditions, mostly by the vacancy mechanism.

At the same time, if, as already mentioned, nickel in the Fe–Cr–Ni system is characterised by a low diffusion coefficient, and $D_{Cr}>D_{Fe}>D_{Ni}$, then in Fe–Cr–Mn alloys, the rate of diffusion of Cr and Mn is similar or higher than that of iron [254], i.e. $D_{Cr} \cong D_{Mn}>D_{Fe}$.

The lower efficiency of manganese as an austenite stabiliser in comparison with nickel and its high diffusion coefficient also predetermine the increased tendency for the instability of the austenite of the Fe–Cr–Mn steels as a result of transformations with the formation of ferritic and martensitic phases not only under the effect of radiation but also in conventional thermal annealing and ageing. In irradiation, these factors should stimulate the destabilisation of austenite both in the vicinity of the grain boundaries and in the matrix at sinks of different type as a result of depletion in manganese and enrichment with iron under the effect of the reversed Kirkendahl effect. The high coefficient of thermal diffusion of manganese, which is increased even further by the effect of irradiation, also supports the radiation-stimulated instability of the steels based on the Fe–Cr–Mn system.

In addition to the previously mentioned factors, the Fe–Cr–Mn system in comparison with Fe–Cr–Ni system is characterised by a lower stacking fault energy and increased susceptibility to mechanical twinning. This increases in these materials the tendency for the $\gamma \to \alpha$ and $\gamma \to \varepsilon$ martensitic transformations, taking place during deformation [271, 272]. The susceptibility to radiation-stimulated spinodal breakdown of steels based on the Fe–Cr–Mn system, which was observed in the experiments in [103] in electron irradiation of EP-838 steel (Fe–0.1C-12Cr–14Mn–4.5Ni–1.5Al–0.5Mo) also supports an increase of their phase instability during irradiation. This factor of instability of the Fe–Cr–Mn system has already been reported in [271].

The previously examined general physical considerations regarding the possible special features and mechanisms of thermal and radiation-stimulated phase stability of the steels based on the Fe–Cr–Mn system are in good agreement with the experimental results.

The phase structure of simple Fe–Cr–Mn ternary alloys with a low carbon content and a low content of other alloying elements after annealing (~1100 °C), cold deformation and subsequent ageing at 300–700 °C, includes (depending on temperature and composition) various combinations of γ- and α-ferrite, martensite and the σ-phase [273–276]. The tendency for the formation of the σ-phase increases with the increase of temperature and ageing time. Its formation during ageing is accompanied by the disappearance of martensite and a decrease of the volume fraction of α-ferrite [276]. An increase in the Cr content stimulated its formation [275,276].

The negative effect of the σ-phase on the properties and, in particular, its strong embrittling effect in localisation at the the grain boundaries, necessitates the restriction of the Cr content in the reduced-activation Fe–Cr–Mn steels. However, in this case it is necessary to take into account that the Cr content of the austenitic stainless steels should not be lower than 10–11%, to ensure the required corrosion resistance [271,276].

In simple ternary Fe–Cr–Mn alloys with the Cr content in the range (10–15)% a stable single-phase austenitic structure cannot be produced even during thermal ageing and in irradiation, even at high manganese concentrations of ~30–35 %. In the annealed and cold-worked Fe–10Cr–30Mn alloy, for example, after ageing in the temperature range 300–500 °C, examination showed the γ-phase and α-martensite, and after ageing at 600 °C there were γ- and σ-phases [276].

Under irradiation at temperatures of $\geq$500 °C, the ternary Fe–12Cr-(15–30)Mn alloys in the completely austenitised condition (quenching in water from a temperature of 1100 °C) showed a tendency for breakdown of the formation of the $\gamma+\alpha+\sigma$ three-phase systems [277,272]. Special investigations of the effect of radiation on the two-phase ($\alpha+\gamma$)

Fe–Cr–Mn alloys showed a number of special features of instability and radiation damage of the system [279–281].

In the two-phase ternary Fe–10Cr–15Mn alloy irradiation in a temperature range results in the radiation-stimulated transformations of austenite into ε- and α-martensite and the formation of the ferrite phase at the voids. The formation of this ferrite band around the voids in the austenitic matrix is in complete agreement with experimental data obtained in the examination of radiation-stimulated segregation of Fe and Mn in Cr–Mn alloys and steels [255,282]. In accordance with the experimental data which are in agreement with the previously examined theoretical assumptions on the diffusion and radiation-stimulated segregation of Fe and Mn in the Fe–Cr–Mn system, these elements deplete and enrich the sinks of different type. This stimulated the destabilisation of austenite and the formation of the ferrite phase in their vicinity, including at the voids.

Irradiation of the Fe–10Cr–15Mn alloy at temperatures of 520 and 620 °C resulted in the formation of ferrite also in the austenitic matrix [279–281]. This reflects the temperature tendency for the breakdown of austenite: at low temperatures, the preferential transformations are the martensitic transformation of austenite, and at high temperatures it is the transformation with the formation of ferrite [281].

A special feature of the experimental results of investigations carried out in [279–281] is that at irradiation temperatures of <600 °C voids were detected only in the austenitic matrix. In principle, this effect is in agreement with experimental data obtained in a large number of investigations in which it was established that the processes of void formation and swelling in austenitic steels are more intensive than in ferritic steels.

In the Fe–Cr–Mn alloys characterised by high instability of austenite in relation to transformations with the formation of ferritic and martensitic phases, the stresses formed during the swelling nonuniform in the volume in the presence of the $\alpha+\gamma$ two-phase structures may intensify both martensitic transformations and the processes of void formation and swelling. The austenitic transformations, observed in [279–281], with the formation of the α- and ε-martensite in irradiation, and also the higher susceptibility of the Fe–Cr–Mn alloys to swelling, do not contradict the given assumptions. The introduction of carbon in the range 0.1–0.25% into the ternary Fe–Cr–Mn system increases the stability of austenite to breakdown with the formation of ferritic and martensitic phases.

One of the two main directions in the development of reduced-activation Cr–Mn steels is based on the examination of phase stability, properties and radiation resistance of steels of the type AMCR-0033 (Table 4.4) and of their modified compositions. The AMCR-0033 steel, as already mentioned, is regarded as a promising reduced-activation material for the first wall of the ITER reactor. In the annealed condition (1030 °C), the steel has an austenitic matrix with the precipitation of $M_{23}C_6$ carbides [283]. In the thermal conditions, the tendency for the formation of α-ferrite is observed in the steel only during ageing after deformation. After irradiation in the temperature range 420–600 °C with a fluence of 60–75 dpa, the formation of α-ferrite in the steel was detected only at the voids and only in the condition after annealing and ageing at 760 °C. In this case, neither the annealed nor the annealed and deformed AMCR-0033 steel showed formation of the ferrite phase at sinks after irradiation. After deformation, deformation and ageing, the structure of the both non-irradiated and irradiated steels showed the formation of ε- and α-martensite. At the same time, the formation of σ-phase in the AMCR-0033 steel was not detected [282].

Further investigations, directed at increasing the phase stability and radiation resistance of the AMCR-0033 steel have been carried out on its modifications, additionally alloyed with low concentrations of nickel [284]. The increase of the nickel content to the maximum permissible concentration in the reduced-activation steels in accordance with the requirements of radiological standards for shallow-land waste disposal will evidently cause, together with alloying with silicon, an increase of the swelling resistance of the steels. For a number of Cr–Mn steels this has been shown, in particular, in [285].

The second direction the development of reduced-activation Cr–Mn steels is based on the examination of the phase stability and properties of steels based on the composition Fe–12Cr–20Mn–(0.1–0.25)C with additional alloying of the steel with different combinations of tungsten, vanadium, titanium, phosphorus and boron (see, for example [218,286–290]).

The main ideology (and, in out view, a promising one) in the development of these materials is based on the principle of the development of Cr–Ni steels of the PCA type with high radiation resistance [286, 287]. These principles have been examined in detail in the previous section. For the Cr–Ni steels, these principles are based on the combination of cold plastic deformation with alloying with titanium, silicon and phosphorus, supporting the formation of a dislocation structure with a high density of sinks resistant to recovery and recrystallisation. The stability of the dislocation network in the Fe–Cr–Ni steels is achieved by the formation of fine-dispersion precipitates of carbide and phosphide phases along the dislocations and in the matrix, resistant to coalescence during ageing and irradiation.

The identical principles are also used in the development of radiation-resistant Cr–Mn steels based on the Fe–12Cr–20Mn–C composition.

As shown in [286,287], austenite in the initial composition Fe–12Cr–20Mn–C is unstable and at increased temperatures in ageing breaks down with the formation of large carbides (mostly $M_{23}C_6$) and intermetallic phases. Additional alloying with tungsten, titanium, vanadium, phosphorus and boron results in the formation of fine-dispersion precipitates of the MC carbides along the dislocation network, increasing its resistance to recovery and recrystallization in subsequent ageing [287].

In the formation of fine-dispersion precipitates of the MC carbides, a special role in the Fe–12Cr–20Mn–C alloys is played by titanium which stimulates the formation of these carbides and decreases the tendency for the breakdown of austenite with the precipitates of the large $M_{23}C_6$ carbides at the grain boundaries and in the matrix [287,288]. The TiC carbides are stable and completely insoluble even after annealing at 1050 °C. The effect of the other previously mentioned alloying elements on the phase stability and radiation resistance of Cr–Mn and Cr–Ni steels, like titanium, should also be identical. Unfortunately, the absence at the present time of the required amount of information on the effect of radiation on the structure and properties of Cr–Mn steels of this type does not make it possible to draw any conclusions on their real phase stability in irradiation. However, taking into account the promising nature of the direction of the development of these alloys by analogy with Cr-nickel steels of the PCA type, on the basis of the examined compositions it is possible to expect the development of reduced-activation austenitic Cr–Mn steels with increased phase stability and radiation resistance.

4.6.4. Vanadium-based alloys

The volume and intensity of investigations of the structure, properties and radiation resistance of vanadium-based alloys have increased greatly in recent years. Primarily, this is associated with the fact that, in addition to the austenitic and ferritic stainless steels, the reduced-activation vanadium alloys are regarded as promising materials for the first wall of fusion synthesis reactors. The currently most extensively examined reduced-activation alloys based on vanadium include various compositions of the vanadium–titanium, vanadium–titanium–silicon and vanadium–titanium–chromium system.

The phase composition of both vanadium and its alloys in the condition with and without radiation depends strongly on the concentration of the interstitial impurities oxygen, nitrogen, carbon and the content of other elements, such as silicon, sulphur, phosphorus, titanium, etc, present in the initial vanadium or introduced during the fabrication of alloys and subsequent treatment. The most frequently recorded phases in the unalloyed and non-irradiated vanadium include phases such as V_6O_{13},VS_4, γ-Ti_3O_5 [291,292]. The precipitates of these phases are non-coherent. Unalloyed vanadium also contains a number of coherent precipitates, whose accurate composition has not been identified [292]. In accordance with the measured planar spacings, these precipitates may be γ-Na_3VO_4, CaV_4O_9, $Ca_5Si_2O_7(CO_3)_2$ and $(NaCa)_2Si_6(O,OH)_{12}(CO_3)_{0.5}$.

The neutron and ion irradiation of unalloyed vanadium results in the formation of precipitates in the matrix [293,294], on the dislocations [293,295], the grain boundaries [293] and also segregations in the vicinity of voids [118,119,295]. In all likelihood, these radiation-stimulated precipitates and segregations represent the compounds and clusters of interstitial impurities. As reported in [118,119], the formation of the radiation zone and voids in the vanadium in the process of neutron irradiation at ~150 °C is accompanied by a decrease of the initial microhardness. This effect is highly likely as a result of the departure of hardening elements from the matrix solid solution (mostly interstitial impurities). In vanadium with a higher carbon content (~≥0.06 at%) the process of formation of V_2C carbides is dominant both during thermal annealing and ageing [294,296,297].

A special role in the development of vanadium-based alloys, including radiation-resistant alloys, is played by titanium. The positive properties of titanium as the alloying element compensate its negative effect as a component increasing the duration of the decrease of induced radioactivity of the irradiated alloys to a biologically safe level as a result of mainly the formation of the long-life of ^{39}Ar radionuclide. In addition, as a result of the recent trends in the selection of vanadium alloys with the optimum composition in which the titanium content does not exceed 4–5%, a decrease in the activation parameters of these alloys may be minimised.

Titanium, as an alloying element in vanadium alloys, improves the strength properties, increases the resistance to corrosion and swelling. In this case, as an element with higher chemical activity than vanadium, titanium bonds the interstitial impurities [298], suppressing the formation of identical compounds of vanadium which may play a negative embrittling role. Since the compounds of titanium are characterised by higher thermal stability, the radiation-stimulated diffusion processes, especially with the participation of interstitial impurities, are also suppressed. For example, the capture by titanium of oxygen atoms decreases the rate of diffusion of oxygen in the V–5% Ti alloy by two orders of magnitude in comparison with pure vanadium [299]. These factors also play a significant role in increasing the phase stability and radiation resistance of the vanadium-based alloys.

As shown in a number investigations (see, for example [291,292,300, 301], depending on the composition and also the type and concentration of impurities, the structure of non-irradiated alloys of the vanadium-titanium, vanadium–titanium–chromium and vanadium–titanium–silicon system may contain a wide spectrum of precipitates, mostly on the basis of titanium. These are mostly oxycarbonitrides of titanium Ti(O,N,C), its oxides, phosphides, sulphides and silicides of different stoichiometry. The precipitates of titanium oxycarbonitrides, according to special investigations [302], are dominant in the initial structure of non-irradiated alloys at the total concentration of the interstitial impurities O, N and C $> 5\times10^{-2}$ %. At the concentration of these impurities of $< 4\times 10^{-2}$ %, the formation of Ti(O,N,C) phase is suppressed. In the V–Ti–Cr alloys with a low content of interstitial impurities, additionally alloyed with yttrium, aluminium and silicon (~1% of each element), the primary precipitates consist of the phases $Ti_5(Si,P)_3$, $(Y,Si)_2O_5$ and Ti_5Si_3 [302]. The detected concentration relationship in the formation of phases makes it possible not only to form the initial structure of the vanadium–titanium, vanadium–titanium–chromium and vanadium–titanium–silicon alloys, but also regulate the phase transformations in irradiation in order to increase radiation resistance.

The purity of the initial components of the vanadium-based alloys may have a very strong effect on their phase composition and properties. Investigations of the ductility properties of the V–4Cr–4Ti, V–5Cr–5Ti and V–5Cr–3Ti alloys show [298] that the ductile–brittle transition temperature of these alloys in the non-irradiated condition depends on the type of titanium master alloy used in the production of alloys. If this temperature in the V–4Cr–4 Ti alloy, which was melted using titanium after argon-arc remelting of titanium sponge, was lower than –190 °C, then in other alloys, where the master alloy was directly represented by titanium sponge, the ductile–brittle transition temperature was correspondingly –60 and –85 °C.

Special investigations carried out in [298] show that the fracture surfaces in V–5Cr–5Ti and V–5Cr–3Ti alloys, in contrast to V–4Cr–4Ti alloys, contained large precipitates (2–17 μm) of vanadium compounds of different composition with C, S, O, Cl, Ca, Na, K, Cr, and Cu and fine-dispersion precipitates (30–70 nm) with high density. These phases, as expected, are also compounds of vanadium enriched with O, C, S and Ca.

The observed changes in the phase composition and ductility properties are associated with the impurity contamination, introduced when using titanium sponge, with a relatively increased content, including Cl, Na and K.

The thermodynamic calculations of the free energy show that the formation of, in particular, vanadium oxycarbochlorides in alloys with increased chlorine content is preferred in relation to the formation of titanium oxycarbonitrides. In the absence of chlorine, the formation of titanium oxycarbonitrides is preferred from the viewpoint of energy [298].

On the basis of analysis of the results of investigations carried out in [118,119,301–306], the most important phase changes in the vanadium-based alloys with titanium in irradiation in a general case (depending on composition, irradiation temperature and fluence) include the processes associated with the formation and evolution of either primary or secondary precipitates of mainly the following phases: Ti(O,N,C), α-Ti, TiO_2, Ti_5Si_3 and $Ti_5(Si,P)_3$.

The precipitates of rutile TiO_2 form most intensively in the process of irradiation at temperatures ≥600 °C [301,304]. In most cases, these precipitates are needle-shaped, localised in the matrix and are characterised by a tendency for precipitation in alloys with a high titanium content [301,304]. The precipitation of this phase in the process of irradiation in the vicinity of the grain boundaries is suppressed by the formation of titanium phosphides at the grain boundaries in the production of alloys. The phosphorus atoms are characterised by a tendency for segregation at the grain boundaries, forming phosphides and depleting the solid solution in titanium [304].

In analysis of the mechanisms of formation of precipitates of TiO_2 and other radiation-stimulated phases in vanadium alloys with titanium it is important, in our opinion, to consider the following experimental facts. The precipitates of the phases TiO_2, Ti_5,Si_3, $Ti_5(Si,P)_3$, etc, were observed in [301–306] after irradiation with high neutron fluences. At the same time, investigations of the kinetics of structural-phase changes in the V–21.5 at.% Ti alloy in the process of electron irradiation at 100 °C showed [118,119] that the given alloy is characterised initially by the formation of the α-Ti phase (at doses of ≤5×10^{17} cm^{-2}). This result is in agreement with the equilibrium diagram of the vanadium–titanium system [307] according to which this transformation is possible at temperatures of <670 °C. Within the framework of the considerations examined in section 4.3.2, the intensification of the breakdown of the β-(V,Ti) solid solution in irradiation with the formation of the α-Ti phase may be interpreted on the basis of the mechanisms of radiation-enhanced diffusion and compensation, by point defects, of coherent strains, accompanying transformation processes.

In particular, it should be mentioned that in further electron irradiation of the V–21.5 at.% Ti alloy, the reflections from radiation-stimulated precipitates no longer corresponded to the reflections of the lattice of the α-Ti phase [118,119]. The α-Ti phase was not detected in the investigations [118, 119] and after irradiation of the given alloy with a neutron flux of 4.7×10^{20} cm^{-2} at a temperature of 150 °C.

Thus, on the basis of the previously examined results, the mechanism of formation of the radiation-stimulated phases in the vanadium–titanium alloys may be regarded as a complex process. The first stage of this process is characterised by the radiation-stimulated breakdown of the solid solution with the formation of precipitates of the α-Ti phase. During further irradiation, the absorption, by this phase, of the diffusion flows of oxygen atoms and other elements also leads to the formation of the compounds TiO_2 and other phases, recorded usually after irradiation with high fluences.

According to the results of a series of detailed investigations (for example [303–306,308]), a special role in the variation of the structural properties of the alloys of vanadium with titanium is played by the radiation-stimulated precipitates of the Ti_5Si_3 phase. In alloys with increased silicon content and low concentration of O+N+C, this phase also forms without irradiation, but in the radiation conditions, the precipitates of this phase are considerably smaller and their concentration is greatly increased.

The processes of precipitation of the Ti_5Si_3 phase are relatively intensive at temperatures of ≥420 °C. Being effective sinks for defects, the particles of the Ti_5Si_3 phase suppress the void formation processes. At the same time, they have no detrimental effect on the ductility properties, because they do not operate as the sources of nucleation of cracks and degradation of grain boundary strength [303]. In addition, the formation of these phases during irradiation makes it possible to compensate the radiation softening, associated with the dissolution of the particles of the initial phase Ti(O,N,C) in the transport of oxygen from the matrix into lithium [306].

As in both annealing and irradiation, the vanadium alloys with titanium may be characterised by the formation of the Ti_5P_3 phase which is isostructural with the Ti_5O_3 phase [301,303,304,306]. These phases have the same hexagonal lattice of the type Mn_5Si_3 with very similar parameters. Phosphorus may partially replace silicon in irradiation-stimulated precipitates of the Ti_5Si_3 phase, modifying this compound to the $Ti_5(Si,P)_3$ phase [301,303,304].

In conclusion of this section, it should be mentioned that after irradiation with a fluence of 1.9×10^{23} cm^{-2} at 420 °C, the V–20Ti alloy contained the transmutation phase γ-LiV_2O_5 [308]. Lithium in the alloy could form by the reaction $^{10}B(n,\alpha)^7Li$.

REFERENCES

1. Bystrov L.N. Ivanov L.I., Platov Yu.M., Fiz. Khim. Obrab. Mater., 1970, No. 2, p.34.
2. Bystrov L.N., Ivanov L.I., Platov Yu.M. Phys.Stat.Sol. (a), 1971, v.7,p.617.
3. Schule W. J.Phys.F: Metal.Phys. 1980, v.10, p.2345.
4. Damask A.C. In: Rend.Scuola Int.Fis. "Enrico Fermi" N-Y, London, 1962, p.763.
5. Lomer W.M. AERE, Techn.Report, 1954, p.1540.
6. Dienes G.I., Damask A.C. J.Appl.Phys., 1958, v.29, no12, p.1713.
7. Bystrov L.N., Ivanov L.I., Platov Yu.M., Dokl. Akad. Nauk SSSR 1969, v. 185, No. 2, p.309.
8. Sharp J.V. AERE Report R6267, 1969, 19p.
9. Sizman R. J.Nucl.Mater., 1978, v.69/70, p.386.
10. Adda Y., Beyeler M., Bregec G. Thin Solid Films, 1975, v.25, p.107.
11. Le Claire A.D. J.Nucl.Mater., 1978, v.69/70, p.70.
12. Barbu A. Acta Met., 1980, v.28, p.499.
13. Bocquet J.L. Rapport CEA., R-5112, Soclay, 1981, 45p.
14. Bocquet J.L. Acta Met.,1986, v.34, no4, p.571.
15. Smirnov A.A. Molecular-kinetic theory of metals, Moscow, Nauka, 1966, 488 pp.
16. Bystrov L.N., Ivanov L.I., Platov Yu. M., in: Diffusion in metals and alloys, Tula 1968, p.254.
17. Bystrov L.N., Ivanov L.I., Platov Yu. M., in: Ordering of atoms and its effect on the properties of alloys, Naukova dumka, Kiev, 1968, p. 250.
18. Bystrov L.N., Ivanov L.I., Platov Yu. M., Fiz. Khim. Obrab. Mater., 1970, No. 12, p. 34.
19. Bystrov L.N., Ivanov L.I., Platov Yu. M., Fiz. Khim. Obrab. Mater., 1970, No. 3, p. 34.
20. Bystrov L.N., Ivanov L.I., Platov Yu.M. Phys.Stat.Sol.(a), 1971, v.8, p. 375.
21. Halbwachs M., Hillairet J., Cost J.R. J.Nucl.Mater., 1978, v.69/70, p.776.
22. Halbwachs M., Stanley J.T., Hillairet J. Phys.Rev.B, 1978, v.B18, p. 4938.
23. Schule W. J.Phys.F: Metal.Phys., 1980, v.10, p.2345.
24. Halbwachs M. J.Phys.F: Metal.Phys., 1978, v.8, p.1053.
25. Halbwachs M., Hillairet J. J.Phys.F: Metal.Phys., 1981, v.11, p.2247.
26. Platov Yu.M., Simakov S.V., Fiz. Khim. Obrab. Mater., 1988, No. 6, p.5.
27. Larsen R.E., Dienes G.J., Damask A.C. Acta.Met., 1964, v.12, p. 1141.
28. Schule W., Lang E., Donner D., Penkuhn H. Rad.Effects, 1970, v.2, p. 151.
29. Shule W., Scholz R. Rad.Effects, 1984, no.1-2, p. 115.
30. Schule W. Mater.Sci.Forum. 1987, v.15/18, p. 1225.
31. Schule W. J.Nucl.Mater., 1988, v. 155/157, p.1132.
32. Schule W., In: Radiation materials science, Khar'kov, 1990, No.1, p.148.
33. Wechsler M.S. In: Symposium on radiation effects on metals and neutron dosimetry, ASTM-STP-341, 1963, p.86.
34. Poerschke R., Wollenberger H. Thin Solid Films, 1975, v.25, p.167.
35. Poerschke R., Wollenberger H. J.Phys.F: Metal.Phys., 1976, v.6, no.1, p.27.
36. Chountas K., Papathanassopoulos K., Andronikos P. Rad.Effects Lett., 1979, v.43, p.249.
37. Poerschke R., Theis U., Wollenberger H. J.Phys.F: Metal. Phys., 1980, v.10, p.67.

38. Ivanov L.I., Volkov M.G., Platov Yu. M., Sadykhov S.I.O., Simakov S.V. Fiz. Khim. Obrab. Mater., 1988, no. 1, p.28.
39. Schule W., Crestoni G. Rad.Effects, 1976, v.29, p. 17.
40. Bartels A., Kemkes C., Lucke K. Acta Met., 1985, v.33, no.10, p. 1887.
41. Bystrov L.N. and Platov Yu.M., Fiz. Met. Metalloved.,, 1968, v..25, No.5, p.944.
42. Lam N.Q., Rothman S.J., Merkle K.L., Nowicki L.J., Dever D.J. Thin Sol.Films., 1975, v.25, p.157.
43. Heitkamp D., Bierman W., Lundy. Acta Met., 1966, v.14, p.1202.
44. Muller A., Naundorf V., Macht M.P. J.Appl.Phys., 1988, v.64, no.7, p.3445.
45. Bonzel H.P. Acta Met., 1965, v.13, no.10, p. 1084.
46. Mach M.P., Naudorf V. Wollenberger H. J.Nucl.Mater., 1981, v.103/104, p. 1487.
47. Mach M. P., Naudorf V. J.Appl.Phys. 1982, v.53, N11, p.7551.
48. Shimizu H., Ono M. J.Appl.Phys., 1982, v.53, no.4, p.3044.
49. Marton D., Fine J., Chambers G.P. Mater.Sci.Engin., 1989, v.A115, p.223.
50. Rotman S.G., In.: Phase transformations in irradiation, Chelyabinsk,Metallurgiya 1989, p. 151.
51. Halbwachs M, Yoshida N. Phil.Mag., 1981, v.43, no.5, p.1125
52. Platov Yu.M., Simakov S.V., Fiz. Met. Metalloved., 1986, v.61, no.2, p.213.
53. Cowley J.M. Phys.Rev. 1950, v.77, no.5, p.669.
54. Houska C.R., Averbach B.L. J.Appl. Phys., 1959, v.30, no.10, p.1225.
55. Bartels A. Mater.Sci.Forum., 1987, v.15/18, p.1183.
56. Schule W. Mater.Sci.Forum., 1987, v.15/18, p.1231.
57. Bartels A., Bartusel D., Kemkes-Sieben Chr, Luke K. Mater.Sci.Forum., 1987, v.15/18, p. 1237.
58. Abromeit C., Poerschke R., Pfeiler W. Mater.Sci.Forum., 1987, v.15/18, p. 1245.
59. Krivoglaz M.A., Smirnov A.A., Theory of ordering alloys, Moscow, Fizmatgiz, 1958, 388pp.
60. Iveronova V.I., Katsnel'son., Short-range order in solid solutions, 1977, 255pp.
61. Wagner K., Thermodynamics of alloys, Moscow, Metallurgiya, 1957, 179pp.
62. Kubaschewski O., Catterall J.A. Thermodynamical data of Alloys. London, N-Y, Pergamon Press, 1956, 200pp.
63. Martin G. Phys.Rev. B, 1984, v.30, no.3, p.1424.
64. Martin G., Bellon P. Mater.Sci.Forum., 1987, v.15/18, p.1337.
65. Bellon P., Martin G. Ann.Chim.Fr., 1991, v.16, p.365.
66. Potter D.I. In: Phase transformations in irradiation, Chelyabinsk, Metallurgiya 1989, p.168.
67. Banerjee S., Kulkarni U.D., Urban K. Acta Met., 1989, v.37, no.1, p. 35.
68. Kosaki M., Smith H.P., Khan J.M., Potter D.I, Worley R.D. J.Appl.Phys., 1967, v.38, no.6, p.2497.
69. Butler E.P., Swann P.R. In: Proc. 5-th Int.Conf.High Voltage Microscopy, Kyoto, 1977, p.555.
70. Butler E.P. In: Inst.Phys.Conf.Ser.,1978, no.41, p.98.
71. Van Tendeloo G., Van Landuy J., Amelinck S. Rad. Eff., 1979, v.41, p.179.
72. Howe L M., Rainville M.H. Phil.Mag.A., 1979, v.39, no.2, p.195.
73. Zee R.H., Guinan M.W., Kulcinski G.L. J.Nucl.Mater., 1983, v.114, p.190.
74. Kirk M.A., Blewitt T.H. J.Nucl.Mater., 1982, v.108/109, p.124.
75. Benerjee S., Urban K. Ann.Chim.Fr., 1984, v.9, p.233.
76. Huang J.S., Guinan M.W., Hanh P.A. J.Nucl.Mater., 1986, v.141/143, p.888.

77. Huang J.S., Guinan M.W., Kirk M.A., Hanh P.A. J.Nucl.Mater., 1988, v.155/157, p.1084.
78. Kulik J., Gratias D., de Fontaine D. Phys.Rev. B, 1989, v.40, no.13, p. 8607.
79. Cheng J., Ardell A.J. Acta Met., 1989, v.37, no.7, p.1891.
80. Zee R.H., Wilkes P. In: Phase Stability during Irradiation. Warrendale, AIME, 1981, p.295.
81. Russell K.S., in:Phase transformations under irradiation, Chelyabinsk, Metallurgiya, p. 65
82. Russell K.C. Progr. Mater.Sci., 1984, v.28, p.229.
83. Bellon P., Martin G. J. Less-Common Metals, 1988, v.145, no.1/2, p.465.
84. Bellon P., Martin G. Phys.Rev.B, 1989, v.39, no.4, p.2403.
85. Bakai A.S. Mater.Sci.Forum., 1992, v.97/99, p.307.
86. Poerschke R., Wollenberger H. Rad.Eff. 1980, v.49, p.225.
87. Wagner W., Poerschke R., Axmann A., Schwahn D. Phys. Rev.B, 1980, v.21, no.8, p.3087.
88. Wagner W., Poerschke R., Wollenberger H. J.Phys.F: Metal.Phys., 1982, v.12, p.405.
89. Kinoshita C., Nakai K., Kitajima S. Mater.Sci.Forum, 1987, v.15/18, p.1403.
90. Cahn J.W. Acta Met., 1961, v.9, no.9, p.795.
91. Cahn J.W. Trans.Met.Soc. AIME, 1968, v.242, p.166.
92. Cook H.E. J.Phys.Chem.Solids, 1969, v.30, p.2427.
93. Cook H.E. Acta Met., 1970, v.18, p.297.
94. Cook H.E. Fontaine D., Hilliard J.E. Acta Met., 1969, v.17, p.765.
95. Krivoglaz M.A., Theory of scattering of x-rays and neutrons by real crystals, Moscow, Nauka, 1967, 336pp.
96. Vrijen J., Aalders J, Dijk C., Radelaar S. Phys.Rev., 1980, v. B22, no.4, p.1503.
97. Mozer B., Kenting D.T., Moss S.S. Phys.Rev., 1968, v. 175, no.3, p.868.
98. Garner F.A., Brager H.R., McCarthy J.M. Report HEDL-SA-3430 FP, 1986, 25p.
99. Russell K.C., Garner F.A. Met.Trans. A., 1992, v.23A, p.1963.
100. Bad'yanov A.B., Volkov M.G., Ivanov L.I., Lazorenko V.M., Platov Yu.M., Fiz. Khim. Obrab. Mater., 1985, no.2, p.136.
101. Brager H.R., Garner F.A. In: Effects of Radiation on Materials. Proc.12th Int.Symp., Philadelphia, ASTM, 1985, p.139.
102. Garner F.A., Brager H.R., Dodd R.A., Lauritzen T. Nucl.Instrum. Meth., 1986, v.B16, p.244.
103. Ivanov L.I., Bystrov L.N., Platov Yu.M., Lazorenko V.M., Bondarenko G.G., Kosjanov O.M., Ustinovshikov V.M. J.Nucl.Mater., 1992, v.191/194, p.733.
104. Ivanov L.I., Lazorenko V.M., Platov Yu.M., Simakov S.V., Fiz. Khim. Obrab. Mater., 1984, no.4, p.27.
105. Scheuer U., Wahi R.P., Wollenberger H. J.Nucl.Mater., 1986, v.141/143, p.767.
106. Wollenberger H. Mater.Res.Symp.Proc.,1989, v.138, p.3.
107. Wollenberger H. Mater.Sci.Forum.,1992, v.97/99, p.241.
108. Konobeevskii S.T., Effect of radiation on materials, 1967, 401pp.
109. Herman H., Kauffman J.W., Fine M.E. J.Appl.Phys., 1965, v.36, no.6, p.1976.
110. Ivanov L.I., Bystrov L.N., Platov Yu.M., Fiz.Metallov Metalloved., 1968, v.25, No.5, p.950.
111. Sklad P.S., Mitchell T.E. Acta Met., 1975, v.23, p.1283.
112. Bertram K., Minter F.J., Hudson J.A., Russell K.C. J.Nucl.Mater., 1978,

v. 75, p.42.
113. Kinoshita C., Hobbs L.W., Mitchell. In: Phase Stability During Irradiation. Warrendale, AIME, 1981, p.561.
114. Knoll R.M., Kulcinsky G.L. J.Nucl.Mater., 1985, v.131, p.187.
115. Erawa T., Sako T., Fujita F.E. Mater.Sci.Forum., 1987, v.15/18, p.1379.
116. Zhung Y.G. J.Nucl.Mater. 1989, v.165, p.278.
117. Kell B., Wollenberger H. J.Nucl.Mater., 1989, v.169. p.126.
118. Ivanov L.I., Lazorenko V.M., Platov Yu.M., Fiz. Khim. Obrab. Mater., 1990, no.3, p.5.
119. Ivanov L.I., Lazorenko V.M., Platov Yu.M., Tovtin V.I. J.Nucl.Mater.,1992, v.191/194, p.928.
120. Solt G., Waeber W.B., Frisins F., Zimmermann U. Mater.Sci.Forum., 1992, v.97/99, p.273.
121. Barbu A., Lorenzelli T.N.Le,N., Maury F., de Novion C.H. Mater. Sci.Forum., 1992, v.97/99, p.351.
122. Maydet S.I., Russell K.C. J.Nucl.Mater., 1977, v.64, p.101.
123. Mruzik M.R., Russell K.C. J.Nucl.Mater., 1978, v.78, p.343.
124. Cauvin R., Martin G. Phys.Rev. B., 1981, v.23, p.3322.
125. Cauvin R., Martin G. Phys.Rev. B., 1981, v.23, p.3333.
126. Russell K.C. Acta Met., 1971, v.19, p.753.
127. Saito K. Phil.Mag. A. 1981, v.43, no.1, p.43.
128. Kaufman L., Watkin J.S., Gittus J.H., Miodownik A.P. In: Phase diagrams in metals and ceramics. Wachington, NBS, 1978, p.1065.
129. Bocquet J.L., Martin G. J.Nucl.Mater., 1979, v.83, no.1, p.186.
130. Yamauchi H., Fontaine D. In: Phase stability during irradiation. Warrendale, AIME, 1981, p.73.
131. Holloway J.P., Stubbins J.F. In: Effects of radiation on materials. Proc.12th int. symp., Philadelphia, ASTM, 1985, p.167.
132. Nastasi M., Mayer J.M. Mater.Sci.Report, 1991, v.6, p.1.
133. Anthony T.R. In: Radiation-induced voids in metals. USAEC, 1972, p.630.
134. Anthony T.R. In: Diffusion in Solids, Acad.Press, 1975, p.353.
135. Manning G., Kinetics of diffusion of atoms in crystals, Moscow, Mir, 1971, 277pp.
136. Marwick A.D. J.Phys.F.: Metal.Phys., 1978, v.8, no.9, p.1849.
137. Lam N.Q., Johnson R.A. J.Nucl.Mater., 1976, v.20, 121.
138. Johnson R., Lam N.Q. Phys.Rev. B., 1976, v.13, no.10, p.4364.
139. Okamoto P.R., Rehn L.E. J.Nucl.Mater., 1979, v.83, no.1, p.2.
140. Martin G. Phil.Mag. (A), 1978, v.32, no.2, p.131.
141. Wiedersich H., Okamoto P.R., Lam N.Q. J.Nucl.Mater., 1979, v.83, p.98.
142. Chen I-Wei. J.Nucl.Mater., 1983, v.116, p.249.
143. Wiedersich H., Okamoto P.R., Lam N.Q., In: Phase transformations in irradiation, Chelyabinsk, Metallurgiya, p.10.
144. Pechenkin V.A., Epov G.A., Fiz. Met. Metalloved., 1988, v. 65, no.2, p.279.
145. Murphy S.M. Phil.Mag. A., 1989, v.59, no.5, p.953.
146. Murphy S.M. J.Nucl.Mater., 1989, v.168, p.31.
147. Pechenkin V.A., Epov G.A., Fiz. Met. Metalloved, 1991, no.3, p.58.
148. Murphy S.M. J.Nucl.Mater., 1991, v.182, p.73.
149. Turkin A.A., Vopr. Atomn. Nauki i Tekhn., Seria Fizika Radiats. Povrezhdenii Radiats. Materialoved., 1991, no.3, p.1.
150. Russell K.C. In: Radiation Effects in Breeder Reactor Structural Materials, AIME, N.Y., 1977, p.821.

151. Potter D.I., Rehn L.E., Okamoto P.R., Wiedersich H. Scripta Met., 1977, v.11, p.1095.
152. Piller R.C., Marwick A.D. J.Nucl.Mater., 1978, v.71, p.309.
153. Rehn L.E., Okamoto P.R., Poter D.I., Wiedersich H. J.Nucl.Mater., 1978, v.74, p.242.
154. Wagner A., Seidman D.N. J.Nucl.Mater., 1979, v.83, p.48.
155. Takeyama T., Ohnuki S., Takahashi H. Scripta Metal., 1980, v.14, p. 1105.
156. Ohnuki S., Takeyama T., Takahashi H. In: Point Defects and Defect Interactions in Metals. Tokio, University Tokio Press, 1982, p.954.
157. Wiedersich H., Okamoto P. In: Phase Stability during Irradiation, AIME, Wrrendale, 1981, p.23.
158. Ardell A.G., Yankhorban K., in: Phase transformations in irradiation, Chelyabinsk, Metallurgiya, 1989, p.226.
159. Ren L.E., Okamoto P.R., In: Phase transformations in irradiation, Chelyabinsk, Metallurgiya, 1989, p.226.
160. Kornblit L., Ignatiev A. J.Nucl.Mater., 1984, v.126, p.77.
161. Schroeder H., Packan N.H., Kesternich W. Mater.Sci.Forum, 1987, v.15/18, p.1391.
162. Kameda J., Bevolo A.J. Acta Met., 1989, v.37, no.1, p.3283.
163. Erck R.A.,Rehn L.E. J.Nucl.Mater., 1989, v.168, p.208.
164. Erck R.A., Rehn L.E. Phil.Mag. 1990, v.62, no.1, p.29.
165. English C.A., Murphy S.M., Perks J.M. J.Chem.Soc.Farad. Trans., 1990, v.86(8), p.1263.
166. Yoshida Y., Fratzl P., Vogl G., Hofer H., Dworschak F. J.Phys.:Condens.Mater., 1992, v.4, p.2415.
167. Ivanov L.I., Lazorenko V.M., Platov Yu.M., Kalinin V.G., Simakov S.V., Smirnov A.V., Tovtin V.I., Toropova L.S., Fiz. Khim. Obrab. Mater., 1985, no.4, p.16.
168. Platov Yu.M., Lazorenko V.M., Simakov S.V., Tovtin V.I., Ivanov V.V. Mater.Sci.Forum., 1992, v.97/99, p.253.
169. King H.W. J.Mater.Science, 1966, v.1, p.79.
170. Lam N.Q., Hoff H.A. Surface Science., 1988, v.193, p.353.
171. Yacout A.M., Lam N.Q., Stubbins J.F. Nucl.Instrum. Meth.Phys.Res., 1992, v.B71, p.148.
172. Nukai T., Mitchell T.E. J.Nucl.Mater., 1982, v.105, p.149.
173. Wahi R.P., Wollenberger H. J.Nucl.Mater., 1983, v.113, p.207.
174. Herschitz R., Seidman D.N. Acta Met., 1984, v.32, no.8, p.1141.
175. Martin G. Phil.Mag., 1975, v.32, no.3, p.615.
176. Martin G. In: Fundamental Aspects of Radiation Damage in Metals. USERDA, 1976, v.II, p.1084.
177. Martin G. Phys.Rev. Lett., 1983, v.50, no.4, p.250.
178. Martin G., Cauvin., Bocquet J.L., Barbu A. In: Phase Stability during Irradiation. Warrendale, AIME, 1981, p.43.
179. Bakai A.S. Vopr. Atomn. Nauki Tekhniki, Seria Fiz. Radiats. Povrezhdenii Radiats. Materialoved., 1983, no.3, p.89.
180. Kirsanov V.V., Kislitsin S.V., Izv. Akad. Nauk Kaz. SSR, 1983, no.1, p.1.
181. Lazorenko V.M., et al.: In: Constructional materials for fusion reactors, 1988, p.167.
182. Ivanov L.I., Lazorenko V.M., Platov Yu.M., Pletnev M.N., Toropova L.S., Dokl. Akad. Nauk SSSR, 1981, v.257, p.1175.
183. Ivanov L.I., Lazorenko V.M., Platov Yu.M., Fiz. Khim. Obrab. Mater., 1989,

no.2, p.11.
184. Platov Yu.M., Simakov S.V., Ivanov V.V., et al., Fiz. Khim. Obrab. Mater., 1990, no.3, p.20.
185. Ivanov L.I., Lazorenko V.M., Platov Yu.M., Fiz. Khim. Obrab. Mater., 1991, no.6, p.50.
186. Hansen M., Anderko K., Structures of binary alloys, 1962, v.1, p.79.
187. Mandolfo L.F., Structure and properties of Al alloys, Moscow, Metallurgiya, 1979, 640pp.
188. King R.T., Jostsons A. Met.Trans., 1975, v.6A, p.863.
189. Nelson R.S., Hudson J.A., Mazey D.J. J.Nucl.Mater., 1972, v.44, p.318.
190. Bleibery M.L., Bajaj R., Bell W.L., Thomas L.E. In: Radiation Effects in Breeger Reactor Structural Materials, AIME, N.Y., 1977, p.667.
191. Gilbert R.W., Griffiths M., Carpenter G.J.C. J.Nucl.Mater., 1985, v.135, p.265.
192. Lin B.X. Phys.Stat. Sol.(a), 1983, v.75, k77.
193. Brimhall J.L., Simonen E.P. Nucl.Instrum. Meth.Phys.Res., 1986, v.B16, p.187.
194. Hung L.S., Mayer J.W. Thin Solid Films, 1985, v.123, p.135.
195. Mori H., Fujita H. Ultramicroscopy, 1991, v.39, p.355.
196. Lifshitz I.M., Slyozov V.V. J.Phys.Chem.Solids, 1961, v.19, p.35.
197. Wagner C.Z. Elektrochem., 1961, v.65, p.581.
198. Wilkes P. J.Nucl.Mater., 1979, v.83, p.166.
199. Liou K.Y., Wilkes P. J.Nucl.Mater., 1979, v.87, p.317.
200. Brailsford A.D. J.Nucl.Mater., 1980, v.91, p.221.
201. Frost H.J., Russell K.C. J.Nucl.Mater., 1982, v.103/104, p.1427.
202. Frost H.J., Russell K.C. Acta Met., 1982, v.30, p.953.
203. Banerjee S., Urban K. Phys.Stat.Sol.(a), 1984, v.81, p.145.
204. Martin G., Bellon P. J.Less.Common Metals, 1988, v.140, p.211.
205. Turkin A.A., Bakai A.S., Buts A.V. Mater.Sci.Forum, 1992, v.97/99, p.343.
206. Salomons E., Bellon P., Soisson F., Martin G. Phys.Rev. B., 1992, v.45, no.9, p.4582.
207. Platov Yu.M., Pletnev M.N. J.Nucl.Mater., 1994, v.211, p.95.
208. Potter D.I., McCormeek. Acta Met., 1979, v.27, p.933.
209. Potter D.I., Wiedersich H. J.Nucl.Mater., 1979, v.83, p.208.
210. Potter D.I.,Ryding D.G. J.Nucl.Mater., 1977, v.71, p.14.
211. Urban K., Martin G. Acta Met., 1982, v.30, p.1209.
212. Baron M. In: Phase Stability during Irradiation. AIME, Warrendale, 1981, p.63.
213. Abuzov A.S., Slezov V.V., Tanatarov L.V., Fiz. Tverd. Tela, 1991, v.33, no.3, p.834.
214. Simonen E.P. In: Optimizing Materials for Nuclear Application, 1985, Warrendale, Metallurgical Soc., p.225.
215. Amaev A.D., Gorynin I.V., Ignatov V.A., Nikolaev V.A., Rjazantzev E.P., Shamardin V.K. Proc.Int.Conf.on Radiation Material Science, 1990, p.106.
216. Akamatsu M., Bourgoin J., Janot C., Moser P., J.C.Van Duysen. Ann.Chim.Fr., 1991, v.16, p.333.
217. Vishkarev O.M., Zvezdin Yu.I., Schamardin V.K., Tulyakov G.A. In: Radiation Embrittlement of Nuclear Reactor Pressure Vessel Steels. 1993, Philadelphia, ASTM, p.218.
218. Ivanov L.I., Platov Yu.M., Demina E.V. J.Advanc.Mater., 1994, 1(2), p.166.
219. Gelles D.S. In: Reduced Activation Materials for Fusion Reactors. Philadelphia, ASTM, 1990, p.113.

220. Abe F., Noda T., Araki H., Okada M. In: Reduced Activation Materials for Fusion Reactors. Philadelphia, ASTM, 1990, p.130.
221. Hiwatashi S., Kohno Y., Asakura K., Kohyama A. J.Nucl. Mater., 1991, v.179/181, p.709.
222. Kimura A., Matsui H. J.Nucl.Mater., 1994, v.212/215, p.701.
223. Gelles D.S. J.Nucl.Mater., 1994, v.212/215, p.714.
224. Kohno Y., Kohyama A., Yoshino M., Asakura K. J.Nucl.Mater., 1994, v.212/215, p.707.
225. Gelles D.S. In: Optimizing Materials for Nuclear Applications. AIME, Warrendale, 1985, p.63.
226. Little E.A., Stoter L.P. In: Effects of Radiation on Materials. ASTM, Baltimore, 1982, p.207.
227. Gelles D.S. J.Nucl.Mater., 1981, v.103/104, p.975.
228. Suzuki M., Sawai T., Mariasz P.J., Hishinuma A. J.Nucl. Mater., 1991, v.179/181, p.718.
229. Kohno Y., Kohyama A., Gelles D.S. J.Nucl.Mater., 1991, v.179/181, p.725.
230. Klueh R.L., Ehrich K., Abe F. J.Nucl.Mater., 1992, v.191/194, p.116.
231. Katoh Y., Kohno Y., Kohyama A., Kayama H. J.Nucl.Mater., 1992, v.191/194, p.1204.
232. Anderko K., Schafer L., Materna-Morris E. J.Nucl.Mater., 1991, v.179/181, p.492.
233. Faulkner R.G., Little E.A., Morgan T.S. J.Nucl.Mater., 1992, v.191/194, p.858.
234. Kimura A., Charlot L.A., Gelles D.S., Baer D.R., Jones R.H. J.Nucl.Mater., 1992, v.191/194, p.885.
235. Little E.A., Morgan T.S., Faulkner R.G. Mater.Sci.Forum, 1992, v.97/99, p.323.
236. Abe F., Noda T., Araki K., Nakazawa S. J.Nucl.Mater., 1991, v.179/181, p.663.
237. Griffin R.D., Gelles D.S., Dodd R.A., Kulcinski G.L. J.Nucl. Mater., 1991, v.179/181, p.714.
238. Kohno Y., Gelles D.S., Kohyama A., Tamura M., Hishinuma A. J.Nucl.Mater., 1992, v.191/194, p.868.
239. Kohyama A., Kohno Y., Asakura K., Kayano H. J.Nucl.Mater., 1994, v.212/215, p.684.
240. Klueh R.L., Alexander D.J., Maziasz J. J.Nucl.Mater., 1992, v.186, p.185.
241. Lee E.H., Rowcliffe A.F., Kenik E.A. J.Nucl.Mater., 1979, v.83, p.79.
242. Garner F.A. In: Phase Stability during Irradiation. AIME, Warrendale, 1981, p.165.
243. Lee E.H., Maziacz P.J., Rowcliffe A.F. In: Phase Stability during Irradiation. AIME, Warrendale, 1981, p.191
244. Yang W.J.S., Brager H.R., Garner F.A. In: Phase Stability during Irradiation. AIME, Warrendale, 1981, p.257.
245. Harries D.R. In: Proc.Conf.Mechanical Behavior and Nuclear Applications of Stainless Steel at Elevated Temperatures. Varese, Italy, London, Metal Society, 1981, p.1.
246. Williams T.M. In: Effects of Radiation on Materials. Philadelphia, ASTM, 1982, p.166.
247. Optimizing Materials for Nuclear Application. Warrendale, Metallurgical Society, 1985, 273 p.
248. Reduced Activation Materials for Fusion Reactors. Philadelphia, ASTM, 1990, 253 pp.

249. Fusion Reactor Materials: Proc.4th Int.Conf. Kyoto, 1989, J.Nucl.Mater., 1991, v.179/181.
250. Fusion Reactor Materials: Proc.5th Int.Conf. Cleawater, 1991, J.Nucl.Mater., 1992, v.191/194.
251. Fusion Reactor Materials: Proc.6th Int.Conf. Stresa, 1993, J.Nucl.Mater., 1994, v.212/215.
252. Lee E.H., Mansur L.K. Metallurgical Transactions A., 1992, p.1977.
253. Thomas L.E. Transactions A.N.S. 1978, v.28, p.151.
254. Garner F.A., Brager H.R., Gelles D.S., McCarthy J.M. J.Nucl. Mater., 1987, v.147, p.294.
255. Kenik E.A., Hojou K. J.Nucl.Mater., 1992, v.191/194, p.1331.
256. Shepherd C.M. J.Nucl.Mater., 1990, v.175, p.170.
257. Hamada S., Suzuki M., Maziasz P.J., Hishinuma A. J.Nucl. Mater., 1991, v.179/181, p.515.
258. Jitsukawa S., Hojou K., Hishinuma A. J.Nucl.Mater., 1991, v.179/181, p.538.
259. Sawai T., Maziasz P.J., Kanazawa H., Hishinuma A. J.Nucl. Mater., 1992, v.191/194, p.712.
260. Kohno Y., Kohyama A., Kayano H. J.Nucl.Mater., 1992, v.191/194, p.742.
261. Suzuki M., Hamada S., Maziasz P.J., Jitsukawa S., Hishinuma A. J.Nucl.Mater., 1992, v.191/194, p.1351.
262. Katoh Y., Kohno Y., Kohyama A. J.Nucl.Mater., 1994, v.212/215, p.464.
263. Shibahara I., Akasaka N., Onose S., Okada H., Ukai S. J.Nucl.Mater., 1994, v.212/215, p.487.
264. Neklyudov I.M., Voyevodin V.N. J.Nucl.Mater., 1994, v.212/215, p.39.
265. Fukuya K., Nakahigashi S., Ozaki S., Shima S. J.Nucl.Mater., 1991, v.179/181, p.1057.
266. Yoshida N., Xu O., Watanabe H., Muroga T., Kiritani M. J.Nucl.Mater., 1992, v.191/194, p.1114.
267. Miroga T., Watanabe H., Yoshida N. J.Nucl.Mater., 1994, v.212/215, p.482.
268. Miroga T., Garner F.A., McCarthy J.M. In: Annual Progress Report for Monbusho-DOE for Collaboration in Fundamental Studies of Irradiation Effects in Fusion Materials. 1989, p.108.
269. Watanabe H., Muroga T., Yoshida N. In Annual Progress Report for Monbusho-DOE for Collaboration in Fundamental Studies of Irradiation Effects in Fusion Materials., 1992, p.217.
270. Watanabe H., Aoki A., Muroga T., Yoshida N. J.Nucl.Mater., 1991, v.179/181, p.529.
271. Garner F.A., McCarthy J.M. In: Reduced Activation Materials for Fusion Reactors. Philadelphia, ASTM, 1990, p.19.
272. Yukawa N., Morinaga M., Nishiyama K., Matsumoto Y., Muraga Y., Ezaki H. In: Reduced Activation Materials for Fusion Reactors. Philadelphia, ASTM, 1990 p.30.
273. Klueh R.L., Maziasz P.J. In: Reduced Activation Materials for Fusion Reactors. Philadelphia, ASTM, 1990, p.7.
274. Maziasz P.J., Klueh R.L. In: Reduced Activation Materials for Fusion Reactors. Philadelphia, ASTM, 1990, p.56.
275. Okazaki Y., Mochizuki M., Miyahara K., Hosoi Y. In: Reduced Activation Materials for Fusion Reactors. Philadelphia, ASTM, 1990 p.80.
276. Cole J.I., Gelles D.S., Hoyt J.J. J.Nucl.Mater., 1992, v.191/194, p.657.
277. Miyahara K., Garner F.A., Hosoi Y. In: Annual Progress Report for Monbusho-DOE for Collaboration in Fundamental Studies of Irradiation Effects in Fusion Materials., 1992, p.119.

278. Miyahara K., Garner F.A., Hosoi Y. J.Nucl.Mater., 1992, v.191/194, p.1198.
279. Takahashi H., Ohnuki S., Kinoshita H., Nakahigashi. J.Nucl. Mater., 1991, v.179/181, p.629.
280. Takahashi H., Ohnuki S., Kinoshita H. J.Nucl.Mater., 1992, v.191/194, p.1183.
281. Takahashi H., Ohnuki S., Kinoshita H., Garner F.A. In: Annual Progress Report for Monbusho-DOE for Collaboration in Fundamental Studies of Irradiation Effects in Fusion Materials., 1992, p.135.
282. Kato T., Takahashi H., Ohnuki S., Kiyotomo N., Kuniya J.J.Nucl.Mater., 1991, v.179/181, p.623.
283. McCarthy J.M. J.Nucl.Mater., 1991, v.179/181, p.626.
284. Piatti G., Boerman D., Marques F.D.S. J.Nucl.Mater., 1992, v.191/194, p.662.
285. Garner F.A., Brager H.R., Takahashi H. J.Nucl.Mater., 1991, v.179/181, p.633.
286. Klueh R.L., Maziasz P.J. In: Reduced Activation Materials for Fusion Reactors. Philadelphia, ASTM, 1990, p.7.
287. Maziosz P.J., Klueh R.L. In: Reduced Activation Materials for Fusion Reactors. Philadelphia, ASTM, 1990, p.56.
288. Klueh R.L., Kenik E.A. J.Nucl.Mater., 1994, v.212/215, p.437.
289. Ivanov L.I., Dyomina E.V., Milicka K., Cadec J. J.Nucl.Mater., 1992, v.191/194, p.681.
290. Schamardin V.K., Bulanova T.M., Neustroev V.S., Ostrovsky Z.E., Kosenkov V.M. J.Nucl.Mater., 1992, v.191/194, p.706.
291. Gazda J., Danyluk S., Loomis B.A., Smith D.L. In: Fusion Reactor Materials, 1992, DOE/ER-0313/13, p.222.
292. Gazda J., Loomis B.A., Nowicki L.J., Smith D.L., Danyluk K.S. In: Fusion Reactor Materials, 1993, DOE/ER-0313/15, p.232.
293. Bradley E.R., Jones R.H. J.Nucl.Mater., 1985, v.133/134, p.671.
294. Ohnuki S., Takahashi H., Kinoshita H., Nagasaki R. J.Nucl. Mater., 1988, v.155/157, p.935.
295. Okamoto P.R., Santhanam A.T., Wiedersich H., Taylor A. Nucl.Technology, 1974, v.22, p.45.
296. Ohnuki S., Takahashi H., Kinoshita H. Ibid.Ref.[248], p.190.
297. Pedraza A.J., Pedraza D. Nucl.Instrum.Meth.Phys.Research, 1986, v.B16, p.270.
298. Chung H.M., Gazda J., Nowicki L.J., Sanecki J.E., Smith D.L. In: Fusion Reactor Materials, 1993, DOE/ER-0313/15, p.207.
299. Nakayima H., Nagata S., Matsui H., Yamaguchi S. Phil.Mag.: A, 1993, v.67, no.3, p.557.
300. Loomis B.A., Smith D.L. J.Nucl.Mater., 1992, v.191/194, p.84.
301. Chung H.M., Smith D.L. J.Nucl.Mater., 1992, v.191/194, p.928.
302. Satou M., Chung H.M. In: Fusion Reactor Materials, 1992, DOE/ER-0313/13, p.227.
303. Satou M., Abe K., Kayano H. J.Nucl.Mater., 1994, v.212/215, p.794.
304. Chung H.M. In: Fusion Reactor Materials, 1992, DOE/ER-0313/13, p.194.
305. Chung H.M., Loomis B.A., Nowicki L., Gazda J., Smith D.L. In: Fusion Reactor Materials, 1993, DOE/ER-0313/15, p.223.
306. Matsui H., Tanno M., Gazda J., Chang H.M. In: Fusion Reactor Materials, 1993, DOE/ER-0313/15, p.240.
307. Binary Alloy Phase Diagrams. Ed.T.B. Massalski. Ohio, ASM, 1986, v.2, p.2134.
308. Chung H.M., Loomis B.A., Smith D.L. J.Nucl.Mater., 1994, v.212/215, p.804.

Chapter 5

RADIATION-RESISTANT STEELS AND ALLOYS WITH ACCELERATED REDUCTION OF INDUCED RADIOACTIVITY

5.1 INTRODUCTION

Further development of nuclear power engineering, accelerator and space technology is greatly determined by the development of materials with the accelerated reduction of induced radioactivity (so-called reduced-activation materials). In particular, this problem is important for nuclear power engineering, in which the application of reduced-activation materials makes it possible to solve more efficiently a number of ecological problems, associated with the radiological aspects of the decommissioning of fuel systems, structures and structural elements of reactors, processing and storage of radioactive waste.

One of the most important directions for the further development of nuclear power engineering is the development of various projects of fusion reactors. In this case, the most developed in both the theoretical and practical viewpoint is the concept of the development of an energy reactor, a TOKAMAK, with the magnetic sustainment of plasma on the basis of the reaction:

$$D+T \rightarrow n(14.1\text{MeV})+\text{He}_4(3.5\text{MeV}) \tag{5.1}$$

The two main projects have been examined in a specific conceptual stage: variants of the hybrid and 'pure' reactor-TOKAMAK. The hybrid reactor is in fact a thermonuclear variant of the fast fission reactor. In addition to the production of electric energy on the basis of heat generation during the reaction of D–T synthesis, in this reactor is necessary to consider the processing of secondary nuclear fuel (plutonium) in heat-generating elements from non-enriched uranium, distributed in the blanket of the reactor. It is evident that this project does not greatly simplify the previously examined ecological problems of utilisation of radioactive waste, typical of fission reactors.

At the present time, the preferred direction is the direction of the development of projects of 'pure' fusion reactors – TOKAMAK's, in which ecological safety is improved by omitting the cycle of processing secondary nuclear fuel. The absence of the traditional fission nuclear fuel in this reactor creates the most suitable conditions for solving ecological problems on the basis of application of reduced-activation structural materials. Their application in the present case not only simplifies service and decreases the ecological risk of nuclear systems, but also decreases the consumption of materials for the processing and storage of nuclear waste.

Thus, the development of the concept of 'pure' thermonuclear power engineering has also determined the development in radiation materials science of the direction towards reduced-activation structural materials. This direction is important also for commercial nuclear engineering where the application of materials with the accelerated reduction of induced reactivity may also result in a large ecological and economic gain, especially when used for the pressure vessels of fission nuclear reactors of different types.

In recent years, the investigations into the radiation-resistant reduced-activation structural materials have been continuing very rapidly. This is also indicated, in particular, by the analysis [1–3] of reports presented at various conferences and seminars and, in particular, for the materials for fusion reactors [4–6], and in the current journal and book publications. Figure 5.1 presents clear information on the rapidly increasing number of published studies on the problem of reduced-activation materials in the period 1975-1994 [3]. Intensive investigations in this direction are also being continued at the present time. The work in this direction is being carried out within the framework of appropriate national and international programmes of Europe, the USA, Japan and Russia (for example, [4,5,7–12]). In these programmes, special attention is given to the development of three classes of materials: austenitic stainless steels, ferritic steels (pearlitic, martensitic and ferritic–martensitic steels) and vanadium-based alloys.

The development of reduced-activation compositions of austenitic and ferritic steels is based on the experience obtained in the service of their prototypes in thermal power engineering, atomic engi-

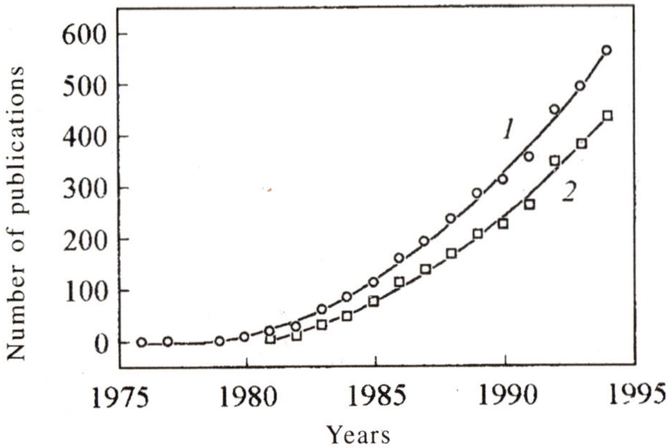

Fig. 5.1 Increase in the number of publications on the problem of reduced-activation materials in 1975–1994. 1) Total number of publications, including articles, theses, conference proceedings, etc; 2) The number of journal articles and books [3].

neering, including for pressure vessels and structural elements of the active zones of nuclear reactors. The direction of research into reduced-activation vanadium alloys is based on experience obtained in development of vanadium alloys for fast reactors. In accordance with the activation requirements, the approach to selecting the alloying elements is changed. A number of traditional components of steels and alloys are replaced by alloying elements, characterised by accelerated reduction of the induced radioactivity without impairing the technological and service properties.

5.2. MAIN DIRECTIONS AND PROBLEMS OF DEVELOPMENT OF REDUCED-ACTIVATION MATERIALS

The scientific base for the development of reduced-activation materials is the determination of the radionuclide composition of the investigated objects on the basis of values of the effective cross-section of nuclear reactions for the known spectra of nuclear particles and radiation doses. From this, the reliability of the calculations of the concentration of radionuclides depends both on the accuracy of determination of these parameters and on the knowledge of the actual chemical and isotope composition of selected materials. The experimental modification of theoretical calculations is hardly possible, especially taking into account long 'cooling' times of the irradiated materials, expressed in tens and hundreds of years. Since the problem of the evaluation of the activation parameters of materials and, primarily, the duration of reduction of radioactivity to a biologically safe level, and also the determination of the conditions of processing and utilisation of radioactive waste are most important for nuclear power engineering, in subsequent sections attention will be given to the problems of development of the reduced-activation materials mainly for this branch.

A number of national computer programmes (for example [13–18]) have been developed the activation calculations. These programmes have been tested by the International Atomic Energy Agency. For example, in Russia, one of the main programmes of calculating the activity of irradiated materials and the kinetics of the subsequent reduction of induced radioactivity is the ACTIVA programme [14,15].

The results of calculations carried out using similar programmes and the currently available databases of nuclear-physical data have been used for the preparation of guidance materials for the activation of hypothetically pure elements for specific neutron spectra. For example, the Culham laboratory (England) has developed a handbook for the activation of all elements of the Periodic table of elements and the kinetics of the subsequent reduction of activity for the neutron spectra of a deuterium–tritium fusion reactor [19]. An identical handbook in the electronic version has also been developed by the Institute of Geochemistry and Analytical Chemistry of the Russian Academy of Sciences.

Figure 5.2 shows the results of one of the first calculations of the activation and kinetics of subsequent reduction of induced radioactivity in a number of hypothetically pure elements for an integral thermonuclear neutron load of 12.5 MV year m^{-2} [20].

These and a number of other similar calculations (see, for example [2,7–9,13,14,17,21,22]) indicate that the unfavourable (in respect of activation) elements and admixture elements of steels and alloys, which can be used as structural materials in thermonuclear power engineering, include primarily Mo, Ni,Co, Al, Nb, Bi and Ag. The permissible concentrations of these and other elements have been published in [2,9,14,19–25]. For example, for the realisation of the radiological criterion permitting hands-on recycling of radioactive waste (28 μSv/h after cooling for 100 years), the maximum permissible concentration of a number of alloying and impurity elements should not exceed the values presented in Table 5.1 [21].

Table 5.1 shows that for the development of optimum reduced-activation materials, it is necessary to inspect accurately the content of not only basic components but also a large number of im-

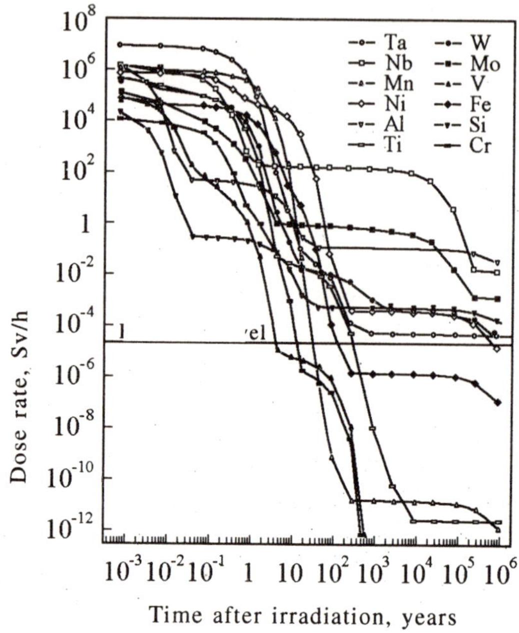

Fig. 5.2 Calculated kinetic dependences of the decrease of induced radioactivity in a number of chemical elements [20].

purity elements. In this case, the accuracy of determination of the concentration of a number of elements should be on the level of 10^{-5}–10^{-6} %.

In the majority of currently available calculation programmes, the role of the reactions (x,n) on secondary particles x (where $x = p,d, t,\alpha$, He the particle, formed by the primary nuclear reaction (n,x)), is usually not taken into account. In the studies by Karlsruhe experts (see, for example [26,27]) it was shown that the these reactions may contribute additionally to the activation of the irradiated chemical elements and materials.

Within the framework of this problem, the authors of [20] carried out special calculations of the parameters of specific activity and contact intensity of γ-radiation for pure elements V, Cr, Ti and Si and alloys of vanadium with Cr, Ti and Si, in relation to the cooling time of these materials after their neutron activation without taking into account and considering the reactions on secondary charged particles. On the whole, the results of the calculation show that the efficiency of the effect of reactions from secondary charged particles on the activation parameters of the vanadium-based alloys is very low. Nevertheless, it should be mentioned that to correct the

Table 5.1 Maximum permissible concentration of elements determining the dose rate after 100 years of "cooling" radioactive waste [21]

Element	Permissible content, %
Ag	10^{-6}
Al	1×10^{-3}–1.7×10^{-2}
Ba	2×10^{-4}–2×10^{-3}
Bi	10^{-5}
Ca	1×10^{-1}
Cd	$(2–5)\times10^{-1}$
Cl	8.5×10^{-1}
Co	1×10^{-5}–2×10^{-4}
Cu	2×10^{-3}
Dy	6×10^{-5}
Er	4×10^{-4}
Gd	4×10^{-4}
Ho	7×10^{-6}
K	8×10^{-3}–1.4×10^{-2}
Lu	3×10^{-5}
Mo	$(2–3)\times10^{-3}$
Nb	1×10^{-5}–1×10^{-6}
Ni	1×10^{-2}
Sm	3×10^{-5}
Tb	2×10^{-6}

activation calculations taking these processes into account, it is necessary to have the data banks with reliable sections of the corresponding nuclear reaction. In fact, this problem remains urgent also for the primary neutron reactions which, in fact, has been mentioned that the International Seminar on Reduced-Activation Materials in Culham [7].

It should also be mentioned that in order to obtain the required accuracy of calculating the activation of materials and changes of their chemical and isotope composition, it is important to have correct information on the energy spectra of neutron radiation. As a a suitable example of the effect of different neutron spectra of a D–T fusion reactor on the variation of the chemical and isotope composition and, consequently, the activation parameters, attention will be given to the results of calculations presented in our article [29]. The estimates of the degree of burn-out of elements in the Cu–10%Al_2O_3 composite have been obtained for two neutron spectra, shown in Fig. 5.3. Calculations of nuclear transmutations were carried out using the 'ACTIVA' programme, with the application of the following reactions (n,γ), $(n,2n)$, (n,np), (n,d), (n,t), (n,nd) and (n,He). The nuclear transformations up to the sixth order were taken into account.

Table 5.2 gives the data on the variation of the isotope composition of the Cu–10%Al_2O_3 composite for the two previously mentioned neutron spectra. It may be seen that for the spectrum with a higher degree of thermalisation (Fig. 5.2b), the degree of burn-out of primary stable isotopes is considerably higher, as a result of the efficient course of the reaction (n,γ).

The experimental results show that even in the same fission reactor or a fusion reactor, the activation of materials, and also the variation of their chemical and isotope composition, may be determined by the position of the irradiated material and, in relation to this, the spectra of neutron radiation may vary. For example, the kinetics of reduction of the activity of hypothetically pure aluminium after irradiation with neutrons with an energy of 14 ≥ MeV (Fig. 5.2) is controlled by the decay of the long-life isotope ^{26}Al (the decay period $T_{1/2} = 2.27 \times 10^{13}$ s), formed by the threshold reaction ($n,2n$) on neutrons with an energy of ≥ 14 MeV. At lower neutron energies, the long-life radionuclides do not form, i.e. the in the given case aluminium may be used as reduced-activation material in the elements of structures of D–T thermonuclear reactors which are not subjected to the effect of neutrons with an energy of ≥ 14 MeV [13,31].

The most important characteristics of radionuclides, determining both the level of induced radioactivity and the rate of its reduction are correspondingly their effective formation cross sections and decay periods. According to appropriate calculations, of the elements, presented in Fig. 5.2, the shortest time of the reduction of induced radioactivity to a biologically safe level is exhibited by hypothetically clear V and Cr (see also [21,32]). Since Cr, because of high brittleness, is not regarded as a promising basic element for development of struc-

Table 5.2 Degree of burn-out of matrix elements in Cu–Al_2O_3 composite irradiated with different neutron spectra

Stable isotope	Content in natural mixture, %	Burn–out, %	
		Spectrum Fig. 5.3a	Spectrum Fig. 5.3b
Cu–63	69.17	4.93	16.46
Cu–65	30.83	4.51	15.37
Al–27	100	5.11	19.60
O–16	99.76	1.01	4.62
O–17	0.048	5.65	13.32
O–18	0.20	1.91	7.49

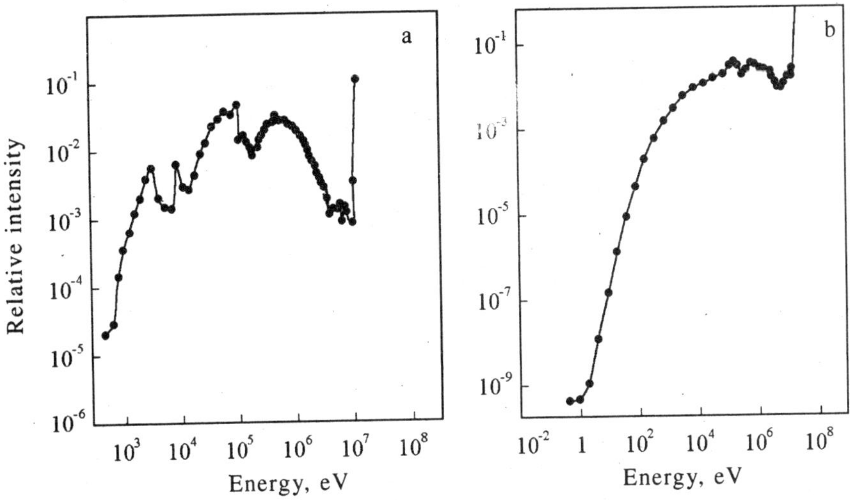

Fig. 5.3 Energy distribution of neutrons used for calculating transmutations in Cu–Al_2O_3 [29].

tural alloys for nuclear power engineering, preference is given to vanadium as the most promising basic element.

Chromium, titanium, silicon and gallium are used as alloying elements for the development of vanadium-based alloys. As regards steels, as the main structural materials of nuclear power engineering, the development of the compositions characterised by accelerated reduction of induced radioactivity is carried out using the following main principles of alloying: in prototypes of the reduced-activation austenitic stainless steels, nickel is partially or completely replaced by Mn, taking into account the nickel equivalent as the criterion of stabilisation of austenite. In prototypes of ferritic reduced-activation steels, molybdenum is replaced by tungsten and vanadium, and niobium is replaced by tantalum. The currently excepted concentration level of the main alloying elements of reduced-activation steels and alloys of vanadium (see, for example [2,7–9,17,32,33]) is presented in Table 5.3 [2].

The theoretical efficiency of the modification alloying of steels aimed at accelerating the reduction of induced radioactivity is clearly confirmed by the results of comparative activation calculations conducted for reduced-activation ferritic and austenitic steels and well-known prototypes of these steels (Fig. 5.4 [34]). At the same time, it should be mentioned that the activation calculations, including those presented in Fig. 5.2 and 5.4, were carried out either for hypothetically pure elements or for hypothetically clean steels and alloys without taking the real impurity composition of these materials into account.

Table 5.3 Content of main alloying elements in reduced-activation steels and vanadium alloys [2]

Element	Concentration of elements, %		
	Austenitic steels	Ferritic steels	Vanadium alloys
C	0.1–0.3	0.1–0.2	–
Cr	5.0–20.0	2.0–12.0	0–15.0
Mn	10.0–30.0	0.3–1.5	–
W	–	0.7–3.0	–
V	–	0.2–0.7	–
Ta	–	0.1–0.3	–
Ti	–	–	3.0–20.0
Si	–	–	0–1.0
Ga	–	–	3.0–5.0 [47]

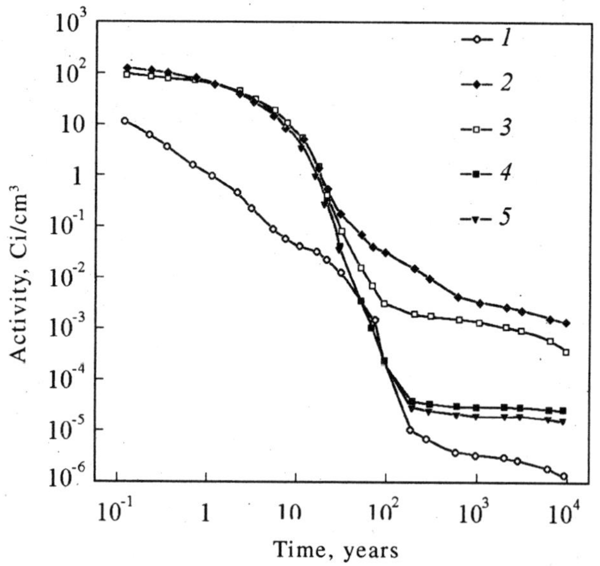

Fig. 5.4 Calculated kinetic dependences of the reduction of reduced radioactivity in reduced-activation austentic and ferritic steels, their prototypes and V–15Cr–15Ti alloy. 1) V15Cr15Ti vanadium alloy, 2) 316 steel (austenitic), 3) modified Fe–9Cr–1Mo steel (ferritic), 4) reduced-activation ferritic steel, 5) reduced activation austenitic steels [34].

In this case, vanadium and its alloys have certain advantages as regards activation parameters in comparison with iron and reduced-activation alloys based on iron (Fig. 5.2 and 5.4). By taking into account the actual impurity composition of the material, it is possible to modify greatly the activation parameters of irradiated materials. In Fig. 5.5

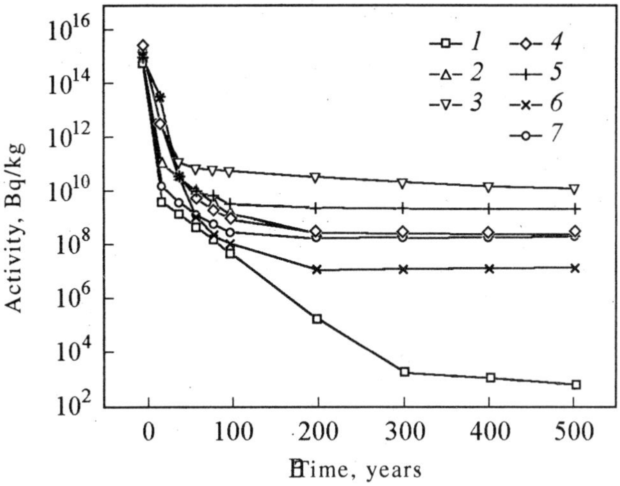

Fig. 5.5 Calculated kinetic dependences of the reduction in induced radioactivity of vanadium and a number of austenitic and ferritic steels: 1) hypothetically pure vanadium, 2) commercial purity vanadium, grade VNM-1 (certificate), 3) steel 316, 4) Fe–12Cr–20MnV steel, 5) Fe–9MoNbV steel, 6) Fe–9CrWVTa steel, 7) commercial purity vanadium, grade VNM-1 (analysis) [3].

this is shown on an example of our calculations of the level of induced radioactivity and kinetics of its subsequent reduction in hypothetically pure vanadium, in industrial vanadium of commercial purity (VNM-1), austenitic Cr–Ni (316SS) and Cr–Mn (10Kh12G20V) steels, and also ferritic steels Cr9MoVNb and Cr9WVTa, taking their actual impurity composition into account [3]. The panoramic composition materials was determined by the methods of mass spectroscopy and radioactivation analysis [3]. Figure 5.5 shows that the reduced activation Cr–Mn steel 10Kh12G20V has even advantages in comparison with technical purity vanadium with an actual impurity composition.

The authors of [21] presented the panoramic analysis of the impurity composition of vanadium, produced from special purity vanadium pentoxide with subsequent operations of electrolytic refining (single and double) and triple electron-beam remelting in high vacuum. Figure 5.6 [21] shows a histogram of the content of a number of critical (in respect of activation) impurity elements in vanadium after the previously mentioned methods of purification in relation to the acceptable concentrations (Table 5.1). The results show that the main contribution to the radioactivation characteristics of vanadium irradiated with D–T netrons is also provided by various metallic impurities, such as niobium, cobalt, molybdenum and aluminium. For comparison, Fig. 5.6 shows the data for specimens of super-pure 'ex-

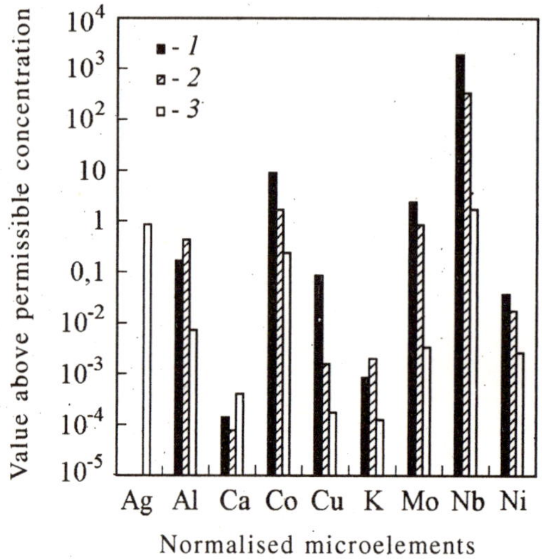

Fig. 5.6 Concentrations of a number of critical elements in specimens of metallic vanadium produced from vanadium pentoxide, grade VNO-0, after additional purification (single and double electrolytic refining with subsequent triple electron beam remelting), and also in vanadium from the Exhibition-Collection of Special Purity Substances of the Institute of Chemistry of the Russian Academy of Sciences (Nizhnii Novgorod), in relation to the maximum possible concentration: ■) single refining, ▨) double refining, □) Exhibition-Collection [21].

hibition' vanadium which is almost impossible to produce in the industrial production conditions.

Figure 5.6 shows that only this vanadium approaches the permissible level of critical elements on the basis of its content of these elements and, in the first approximation, it is close to hypothetically clean vanadium which is often used as the main component in appropriate activation calculations. However, even in this case, as shown by calculations [19, 21], after irradiation of this type of vanadium with D–T synthesis neutrons and holding for 300 years, its residual radioactivity is ~4.3×10^3 Bq kg^{-1} as a result of the retention of a relatively high content of long-life radionuclide ^{39}Al ($T_{1/2}$ = 289 years).

The results of experimental investigations into the effect of impurities on the activation parameters of a V–21.5 at% Ti alloy in neutron irradiation in an SM-2 reactor (neutron fluence with the energy of greater than 0.1 MeV was $4.7 \times 10^{20} cm^{-2}$), carried out after holding irradiated specimens for six years, have been published in [35, 36]. The results show that, in addition to other radionuclides, the residual radioactivity in the case of long cooling times will be controlled

Table 5.4 Activity of a number of materials ($Bq \cdot kg^{-1}$) irradiated with neutrons with a fluence of 1×10^{23} cm^{-2} in the spectrum of DEMO nuclear reactor, after holding for 100 years

Material	Activity, $Bq \cdot kg^{-1}$		
	Total	^{3}H	^{14}C
Vanadium	8.1×10^{7}	8.1×10^{7}	–
VNM–1	1.2×10^{9}	9.7×10^{8}	1.4×10^{8}
Steel 316SS	5×10^{10}	1.5×10^{8}	1.6×10^{5}
Steel 10Kh12G20V	7.9×10^{8}	4.7×10^{8}	1.5×10^{8}

mainly by the decay of various long life radioactive isotope at such as ^{60}Co and ^{94}Nb. The results are in complete agreement with the calculated data and the results obtained in mass spectroscopic analysis, published in [21] (see Table 5.1 and Fig. 5.6).

When evaluating the activation characteristics of reduced-activation steels and alloys based on vanadium, irradiated with neutrons of different spectra, it is also necessary to consider the formation of tritium and ^{14}Ca in them on interstitial impurities (nitrogen, oxygen, carbon), which are always present in industrial materials. Their concentration in commercial vanadium is, for example ~10^{-2}%. Table 5.4 presents our calculated data on the content of tritium and isotope ^{14}C in hypothetically pure vanadium, technical vanadium of VNM-1 grade and two austenitic steels after neutron irradiation with a fluence of $1 \times 10^{23} cm^{-2}$ in the spectra of a DEMO thermonuclear reactor followed by holding for 100 years. Calculations were carried out on the basis of the actual content of nitrogen, carbon and oxygen in these materials.

For comparison, the concentration of ^{14}C in natural carbon gives a radioactivity level of ~10^{2} $Bq \cdot kg^{-1}$, which is considerably lower than in irradiated industrial vanadium or 10h12G20V reduced-activation steel even after holding for 100 years after irradiation.

Within the framework of the problem of development of materials with accelerated reduction of induced radioactivity, in addition to the actual aspects of their purification in removing impurities unfavourable from the viewpoint of activation, attention is also given to the problem of the effect of the generation of heat of the thermal operating regime of thermonuclear reactors and possible consequences of the emission of radionuclides into the environment in emergency situations [7,37]. In the latter case, if the main factor is considered to be only the time period of the effect of radioactivity, it is evident that the problem of short-life radionuclides will be ecological less dangerous, although in this case it is also necessary to take into account the important role of the intensity of the radiation energy of radioactive emission products.

Examining the problem of large-scale industrial production of reduced-activation austenitic and ferritic steels, the authors of [3,38] proposed a number of possible methods based on conventional metallurgical technologies. It is shown that the selection of the starting material plays an important role in the production of steels. For example, in [38–40], these problems were examined on the basis of panoramic impurity analysis and recommendations were given for the use, as a base and a master alloy, of the most acceptable types of chemical elements and semifinished products (ferrochromium, ferromanganese, ferrovanadium), produced by the domestic industry.

Previously, on an example of aluminium alloys, we already examined the possibility of using some not low-activation materials in structures and structural elements of thermonuclear reactors not subjected to the effect of neutrons with an energy of $\geq$ 14 MeV. These materials, like aluminium, are characterised by the fact that the long-life radionuclides form in them by threshold reactions only in irradiation with neutrons with an energy of $\geq$ 14 MeV. In subsequent stages, we also considered the results of many of our investigations [2,3,14,35,36,38,39–41], concerned with the technological methods of production and treatment of reduced-activation materials in order to accelerate the reduction of their induced activity after irradiation.

One of the most effective and promising methods is, in our view, the method of isotope separation [2,3,14,38,39–41]. The main principle of the method is the application, as both main matrix elements and alloying components of steels and alloys, of their certain isotopes from a natural mixture. During irradiation these isotopes form radionuclides characterised by the highest rate of reduction of induced radioactivity.

Figure 5.7 and 5.8 show the results of our detailed evaluations of the activation and reduction of induced radioactivity for the isotopes of the elements Fe, Cr, Ni, Ti, W, Mo, V and Zr, which are the main components in the development of structural metallic materials for atomic and thermonuclear engineering. These calculations were carried out for the conditions of irradiation in a DEMO thermonuclear reactor, with a neutron fluence of $1 \times 10^{23} cm^{-2}$. In the calculations, carried out using the ACTIVA programme, we considered nuclear reactions up

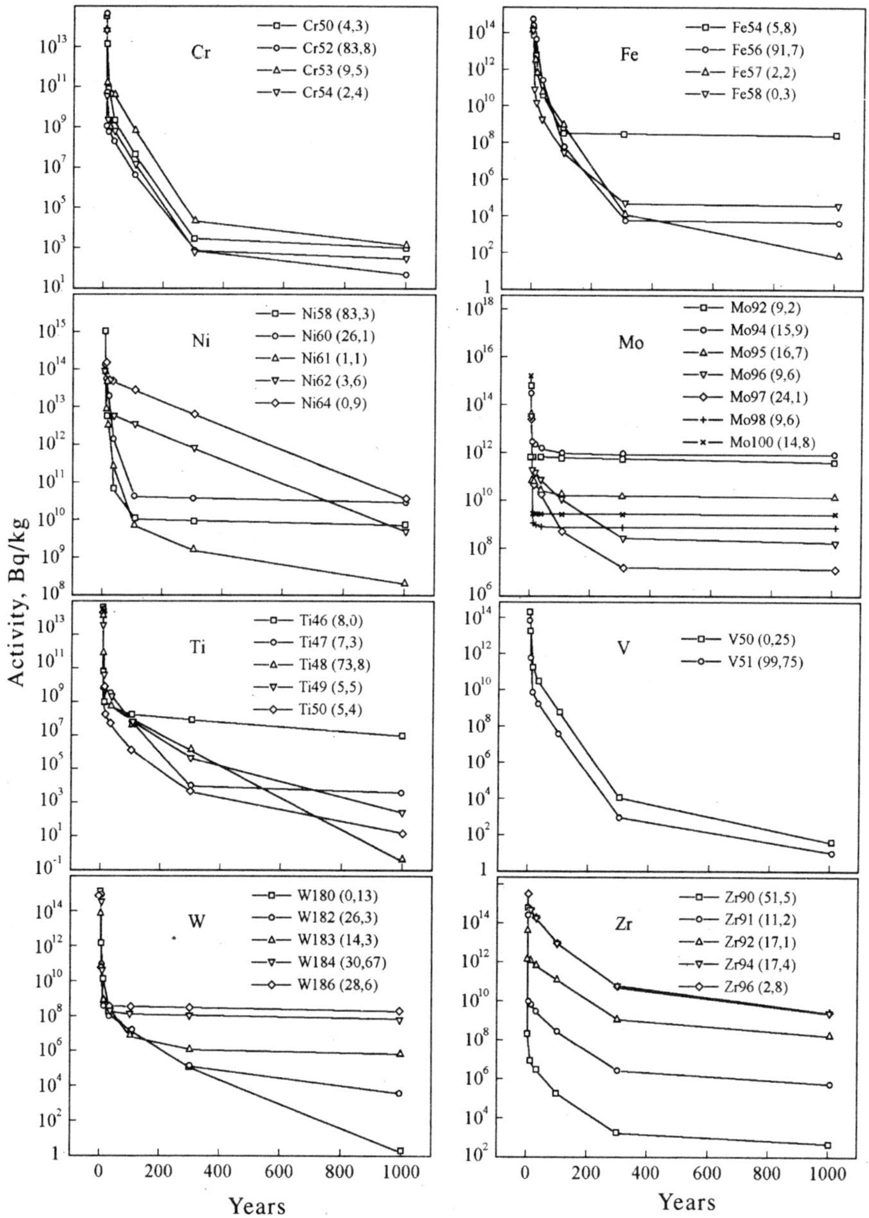

Fig. 5.7 Calculated kinetic dependences of the reduction of radioactivity in isotopes of a number of chemical elements [3].

to the sixth order, of the following types: (*n,np*), (*n,p*), (*n,d*), (*n,n*He), (*n,nd*), (*n,t*), (*n,*2*p*), (*n,*^{3}He), (*n,*γ), (*n*2*n*), (*n*3*n*), (*n,n*'), (*n,*^{2}He), (*n,p*He), (*n,pd*), (*n,pt*), (*n,p*^{3}He). When analysing the results presented in Fig. 5.7 and 5.8, attention will be given initially in detail to iron ow-

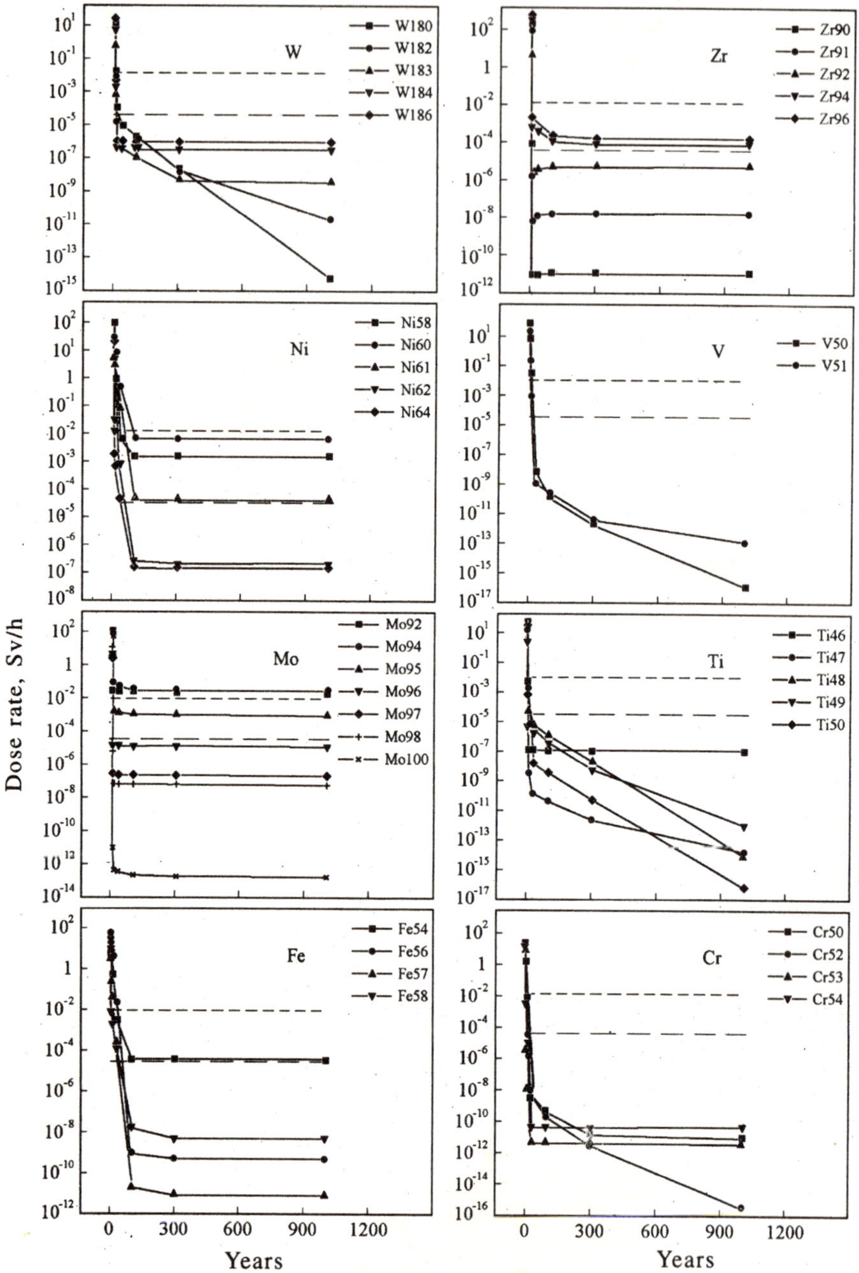

Fig. 5.8 Calculated kinetic dependences of reduction in the dose rate for isotopes of a number of chemical elements [41].

ing to the fact that iron is the main matrix element of reduced-activation austenitic and ferritic steels.

Natural iron consists of a natural mixture of stable isotopes ^{54}Fe(5.8%), ^{56}Fe(91.7%), ^{57}Fe(2.7%) and ^{58}Fe(2.7%). In Fig. 5.7 and 5.8, the reduction of induced radioactivity in iron is controlled mainly by the decay of long-life radionuclide ^{53}Mn, formed on the ^{54}Fe isotope by the reaction $^{54}Fe(n,np)^{53}Mn$. A considerably faster reduction of induced radioactivity is observed for other isotopes of iron and, most importantly, for the ^{56}Fe (91.7%) isotope, which is the main component in the natural mixture.

The results show that the ^{56}Fe isotope may be used in principle for the development of steels with improved activation parameters. The production of iron, enriched with large quantities of the ^{56}Fe isotope is possible, for example, when using the method of centrifugal separation of the isotopes [42]. A significant factor for the efficient realisation of the given project is a reduction to the minimum of the amount of ^{54}Fe isotope in enriched iron. The complete removal of the ^{54}Fe isotope makes it possible to decrease the radioactivity by four orders of magnitude.

For other elements, it is possible to make the following main conclusions:

Chromium – the application of ^{54}Cr and ^{52}Cr isotopes results in a reduction of induced radioactivity by a factor of 10–100. ^{53}Cr is the most undesirable isotope from the viewpoint of activation.

Nickel – regardless of the fact that the application of ^{61}Ni isotope makes it possible to decrease the induced radioactivity and accelerate its reduction, but the still a high level of radioactivity of irradiated isotope ^{61}Ni and its low content in the natural mixture (1.1%) make the process of isotope separation ineffective in this case.

Molybdenum – isotope ^{97}Mo makes it possible in comparison with two lighter isotopes ^{92}Mo and ^{94}Mo to decrease the activation by five orders of magnitude whilst retaining, however, a relatively high level of radioactivity during the long cooling period. The application of the method of isotope separation in the given case is evidently effective for the application of ^{97}Mo isotope in limited concentrations, for example, as an alloying element.

Titanium – the application of isotope ^{50}Ti makes it possible to decrease the level of induced radioactivity by several orders of magnitude.

Vanadium – the application of the method of isotope separation has no obvious advantages in this case.

Tungsten – the application of the method of isotope enrichment has certain advantages with respect to the rate of reduction of induced radioactivity. The efficiency of isotopes ^{180}W, ^{182}W and ^{183}W

in this respect decreases with increase of their atomic weight.

Zirconium – the application of isotope ^{90}Zr greatly accelerates the rate of reduction of induced radioactivity.

Figure 5.8 show the accepted permissible levels of the dose rate for hands-on recycling of radioactive materials (28 × 10^{-6} Sv·h^{-1}) and for remote operation with special manipulators (1×10^{-2}Sv·h^{-1}). These data show that only two hypothetically pure elements, vanadium and chromium, irradiated with a fluence of 1 × 10^{23} cm^{-2} of neutrons of D–T thermonuclear synthesis, after holding for 30 years, permitted hands-on recycling. Close to these elements is titanium. However, the formation in titanium of longlife ^{41}Ca, ^{39}Ar and ^{42}Ar radionuclides in neutron irradiation on the stable isotope ^{46}Ti results in a relatively high level of the dose rate (~10^{-7} Sv·h^{-1}), even after 'cooling' for more than 100 years.

On the whole, analysis of the results show that in the application of the method of isotope separation it is possible to improve the activation parameters of the currently available reduced-activation alloys, i.e. reduction of the level of induced radioactivity and cooling time. In addition to this, isotope enrichment is promising for the development of reduced-activation alloys also on the basis of other elements (possibly titanium, tungsten and zirconium). In this respect, special attention is given to the possibility of development of reduced-activation alloys based on isotope ^{90}Zr. In addition to its activation characteristics and a relatively high content of isotopes in thc natural mixture (15.5%), this is also determined by the fact that the zirconium-based alloys are used efficiently as irradiation-resistant structural materials for atomic power engineering.

The application of isotopes producing, after irradiation, radionuclides with accelerated reduction of induced radioactivity, is also promising when they are used as alloying elements of reduced-activation alloys. This widens the concentration level of permissible alloying elements and also their range a result of the application of the appropriate isotopes of those natural elements whose content in the reduced-activation materials is (because of the radiological criteria) permissible in limited concentrations. For example, as already mentioned, the permissible concentration of molybdenum as a natural element in regular reduced-activation steels should not exceed ~3×10^{-3} % (Table 5.1). In accordance with the results of the calculations, examined previously, the application of isotope ^{97}Mo and alloying elements is already permissible within the range of several per cent [3].

Figure 5.9 shows a characteristic example of the efficiency of the

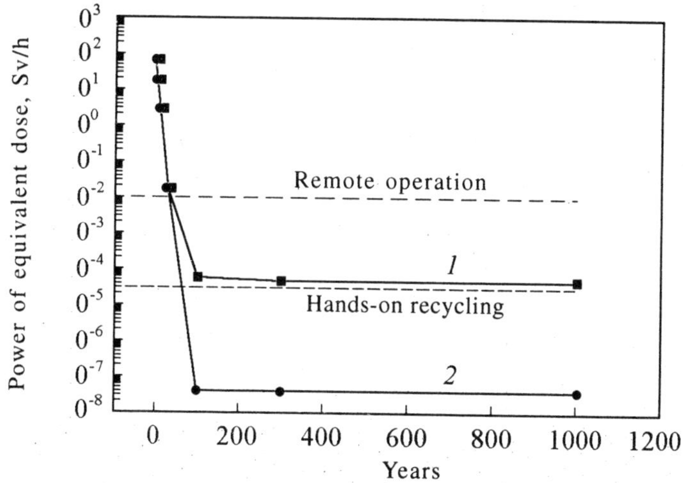

Fig. 5.9 Decrease in the dose rate with time in austenitic steels, 1) Fe–20Mn–12Cr–1W commercial steel, 2) hypothetically pure steel Fe–20Mn–12Cr–1W with isotope components ^{56}Fe, ^{52}Cr, ^{184}W [41].

method of isotope separation in comparison of the kinetics of reduction of the dose rate for the irradiated Fe–20Mn–12Cr–1W steel of the actual composition and identical hypothetical steels with the isotope components ^{56}Fe, ^{55}Mn, ^{55}Cr and ^{184}W, forming radionuclides with the highest rate of reduction of induced radioactivity [41].

In [43] and [44], the efficiency of the method of isotope separation was demonstrated on respectively V–Cr–Ti alloy and SiC/SiC composite, represented by a matrix of SiC, reinforced with SiC fibres. In the first case it was shown that alloying of vanadium not with a natural mixture of isotopes of Ti and Cr and only with isotopes ^{50}Ti and ^{54}Cr makes it possible to accelerate by two orders of magnitude, the reduction of induced radioactivity.

In the development of SiC/SiC composites, regarded recently as a promising reduced-activation material for the first wall and the blanket of thermonuclear reactors, its biological safety may be increased by preventing the formation of the long-life isotope ^{26}Al, using only stable isotopes ^{29}Si and ^{30}Si in the production of composites.

On the whole, the analysis results show that the method of isotope separation makes it possible to re-examine the currently available criteria for the development of reduced-activation materials both from the viewpoint of selection of their base and alloying elements. The technological possibilities of practical realisation of the given method for the development of alloys with accelerated reduction of induced radioactivity are available at the present time about it is obvious that the economic factor will be controlling in this case.

At the same time, it should be mentioned that, regardless of the obvious promising nature of the method of isotope separation in the development of reduced-activation irradiation-resistant alloys, one of the main problems is still the purification of initial materials to remove impurities resulting in the formation of long-life radionuclides and the development of methods for preventing their appearance in the process of production of alloys.

The second method for decreasing the time to establishment of the biologically safe level in irradiated vanadium–titanium alloys was proposed in [14,35,36]. In the given materials, it is limited by the decay of long-life radionuclides ^{39}Ar and ^{42}Ar (mainly as a result of their formation in titanium), but may greatly decrease in degassing in the process of subsequent annealing. This method is technically fully realistic if we consider the currently available experience with the annealing of radiation damage in the heating of pressure vessels of nuclear reactors, and also the technical aspects of the generation of tritium from breeder materials. In fact, this method is also highly promising for the generation of tritium from irradiated vanadium-based alloys where it is intensively formed on interstitial elements (nitrogen, oxygen, carbon) and provides a significant contribution to the activation characteristics of vanadium (Table 5.4). This method is especially important if different methods of purification do not make it possible to decrease thc content of these interstitial elements to the required level.

It should also be mentioned that for the vanadium-based alloys, this method is obviously more promising and technically more feasible in comparison with the method of isotope separation examined previously for the V–Ti–Cr alloys. The method was proposed in [43]. In this case, it is also important to take into account the fact that in the natural mixture of isotopes of Ti and Cr, the content of ^{50}Ti and ^{54}Cr is only 5.4 and 2.4%, respectively.

The process of activation of materials is accompanied by the formation of transmutation products of nuclear reactions, stable and radioactive isotopes, and also by changes of their isotope composition.

The strong negative effect on the properties of irradiated materials has been established for the transmutation gas elements – helium and hydrogen. The effect of non-gas products of nuclear reactions on the phase composition, structure and properties of metals and alloys has not been studied. As shown in Chapter 4 (section 4.4.4), on the basis of the theoretical results of [45], the burn-out of alloying elements of the alloys may not only change their chemical composition but also cause significant phase changes and, consequently, extensive degradation of the properties, especially at high neutron fluxes. The results of the calculations of the variation of the chemical composition of W and W–26Re alloy as a result of the formation of transmutation elements (mostly Re and Os) in the conditions of thermonuclear neutron irradiation [46] also lead to conclusions on the phase instability of these systems with a tendency for the formation of embrittling χ- and σ-phases.

The formation of transmutation products of nuclear reactions and the activation of materials are of the same nature and the current tasks of the further development of these aspects are virtually identical.

Considering the results of analysis, it is useful to indicate the tasks whose solution will control further development of the investigations into the problem of reduced-activation materials and the effect of transmutants on the properties of metals and alloys. We believe that they include:

1. The development of methods of purification of materials in order to remove impurities resulting in the formation of long-life radionuclides, the development of technologies of production of high-purity elements, alloys and components using panoramic inspection and the actual impurity composition.

2. The development of new methods of improving the activation parameters of reduced-activation materials (isotope separation, degassing, etc.).

3. The realisation of a complex approach to the application of reduced-activation materials in different elements of structures and functional systems, operating in radiation fields.

4. The development of methods of modelling structural-phase changes in metallic and other materials during the formation in them of transmutation products of nuclear reactions.

REFERENCES

1 Platov Yu.M., Fifth International Conference on Materials for thermonuclear reactors, November 17–22, 1991, Clearwater, USA.

2 Ivanov L.I., Platov Yu.M., Demina E.V., Perspektivnye Materialy, No.2, 37, 1995.

3 Lyakischev N.P., Demina E.V., Ivanov L.I., Kolotov V.P., Kashin V.I., Platov Yu.M., Savvateev N.N., Vinogradova N.A. J.Nucl.Mater. 1996, v.233/237, p.1516.

4 Fusion Reactor Materials: Proc.4th Int.Conf., Kyoto, 1989, J.Nucl.Mater. 1991, v.179/181.

5 Fusion Reactor Materials: Proc.5th Int.Conf., Cleawater, 1991, J.Nucl.Mater. 1992, v.191/194.

6 Fusion Reactor Materials: Proc. 7th Int.Conf., Obninsk, 1995, J.Nucl.Mater. 21996, v.233/237.

7 Proc.workshop on low activation mater. Culham, 1991, v.1.

8 Reduced activation mater.for fusion reactors. ASTM-STP 1047, PCNO4-01047035 Philadelphia, ASTM, 1990, 253pp.

9 Hosoi Y., ISIJ Int. 1990, v.30, no.11, p.917.

10 Bannich O.A., Zavialsky L.P., Ivanov L.I., Lyakishev N.P. Abstracts 5th Int.Conf.on Fusion Reactor Mater., Cleawater, 1991, p.140.

11 Mattas R.F., Smith D.L., Wu C.H., Kuroda T., Schatalov G. J.Nucl.Mater. 1992, v.191/194, p.139.

12 Solonin M.I., Chernov V.M., Gorokhov V.A., Ioltukhovskiy A.G., Shikov A.K., Blokhin Aa.I. J.Nucl.Mater., 2000, v.283/287, p.1468.

13 Forrest R.O., Endacott D.A.J. Report AERE R-13402, 1989, 117p.

14 Ivanov V.V., Kolotov V.P., Atrashkevich V.V., Platov Yu.M., Preprint GEOCHI, No.134, 1991, 33pp.

15 Ivanov V.V., Kolotov V.P., Atrashkevich V.V., Nagy P. Mater.Sci.Forum, 1992, v.97/99, p.769.

16 Grudzevich O.T., Zeleneskii A.V., Pashchenko A.B., Preprint of FEI no.2058, 1989, 20pp.

17 Harries D.R., Butterworth G.J., Hishinuma A., Wiffen F.W. J.Nucl.Mater., 1992, v.191/194, p.92.

18 Grudzevich O.T., Zelenskii A.V., Ignatyuk A.V., Pashchenko A.B. Atomnaya Energiya, 1994, 129.

19 Forty C.B.A., Forrest R.A., Compton D.J., Rayner C. Handbook of Fusion Activation Data. AEA Technology, 1992, 117pp.

20 Cierjacks S., Ehrling K., Cheng E.T., Conrads H., Ulmaier H. Nucl.Sci.and Eng., 1990, v.106, p.99.

21 Lyakishev N.P., Ivanov L.I., Kolotov V.P., Savvateev N.N., Davydov V.V., Perspektivnye Materialy, 1995 No.6, 43.

22 Butterworth G.J., Keown S.R. Preprint of paper presented at 4th Int.Conf.on Fusion Reactor Mater. Kyoto, 1989, 12pp.

23 Murphy D., Butterworth G.J. J.Nucl.Mater., 1992, v.191/194, p.1444.

24 Sanz J., Perlado J.M., Perez A.S., Guerra D. J.Nucl.Mater., 1992, v.191/194, p.1450.

25 Piatti G., Boerman D., F.Dos Santos Margues. J.Nucl.Mater., 1992, v.191/194, p.662.

26 Cierjacks S., Oblozinsky P., Rzeholz B. KFK Report-4867, 1991.

27 Cierjacks S., Oblozinsky P., Ehrlich K. KFK Report-4944, 1991
28 Butterworth G.J., Forty C.B.A. J.Nucl.Mater., 1994, v.212/215, p.628.
29 Ivanov V.V. , Ivanov L.I. Kolotov V.P., Fiz. Khim. Obrab. Mater., 1993, No.5, p.7.
30 Ivanov L.I., Ivanov V.V., Lazorenko V.M., Platov Yu.M., Tovtin V.I., Toropova L.S. J.Nucl.Mater., 1992, v.191/194, p.1075.
31 Ivanov L.I., Ivanov V.V., Lazorenko V.V., Platov Yu.M., Tovtin V.I., Tekhnologiya Legkikh Splavov, 1990, No.12, p.46.
32 Loomis B.A., Smith D.L., J.Nucl.Mater., 1992, v.191/194, p.84.
33 Zuechetti M. Fusion Technology, 1991, v.12, p.294.
34 Mann F.M. In: U.S. Program on Reduced-Activation Ferritic Steels. Clearwater, 1991.
35 Ivanov L.I. Ivanov V.V., Kolotov V.P., Platov Yu.M., Tovtin V.I., Metally, 1992, No.1, p.141.
36 Ivanov L.I., Ivanov V.V., Lazorenko V.M., Platov Yu.M., Tovtin V.I. J.Nucl.Mater., 1992, v.191/194, p.928.
37 Butterworth G.J. J.Nucl.Mater., 1991, v.179/181, p.135.
38 Lyakishev N.P., Dashevsky V.Ya., Dyomina E.V., Ivanov L.I., Platov Yu.M., Prusakova M.D., Kolotov V.P., Alenina M.V. J.Nucl.Mater., 1998, v.258/263, p.1300.
39 Demina E.V., Ivanov V.V., Ivanov L.I., Kolotov V.P., Platov Yu.M., Fiz. Khim. Obrab. Mater., 1992, No.4, p.5.
40 Dyomina E.V., Ivanov V.V., Ivanov L.I., Kolotov V.P., Platov Yu.M. J.Nucl.Mater., 1992, v.191/194, p.1461.
41 Alennina M.V., Ivanov L.I., Kolotov V.P., Platov Yu.M., Fiz. Khim. Obrab. Mater., 1998, No.4, 67.
42 Kruglov A.K., How has the atomic industry in the USSR developed, TsNIIATOMINOFORM, Moscow, 1991, p.5.
43 Piet S. et.al. Fusion Technology, 1991, no.19, p.146.
44 Scholz H.W., Zuechetti M., Kosteleyn K., Adelhelm C. J.Nucl.Mater., 1994, v.212/215, p.655.
45 Platov Yu.M., Pletnev M.N. J.Nucl.Mater., 1994, v.211, p.95.
46 Forty C.B.A., Butterworth G.J., Sublet J. J.Nucl.Mater., 1994, v.212/215, p.640.
47 Ivanov L.I., Platov Yu.M., Dediurin A.I., Borovitskaia I.V., Lazorenko V.M., Savvatev H.H., Sidorova V.V., Zakharova M.I., Artemov N.A. J.Nucl.Mater., 1996, v.233/237, p.395.

Chapter 6

MAIN PRINCIPLES AND MECHANISMS OF RADIATION DAMAGE OF STRUCTURAL METALLIC MATERIALS

6.1. INTRODUCTION

The processes of buildup of radiation defects, phase changes and general evolution of the microstructure of metals and alloys are accompanied by changes in the properties of irradiated materials. In structural materials of atomic and thermonuclear power engineering, the most important problems are associated with radiation hardening and emrittlement, swelling and changes of the characteristics of creep and long-term strength. The problems of low-temperature hardening and embrittlement are also very important for stabilising and superconducting materials of the magnetic systems of thermonuclear equipment and also structural materials for space technology. The solution of these problems depends greatly on the correct interpretation of the mechanisms of radiation damage and the development of methods of suppressing the negative effect of radiation.

The mechanisms of degradation of the properties of metallic materials in irradiation are linked with special features of the mechanisms of structural-phase changes, examined in previous chapters. It was shown that they are determined by the type of crystal lattice, chemical and phase composition, dislocation structure and irradiation parameters (temperature, type of radiation, irradiation energy, intensity and dose). In the actual operating conditions of materials in elements of sections and structures of nuclear power systems and other systems, the intensity and extent of degradation of the properties of irradiated materials and, on the whole, their service life, depend greatly on the nature and level of applied stresses, in addition to the previously mentioned factors. This problem is especially important for structural materials of the first wall of fusion reactors, operating in the most severe service conditions, and also stressed elements of atomic power stations.

In this chapter, special attention will be given to the main experimental relationships and considerations of the mechanisms of the previously mentioned processes.

6.2. THE MAIN PRINCIPLES AND MECHANISMS OF THE VARIATION OF THE MECHANICAL PROPERTIES OF METALLIC MATERIALS DURING IRRADIATION

6.2.1. The mechanical properties in active tensile and impact loading

6.2.1.1. Pure metals and diluted solid solutions

Within the framework of the general problem of examination of the physics of radiation damage, the main bulk of the initial experimental investigations into the effect of radiation on the strength and ductility properties of metals was carried out mainly on single crystals and polycrystals of copper [1–6]. The experimental data, presented in this and other investigations, make it possible to specify the following general relationships governing the variation of the mechanical properties of pure FCC and BCC metals.

1. The increase of the strength characteristics of irradiated metals (flow stress, hardness, tensile strength, yield stress) (Fig. 6.1–6.10).
2. The more intensive increase of the yield stress in comparison with the ultimate tensile stress with a tendency for a reduction of the difference between these parameters in the case of high radiation doses (Fig. 6.6 and 6.7).
3. The decrease of radiation hardening with increasing irradiation or testing temperature (Fig. 6.2, 6.3, 6.7, 6.10).
4. A decrease of the ductility of irradiated metals, including total and, in particular, uniform elongation (Fig. 6.1, 6.5–6.7).
5. The shift of the temperature of the ductile–brittle transition to higher temperatures (Fig. 6.11).
6. A tendency for the saturation of the variation of the mechanical properties of metals with increasing irradiation dose (Fig. 6.2, 6.4, 6.6, 6.8, 6.9).

The above-mentioned relationships governing the variation of the mechanical properties of irradiated metals reflected the temperature and dose dependence of the evolution of the interaction of mobile dislocations with radiation defects, i.e. effective obstacles to the movement of dislocations. Depending on the temperature and irradiation

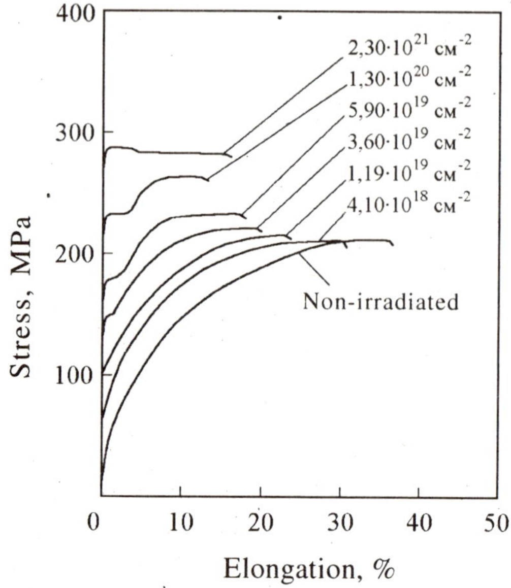

Fig. 6.1. Stress–strain curves of polycrystalline copper at 20 °C after irradiation with different neutron fluences (thermal and fast neutrons) at 80 °C [4].

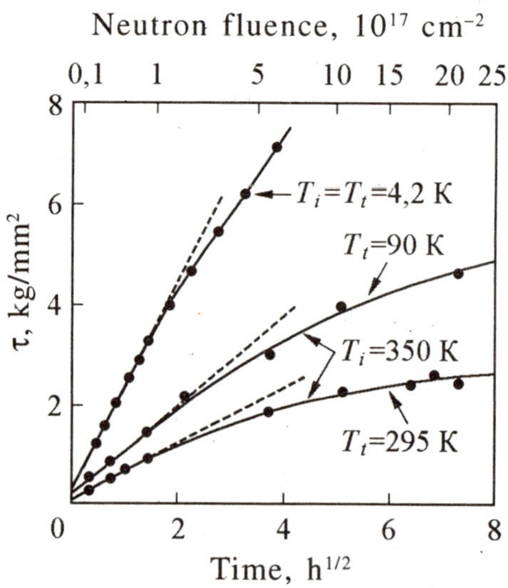

Fig. 6.2. Critical shear stress of copper in relation to neutron fluence and irradiation temperature T_i and test temperature T_t [2].

parameters, the degradation of the mechanical properties may be determined by the controlling effect of the radiation defects of different type, i.e. complexes or clusters, dislocation loops or voids, and also, possibly, their combined effects. The variation of the concentration and size of the defects also determine the kinetics of radiation hardening. The network of the dislocations, which may form in the process of irradiation as a result of the interaction of radiation defects with the initial dislocation structure, may also be an effective component of radiation hardening.

Analysis of the temperature dependence of radiation hardening in the BCC and FCC metals may explain the nature of radiation defects

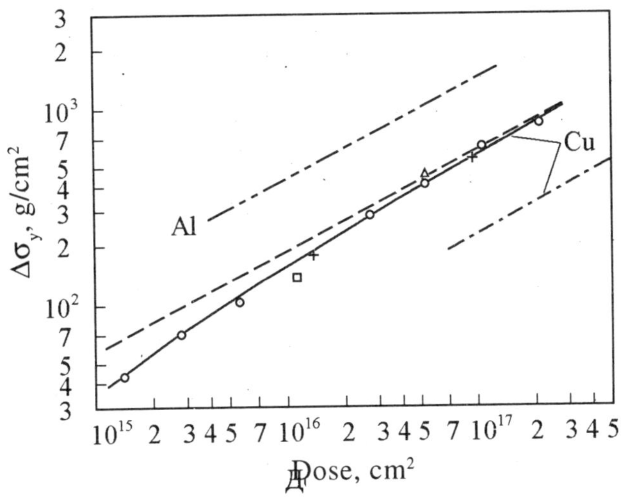

Fig. 6.3. Change in the yield stress of single crystals of aluminium (dot-and-dash-line) [5] and copper (broken line [5] and dotted line [1]) in relation to the dose of electron irradiation at temperatures of 23, 31 and 78 K, respectively. Electron energy 1, 2 and 4 MeV.

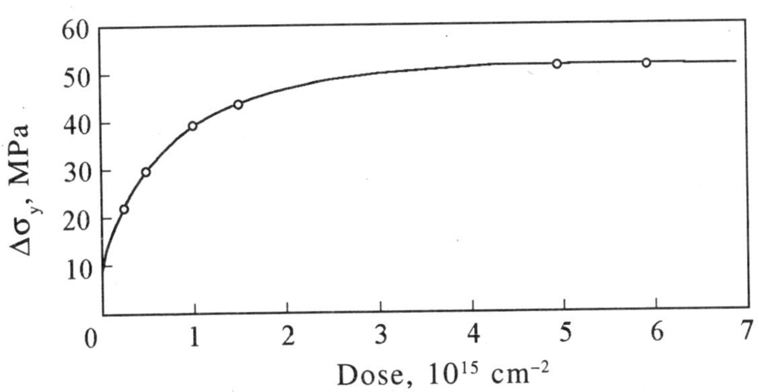

Fig. 6.4. Variation of the yield stress at 20 °C in copper in relation to the dose of irradiation with Ne ions with an energy of 233 MeV at <100 °C [7].

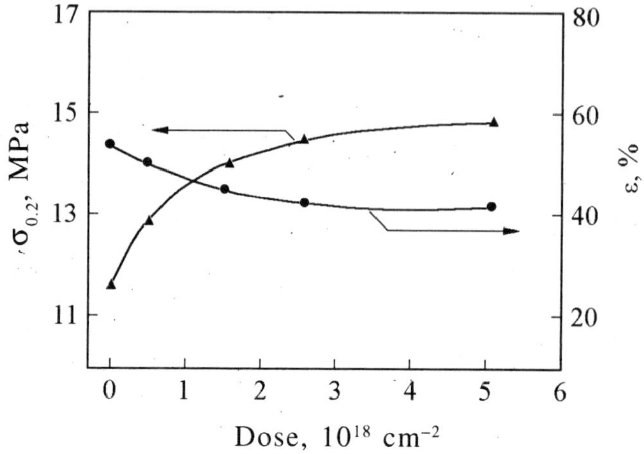

Fig. 6.5. Variation of the yield stress and relative elongation of aluminium in irradiation with electrons with an energy of 23 MeV at room temperature [8].

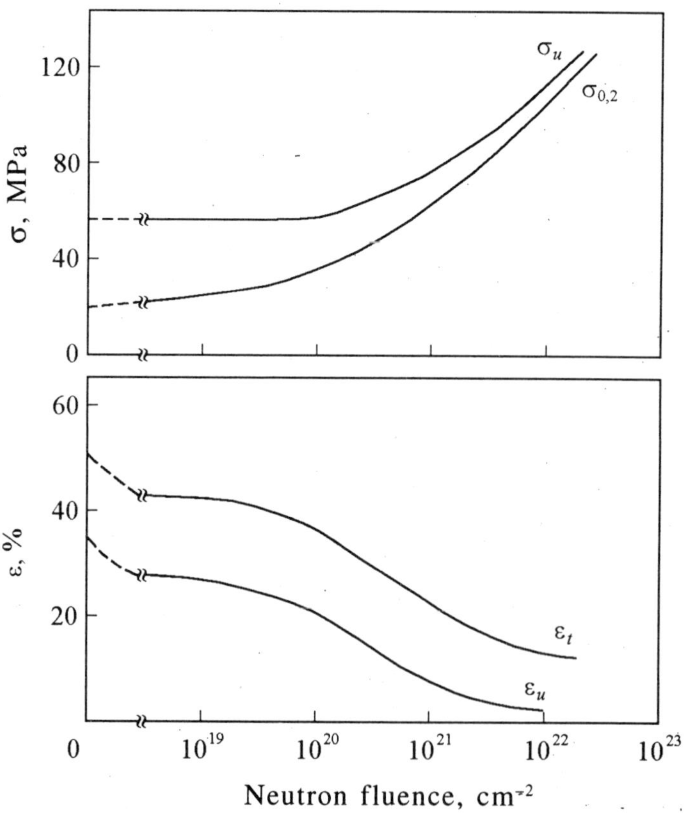

Fig. 6.6. Variation of mechanical properties of aluminium during neutron irradiation at 55 °C [9].

as barriers for the movement of dislocations. The following results obtained [14–17]. In the BCC metals, characterised by the strong temperature dependence of the yield stress in the non-irradiated condition, the controlling mechanism of hardening after irradiation does not change. In the FCC metals, the weak temperature dependence of the yield stress in the non-irradiated condition becomes stronger after irradiation. This indicates that in both the BCC and FCC metals, the radiation defects provide the main contribution to the thermally activated component of the flow stress, i.e. for mobile dislocations they are mainly barriers with short-range stress fields. The overcoming of these barriers by dislocations is carried out both as a result of applied stresses and as a result of thermally activated processes. This also determines the observed temperature dependence of the flow stress of irradiated metals.

The theoretical and practically important relationships of radiation hardening reflect the dose dependences of the saturation of the mechanical properties of irradiated metals. The most convincing

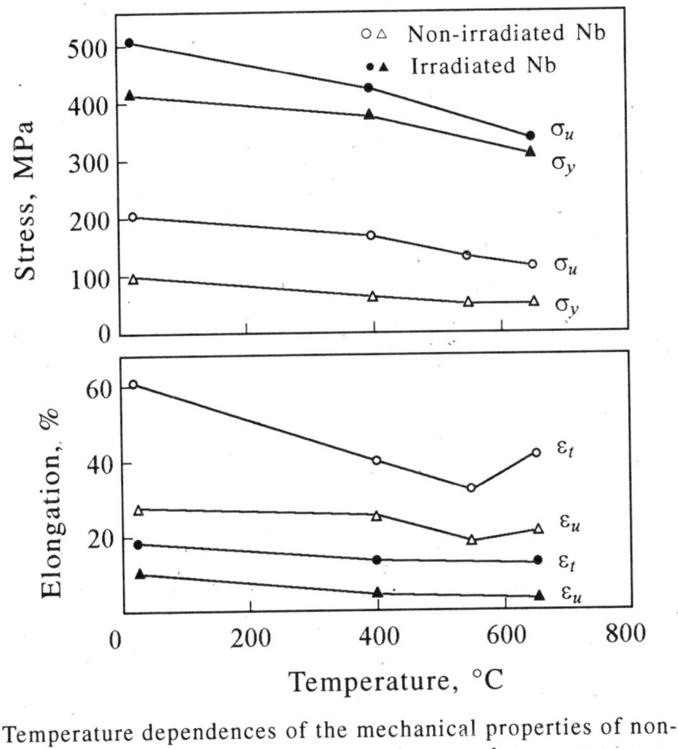

Fig. 6.7. Temperature dependences of the mechanical properties of non-irradiated and irradiated niobium. For irradiated specimens where test temperature was approximately equal to irradiation temperature [10].

examples of these phenomena are presented in Fig. 6.4, 6.5, 6.8 and 6.9 for the irradiation of aluminium, vanadium and copper with ions, electrons and neutrons.

Analysis of the appropriate experimental data and modelling representations shows that the kinetics of the process is controlled by the mechanism of approach to the quasi-equilibrium of the structure of radiation defects, determining radiation hardening. The quasi-equilibrium condition is retained until the concentration and size of the defects start undergoing significant changes as a result of some other processes. In the case of electron irradiation of vanadium, for example, we established a distinctive correlation between the kinetics of changes of microhardness with a tendency for saturation (Fig. 6.8) and the kinetics of increase in the size of the loops [11] (in the process

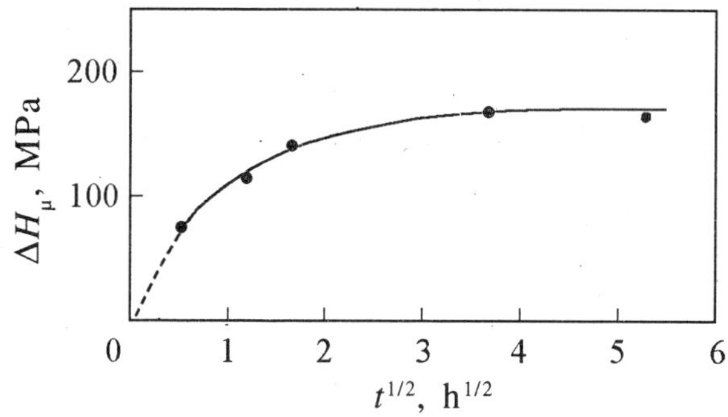

Fig. 6.8. Kinetic dependence of the change of microhardness in irradiation of vanadium with electron with an energy of 2.2 MeV at 100 °C; measurements were taken at 20 °C [11].

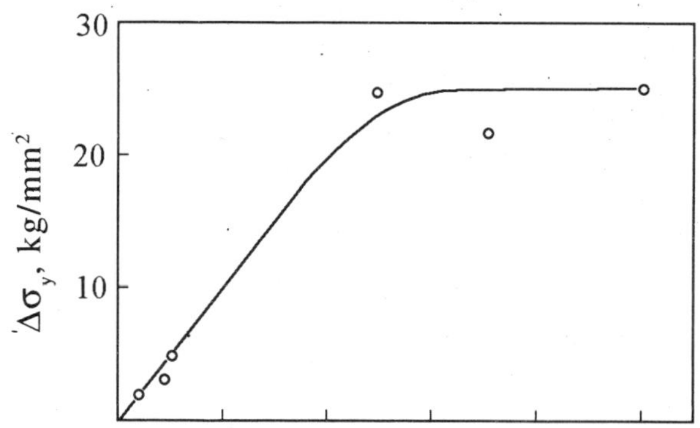

Fig. 6.9. Change of the yield stress of vanadium in relation to neutron fluence at irradiation temperature of 50–70 °C [12].

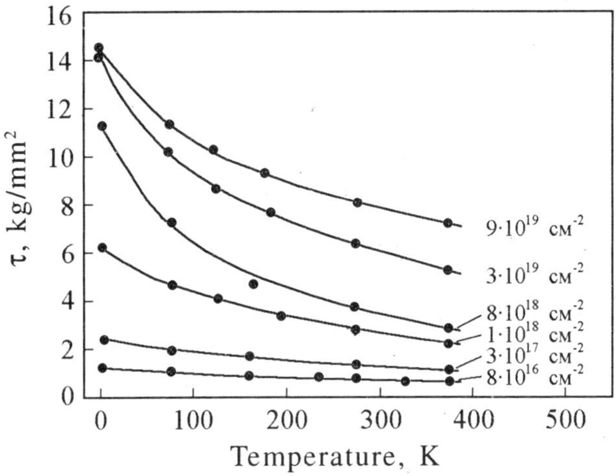

Fig. 6.10. Change of the flow stress in copper crystals irradiated with different neutron fluences in relation to test temperature; irradiation temperature 60–100 °C [3].

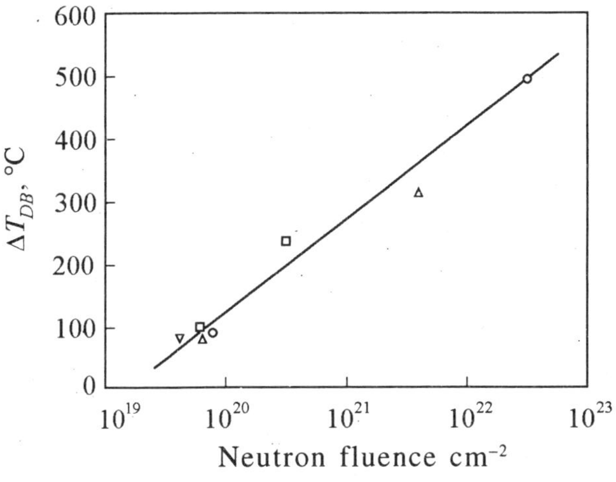

Fig. 6.11. Generalised data for the variation of the ductile–brittle transition temperature in single and polycrystals of molybdedum in relation to the fluence of neutron irradiation at 700–750 °C [13].

of growth of the loops, their density remained unchanged).

In the course of the given experiments, after the application of specific radiation doses, simultaneous measurements were taken of microhardness, and transmission electron microscopy was used to determine the density and size of the loops and evaluate the total concentration of the interstitials in loops. The appropriate dependences presented in Fig. 6.12 [11].

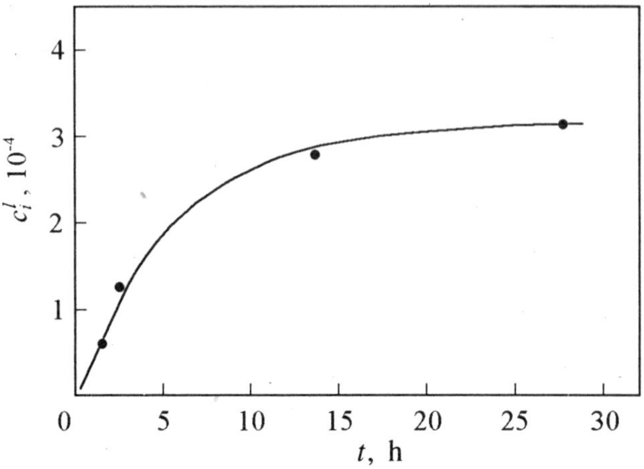

Fig. 6.12. Kinetic dependence of the change of the total concentration of interstitials in dislocation loops in irradiation of vanadium with electrons with an energy of 2.2 MeV at 100 °C [11].

The kinetic dependences of the type of the dependence shown in Fig. 6.12 for vanadium are typical of the process of growth of the dislocation loops during irradiation in other metals. Similar experimental data for aluminium and its diluted alloys, and also for silver, are presented in Fig. 3.13 and 3.16. This case is also characterised by a distinctive tendency for the saturation of imperfect structures. The modelling representations [18], used for the interpretation of similar dependences, have been examined in section 3.7.2.

The quasiequilibrium condition, to which the saturation of the properties and the imperfect structure in Fig. 6.4, 6.5, 6.8, 6.9, 3.13 and and 3.16 corresponds may, as already mentioned, be retained for a specific period of time until other processes disrupt this quasiequilibrium. The duration of quasiequilibrium and of the transition period to a new quasiequilibrium condition will be determined in the first instance by the parameters of radiation-enhanced self-diffusion, i.e., by the temperature and rate of introduction and annealing of radiation defects, and also by the nature and level of applied stress. In this case, the stress factor may play a significant role both as a result of the intensification of thermally activated diffusion processes of evolution of the imperfect structure and as a result of the direct interaction of the active dislocations with the radiation defects. For the previously examined processes of evolution of the loops, further transformation of their structure is possible both as a result of radiation-stimulated coalescence (Fig. 3.5) and their interaction with the initial dislocation structure, active dislocations (in the case of applied stresses) and other radiation defects with subsequent formation of a new quasiequilibrium state – a network of dislocations. Further transition processes and the quasiequilibrium condition in relation to temperature and the radiation dose may be controlled by the mechanisms of void formation and radiation-stimulated recovery of the dislocation structure. It is evident that the corresponding evolution would also be observed in the case of the properties of the irradiated material.

The substitutional and interstitial impurities in both the BCC and FCC metals and also the alloying elements in diluted alloys usually lead to additional radiation hardening [9,16,17,19–22], especially in the case of not too high doses and radiation temperatures. Figure 6.13 shows this on an example of the variation of microhardness in aluminium and its non-saturated solid solutions with magnesium (0.06, 0.54 and 2.1 at.%) after irradiation with a neutron fluence of 6×10^{20} cm^{-2} ($E > 0.1$ MeV) at a temperature of 100 °C [22]. A special feature of the results presented in Fig. 6.13 is the different nature of concentration dependences of the hardening of irradiated and non-irradiated alloys. The non-irradiated alloys are characterised by the linear dependence of microhardness on concentration, reflecting the additive nature of hardening, typical of the diluted solid solutions. In the irradiated alloys, the concentration dependence of hardening has a tendency for saturation.

The results of subsequent electron microscopic analysis of the irradiated aluminium and aluminium–magnesium alloys show [23] that the formation of voids directly during irradiation takes place only in pure aluminium. At the same time, in both aluminium and in aluminium alloys, examination showed dislocation loops whose

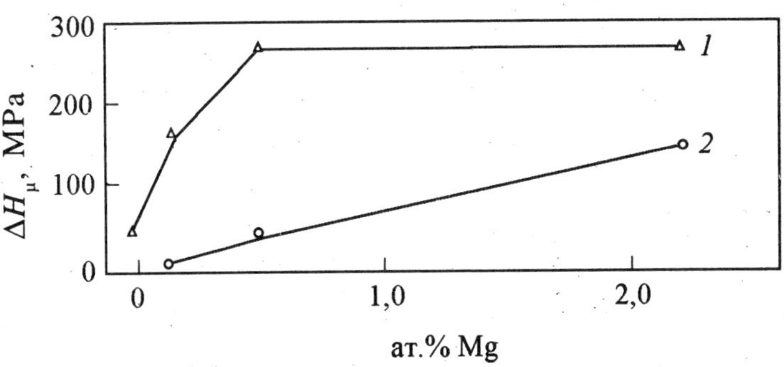

Fig. 6.13. Variation of the microhardness in non-irradiated (○) and irradiated, with a nuetron fluence of 2.6×10^{20} cm^{-2}, at a temperature ~100 °C (△) aluminium-magnesium alloys in relation to magnesium concentration [22].

concentration increased with an increase in the concentration of magnesium and the size decreased. Consequently, the concentration dependence of radiation hardening in Fig. 6.13 can be related to the density of dislocation loops. It is characteristic that the dependence of radiation hardening, presented in Fig. 6.13, is identical with the concentration dependence of the parameters of the nucleation and growth of dislocation loops in the aluminium–magnesium alloys of the same composition, irradiated in a high-voltage electron microscope [18] (Fig. 3.14). As in neutron irradiation, the concentration of the loops also increased with an increase of the magnesium content, and the mean size of the loops decreased, tending to saturation with increasing concentration.

Taking into account the barrier mechanism of radiation hardening, the electron microscopic data [23] and the form of the dependences of the density and dimensions of the dislocation loops on the magnesium content (Fig. 3.14), it may be assumed that the concentration dependence of the variation of microhardness, presented in Fig. 6.13, reflects primarily the tendency for the saturation of density and the dimensions of the loops, i.e. effective obstacles to the movement of dislocations.

As already mentioned, voids were detected immediately after neutron irradiation in [23] only in pure aluminium. It is evident that in addition to the dislocation loops, they also provided a specific contribution to the radiation hardening of aluminium.

An identical conclusion also results from the analysis of the data on the variation of mechanical properties in irradiation of 99.999% aluminium with electrons with an energy of 23 MeV, Fig. 6.5 [8]. Although special investigations of the microstructure were not carried out in [8], on the basis of the investigations [11,24] it may be assumed that, in this case, the radiation hardening of aluminium is caused not only by the formation of dislocation loops but also of voids, as in the case of neutron irradiation [22,23]. As reported in [11,24], in the microstructure of 99.52% vanadium, irradiated with electrons with an energy of 21 MeV at a temperature of 100 °C (dose 1×10^{19} cm^{-2}), in addition to the interstitial dislocation loops there were also large vacancy clusters (microvoids) with a size of ~3 nm and a density of 1.5×10^{16} cm^{-3}. Appropriate theoretical estimates show [24] that the formation of nuclei of these microvoids may take place in depleted zones during the development of cascades of atomic collisions.

In fact, in this connection, it should be mentioned that the cascade mechanism of the formation of nuclei of voids in depleted zones in irradiation of metals with electrons with an energy of several megaelectronvolt does not take place. This is also confirmed by the previously examined experimental data in [11], Fig. 6.12. In this case, in irradiation of vanadium of the same purity and at the same temperature with electrons with an energy of 2.2 MeV, the structure of vanadium contained only interstitial dislocation loops.

Additional radiation hardening and embrittlement of both the diluted undersaturated solid solutions and also of pure metals (during the formation of transmutation products of nuclear reactions), in addition to increasing the concentration of barrier–complexes of defects, dislocation loops, voids, etc, may also be the result of radiation-stimulated and radiation-enhanced phase changes. The physical fundamentals of similar mechanisms of phase stability and the appropriate experimental data are examined in sections 4.2 and 4.4. Characteristic examples of phase changes in the irradiated undersaturated solid solutions on the basis of aluminium are presented in Table 4.2.

The results of examination of the recovery of radiation hardening after neutron irradiation in a series of aluminium-based alloys, taken from the previously cited study in [22], are in good agreement with the results of electron microscopic examination, presented in Table 4.2. In the process of annealing of aluminium and its alloys with magnesium (0.06, 0.54 and 2.1 at.% Mg), up to 450 °C, at temperatures of >~350 °C, the curves of recovery of microhardness reached saturation. In this case, residual microhardness increased with an increase of the magnesium content. As indicated by Table 4.2, the residual radiation hardening is most probably associated with the formation of Mg_5Al_8 and Mg_2Si phases. The formation of the transmutation phase Mg_2Si is the result of formation of silicon by the nuclear reaction $^{27}Al(n,\gamma)^{28}Si$. The distinctive correlation of radiation hardening and embrittlement of aluminium and its alloys with magnesium as a result of the formation of transmutation silicon and the Mg_2Si phase after irradiation with high fluxes of thermal ($\sim 3\times10^{23}$ cm^{-2}) and fast ($\sim 1.8\times10^{23}$ cm^{-2}) neutrons at a temperature of ~55 °C, has been established in [9,25,26].

It should be mentioned that the solubility of silicon in aluminium is very low ($<10^{-2}$ % at a temperature of ~55 °C [25]). Therefore, the effect of transmutation silicon on radiation hardening and embrittlement of aluminium and its alloys is determined mainly by the processes of its radiation-stimulated segregation at sinks of different type, including grain boundaries, and also by phase formation. At the same time, more soluble transmutants may result in additional hardening due to dilation distortion of the matrix solid solution or

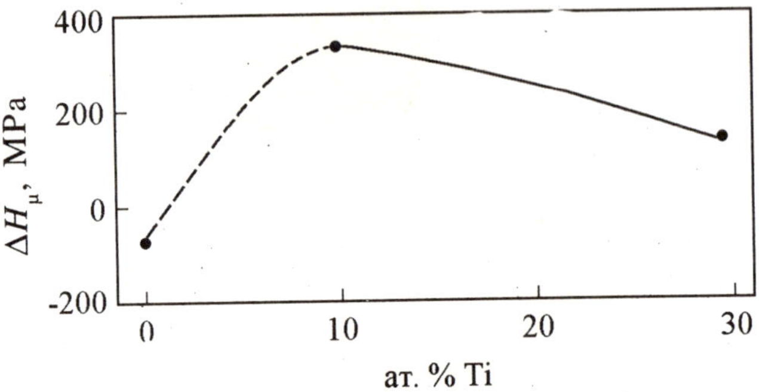

Fig. 6.14. Radiation of microhardness of vanadium and vanadium–titanium alloys after neutron irradiation with a fluence of 4.7×10^{20} cm^{-2} at a temperature of ~150 °C [27,28].

the initial pure metal. For example, in the same aluminium, a similar effect is principally possible as a result of formation of magnesium by the nuclear reaction $^{27}Al(n,\alpha)^{24}\xrightarrow{\beta} Na^{24}Mg$.

In certain conditions, the impurities may also control the opposite effect, i.e. radiation-enhanced softening. A similar effect was detected, in particular, in [27, 28] in neutron irradiation at ~150 °C of 99.52% vanadium annealed at 1000 °C (Fig. 6.14). After irradiation with a fluence of 4.7×10^{20}cm^{-2}, the microhardness of vanadium decreased by 100 MPa. This result is in contrast with the experimental data obtained in electron irradiation at 100 °C of annealed vanadium of the same purity (Fig. 6.8). In this case, after irradiation with a dose of 1×10^{19} cm^{-2}, the microhardness increased by 170 MPa as a result of the formation of interstitial dislocation loops.

The characteristic special feature of the radiation damage of vanadium in neutron irradiation in [11] was that the structure contained, in addition to the dislocation loops and 'pure' voids, also voids with a segregation band. The concentration and size of the 'pure' voids and voids with segregations is presented in Table 6.1. For the voids with the segregations the values of d_1 and d_2 correspond to the mean diameter of the voids without taking into account (d_1) and taking the segregation band into account (d_2).

The density of the vacancy dislocation loops of vanadium, irradiated with neutrons, was almost identical with the concentration of interstitial loops in vanadium irradiated with electrons (~3.0×10^{14} and 2.1×10^{14} s^{-2}, respectively). However, their mean size was considerably smaller (50 and 90 nm, respectively).

The analysis of the previously mentioned experimental facts makes

Table 6.1 Parameters of voids in vanadium irradiated with a neutron fluence of 4.7×10^{20} cm^{-2} with an energy of >0.1 MeV at ~150 °C [11]

Type of voids	ρ, 10^{14} cm^{-3}	d_1, nm	d_2, nm
Voids withont segregations	3.5	25	–
Voids with saegregations	0.7	50	100

it possible to interpret that the softening of vanadium in neutron irradiation on the basis of the radiation-stimulated precipitation of hardening impurity elements from the matrix solid solution. The dominant process mechanism is the segregation of dissolved elements on the voids and, primarily, interstitial impurities (carbon, nitrogen and oxygen), i.e. effective hardening agents of the BCC metals. It is characteristic that in the examined case, the effect of segregation softening is dominant in relation to radiation hardening as a result of the formation of loops and voids, although in comparison with electron irradiation, accompanied by hardening, the total concentration of the planar and volume radiation defects was considerably higher (~3.5 times). An estimate of the increase of the microhardness of vanadium in neutron irradiation as a result of the formation of loops and voids within the framework of the barrier model of radiation hardening [16] (for more detail, see section 6.2.2) gives the values of $\Delta H_m \cong 330$ MPa. Therefore, the real decrease of microhardness as a result of the segregation mechanism of softening gives a very high value: $\Delta|H_m|=330+100=430$ MPa.

In contrast to electron irradiation, the extensive segregation in neutron irradiation is caused primarily by the large increase of the diffusion length of the impurities $l=(2Dt)^{1/2}$, as a result of the increase of temperature and irradiation time and, correspondingly, the probability of their segregation at sinks.

The intensive hardening of irradiated metals as a result of the barrier inhibition of the dislocations by radiation defects results in the formation of local sources of high internal stresses and prerequisites for heterogeneous plastic deformation and for increasing the probability of failure. A characteristic example of the extensive localisation of plastic yielding, typical of both irradiated and quenched BCC, FCC and hexagonal metals, is the mechanism of dislocation channelling (see, for example [16,17,29–31]). The movement of dislocations in this case is restricted by narrow deformation channels in which the concentration and size of defects after dislocation slip rapidly decrease. This process is accompanied by the formation of wide slip bands and by a large decrease of plasticity, especially uniform elongation.

The decrease of plasticity and the increase of the tendency of irradiated metals for both ductile and brittle fracture at relatively low temperatures ~<0.4 T_m (low-temperature irradiation) is caused primarily by the increase of the flow stress (or yield stress). The most convincing example of low-temperature irradiation embrittlement as a result of radiation hardening is the shift of the temperature of the ductile–brittle transition (T_{DB}) of irradiated metals to the range of higher temperatures. This is shown schematically in Fig. 6.15 [16, 17], where σ_y and σ_f denote respectively the yield stress and fracture stress.

The phenomenological interpretation of radiation embrittlement on the basis of the schema in Fig. 6.15 is based on the following criteria: the irradiated metal is brittle during deformation if $\sigma_f \leq \sigma_y$ and is ductile if $\sigma_f > \sigma_y$. In this case, it is taken into account that σ_f is almost independent of temperature, and σ_y increases with a decrease of temperature and also as a result of irradiation. Thus, the shift of T_{DB} of irradiated metals to higher temperatures is a direct consequence of low-temperature irradiation hardening.

The mechanisms of radiation embrittlement and fracture of pure metals and diluted solid solutions are determined mainly by the radiation temperature. At temperatures of ~≤0.4 T_m the variation of the plasticity properties is determined, as already mentioned, by radiation hardening, and fracture is usually of the transcrystalline nature. With increase of radiation temperature >(0.4–0.5)T_m there is a transition to intergranular or mixed fracture. In this case, it is characteristic that in comparison with non-irradiated materials, the temperature of transition from one

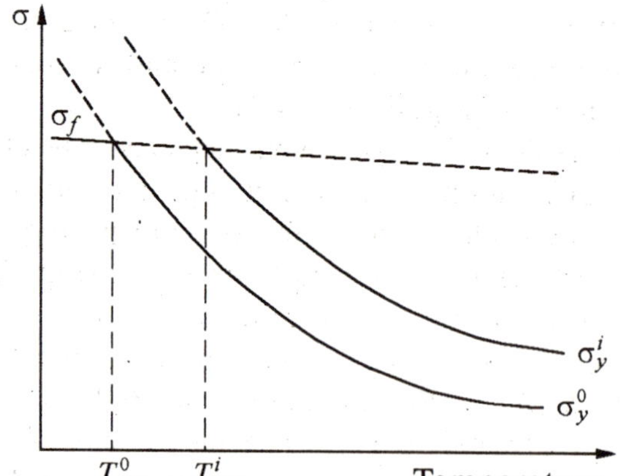

Fig. 6.15. Schematic diagram of the shift of the ductile–brittle transition temperature of irradiated metals [17].

fracture mechanism to another in irradiated metals is displaced to lower temperatures. The shift is caused by two main reasons: hardening of the grains by radiation defects and simultaneous weakening of their boundaries as a result of radiation-stimulated segregation of impurity, alloying and transmutation elements.

The radiation-enhanced diffusion of these elements to their boundaries result in the so-called phenomenon of high-temperature radiation embrittlement which is reflected in a decrease of the total and uniform elongation during short term and long-term loading and preferential failure of material at the same boundaries (see, for example, [32, 34]). It is characteristic that in contrast to low-temperature irradiation embrittlement, the phenomenon of high-temperature radiation embrittlement is not controlled by the mechanism of radiation hardening and is determined by radiation-stimulated processes, taking place at the grain boundaries. In this case, as shown in a large number of investigations (for example, [32–35]), one of the main factors of the loss of high-temperature plasticity and intensification of the process of intergranular failure of neutron-irradiated metals is the formation of clusters and bubbles of transmutation helium at the grain boundaries.

On the whole, the relationships of radiation embrittlement and hardening, examined in this section, are characteristic not only of pure metals and diluted solid solutions but also of two-phase and more complicated metallic systems. However, in the latter case, a significant additional factor of the variation of the mechanical properties of irradiated metallic materials are the processes of radiation-enhanced and radiation-stimulated phase changes. They may play a specific role also in the irradiation of diluted undersaturated solid solutions and even pure metals as a result of the formation of transmutation products of nuclear reactions. In the two-phase and multiphase alloys, the effect of phase changes on the mechanical properties and, in particular, plasticity in specific irradiation conditions may also be controlling as a result of, for example, coalescence or formation of embrittling phases in the matrix, at the grain boundaries and other interfaces. Taking these special features into account, in further discussion we shall pay special attention to the changes of the mechanical properties of industrial or promising alloys for application in nuclear engineering based on vanadium and aluminium, and austenitic and ferritic steels will also be considered.

6.2.1.2. Aluminium-based alloys

Aluminium-based alloys are used in low-temperature thermal reactors with a water or gas carrier for containers of active zones, shells of heat-generating elements, technological channels and other design elements (see, for example, [9,36–38]). In particular, the container of the first Soviet heavy water nuclear reactor, constructed in 1949, was made of an aluminium alloy. In a certain period, aluminium alloys were regarded as promising materials for the first wall of fusion reactors, in particular, the INTOR reactor [39–41]. This approach was abandoned mainly because of two main reasons: the transition from the concept of the development of the INTOR experiment reactor to the ITER reactor with a higher temperature of the first wall, and ecological problems, associated with the formation of long-life radionuclides ^{26}Al and ^{22}Na with respect to the threshold reactions on neutrons with an energy of $\geq$14 MeV. Nevertheless, as mentioned in the previous chapter, this does not exclude the application of aluminium based alloys as reduced-activation materials outside the limits of the first wall in the conditions of softened neutron spectra. In this case, aluminium alloys remain one of the most promising reduced-activation materials for structural members of thermal neutrons reactors.

In the group of the industrial aluminium-based alloys, the low-alloyed alloys based on systems Al–Mg–Si (avials) and Al–Mg–Mn (magnals) have been used on the largest scale. According to the Russian and USA classifications, they belong to the grades A (SAB) and AMG and series 6000 and 5000. The alloys of the Al–Mg–Si system are used normally in the condition dispersion-hardened by Mg_2Si particles. The Al–Mg–Mn alloys are not hardened in heat treatment and are used in the annealed condition.

Figure 6.16 and 6.17 show characteristic tendences of the variation of the mechanical properties of the 6061 (Al–Mg–0.6Si) [26] and 5154 (Al–3.9Mg–0.3Fe–Mn,Si,Zn) alloys [38] in relation to the neutron flux and test temperatures close to their service temperatures. The 6061 alloy was irradiated at a temperature of 328 K to a maximum fluence of 1.8×10^{23} cm^{-2} (E>0.1 MeV) and 3×10^{23} cm^{-2} (E<0.0025 eV). The 5154 alloy was irradiated at a temperature of 360 K. The maximum fluence in this case was 6×10^{22} cm^{-2} ($E > 0.1$ MeV) and 5.7×10^{22} cm^{-2} ($E < 0.414$ eV).

The dose dependences, presented in Fig. 6.16 and 6.17, reflect almost completely the main relationships of the variation of the mechanical properties, examined previously for the pure metals: the radiation hardening (with the dominant change of the yield stress)

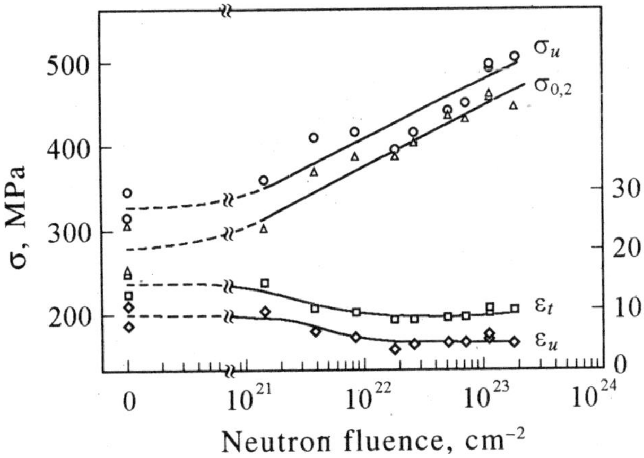

Fig. 6.16. Variation of the mechanical properties of 6061 alloy irradiated at ~328 K in relation to neutron fluence, test temperature 323 K [26].

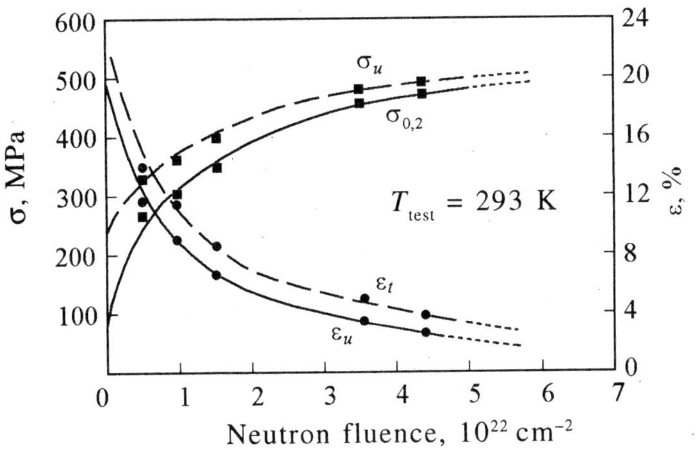

Fig. 6.17. Variation of the mechanical properties of 5154-0 alloy irradiated at 360 K in relation to neutron fluence, test temperature 293 K [38].

and embrittlement with a tendency for saturation of these characteristics at relatively high radiation doses. At the same time, in contrast to pure metals, the ductility of the irradiated 6061 and 5154 alloys, decreases with increasing test temperatures. One of the possible reasons for this effect may be associated with the additional formation, during heating to test temperature, of precipitates of the Mg_2Si phase from the supersaturated solid solution of transmutation silicon in aluminium [38]. This interpretation is in agreement with the results of our investigations [18,23] into the formation of Mg_2Si and Mg_5Al phases in the process of annealing of undersaturated aluminium–magnesium solid solutions (Table 4.2). The precipitates of the Mg_2Si phase were detected by us in particular at the grain boundaries (Table 4.2). Evidently, this increases the probability of intergranular brittle fracture. In accordance with the result of investigations carried out in [18,23], it may be assumed that the decrease of the ductility of the irradiated 6061 and, in particular, 5154 alloy in [26,38] is also associated with the formation of the Mg_5Al_8 phase.

The new promising radiation-resistant alloys based on aluminium and proposed in [40,42–45] include alloys based on the Al–Mg–Sc (Al–(1.8–2.5)Mg(0.3–05)Sc–(0.1–0.2)Zr system. As shown initially in [42], neutron irradiation of the Al–2.5Mg–0.4Sc alloys at ~100 °C with a fluence of 1.1×10^{21} cm^{-2} ($E > 0.1$ MeV) has almost no effect on the mechanical properties of the alloy in the entire test temperature range (20–300 °C). At the same time, the identical irradiation of AMG-2 industrial alloy with the same magnesium content results not only in a decrease of the absolute values of relative elongation but also in a change of the nature of the temperature dependence of this parameter (Fig 6.18).

Figure 6.19 shows the temperature dependences of the variation of the mechanical properties of the Al–2Mg–0.38Sc–0.15Zr alloy irradiated at a higher temperature (~150 °C) with a neutron fluence of 2.5×10^{20} cm^{-2} ($E > 0.1$ MeV). Also in this case, the Al–Mg–Sc–Zr alloy, subjected to preliminary annealing at 400 °C, almost completely retained its initial mechanical properties in the entire test temperature range.

Special investigations, whose results are presented in [42,46], show that the high radiation resistance of the alloys of the aluminium-magnesium–scandium system is caused mainly by the following reasons:

1. The high efficiency of scandium as an alloying element with respect to intensification of the mutual recombination of interstitials and vacancies in the solid solution.

2. The stability of the recombination effect of scandium in the matrix solid solution as a result of its low susceptibility to radiation-stimulated segregation.

3. The efficiency of coherent precipitates of the $ScAl_3$ phase as a neutral sink, causing the recombination of vacancies and interstitials.

4. The stability of the particles of the $ScAl_3$ phase to radiation-stimulated coalescence.

In addition to the relatively high radiation resistance of the alloys of the Al–Mg–Sc system, it is also important to note a number

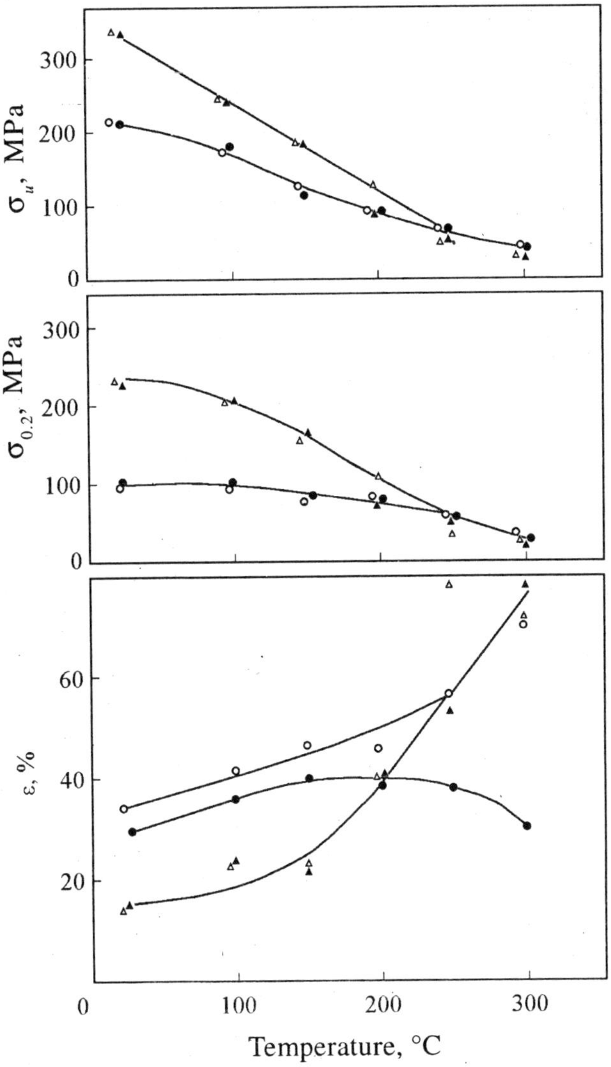

Fig. 6.18. Temperature dependence of the relative elongation of alloys AMG-2 (○,●) and Al–Mg–Sc (△, ▲). ○, △ – prior to irradiation; ●, ▲ – after irradiation [42].

of other positive properties, including in comparison with the traditionally employed avials, magnals and alloys of the 6000 and 5000 series.

The alloys of the Al–Mg–Sc system belong to the group of weldable aluminium alloys, not hardened by heat treatment. In contrast to the alloys of the series 6000 (AB, CAB), to obtain the required properties, these alloys do not require long-term heat treatment. The required level of the properties is obtained after short-term annealing and tem-

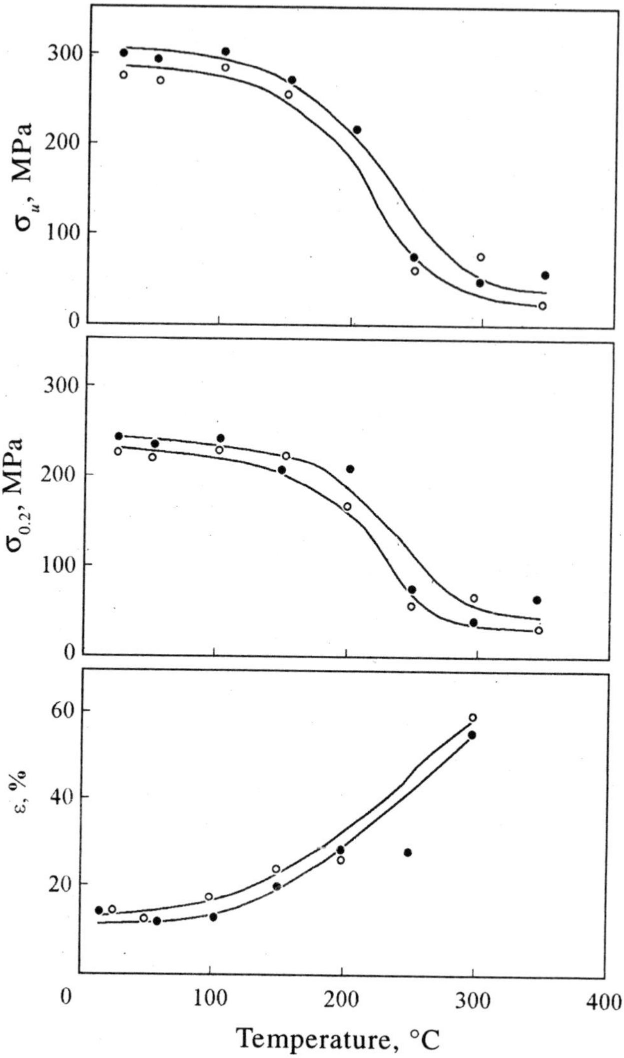

Fig. 6.19. Temperature dependences of the change of the mechanical properties of Al–2Mg–0.38Sc–0.15Zr alloy prior to (○) and after irradiation (●) [43–45].

peratures of ~320–420 °C. As regard the mechanical properties, these alloys in the initial condition are superior to those of the alloys of the 5000 series (AMG-2), including the previously examined 5154 alloy. As regards their strength, the alloys of the Al–Mg–Sc system only slightly inferior to a number of dispersion-hardened alloys of the series 6000 and, in particular, alloy 6061. The alloys of the given system are also characterised by high corrosion-resisting properties.

6.2.1.3. Ferritic steels

The low-alloy ferritic–pearlitic and pearlitic steels are used extensively in Russian and foreign atomic energy systems for pressure vessels of water-water reactors and steam generators, volume compensators, pipelines and structural elements of other functional systems [37,47]. Of special interest for practice is the efficiency of vessel steels of reactors with pressure water which are used as a basis for the majority of currently operating atomic electric power stations. The variation of the mechanical properties of the vessels steels under irradiation restricts greatly the service life of reactors and affects the economic parameters of atomic power stations.

In Russia, low-alloy ferritic–pearlitic steels of the 22K type: (0.19–0.26)C–(0.75–1.0)Mn–(0.2–0.4)Si–<0.3Ni–<0.4Cr are used mainly as structural materials for auxiliary equipment (the vessels of steam generators, etc.), whereas in reactors construction abroad, steels of the same type (for example, A212B, A302B) were initially applied also directly for the vessels of water–water reactors [47]. Abroad, this was followed by the development of steels of the type A553B and A508 of different grades for the construction of vessels of industrial power units of atomic electric power stations. In comparison with steels A212B and A302, depending on the grade, they are partially or completely modified with monitor Mo, Ni and Cr and small concentrations of vanadium (~0.03–0.05%). These materials are used at present in vessel structures in the USA, Japan and France. The detailed chemical composition of these materials, together with the composition of the vessel steels 22NiMoCr37 and 20MnMoNi55, developed in Germany, has been published in literature sources [47-52]. They are used in the condition after quenching and high-temperature tempering with the structure of tempered bainite.

For the vessels of Russian reactors VVER-440 and VVER-1000, the main material used are heat-resisting steels of the pearlitic class 15Kh2MFA (15KhMFAA) and 15Kh2NMFA (15Kh2NMFAA) [47,53,54] with the structure of fine-dispersion sorbite or tempered bainite.

The nature of variation of the mechanical properties of vessel steels for the VVER reactors under irradiation is practically identical with the appropriate relationships, presented in section 6.2.1.1. for pure metals. The working temperatures of the vessel materials are ~260-320 °C ($< 0.4T_m$), i.e. they are used in the conditions in which the main role is played by the processes of low-temperature radiation hardening and embrittlement. A decrease of plasticity and fracture toughness and an increase of the temperature of the ductile–brittle transition are the main factors restricting the service life of the metal of vessels to the fluxes of ~< 3 × 10^{20} cm^{-2}. In the case of the metal of welded joints, the service life with respect to fluence is approximately half this value.

The calculated estimates and inspection of the efficiency of the vessel steels in the process of service of reactors on irradiated reference specimens (inspection experiments) have been carried out mainly on the basis of measurements of the characteristics of impact toughness, fractography, determination of the fracture toughness in static and dynamic tests and the ductile–brittle transition temperature. Figures 6.20 and 6.21 show the temperature dependences of the variation of the impact toughness (in this case, fracture energy) for different neutron fluxes of the base metal (steel 15KhMFA) and welded joints (steel 10CrMVMnA), irradiated in a VVER-440 reactor, at a temperature of 265 °C [55]. They reflect the typical relationships of the degradation of the ductility of irradiated vessel steels: a decrease of impact toughness, corresponding to the upper plateau of the *S*-shaped curves, and the shift of the temperature of the ductile–brittle transition to higher temperatures [56]. The identical temperature shift is also recorded for the temperature dependence of fracture toughness K_{IJ} of irradiated vessel steels in cracking resistance tests (see, for example [57,58]).

As shown in a large number of investigations, a significant role in the radiation embrittlement of vessel ferritic steels is played not

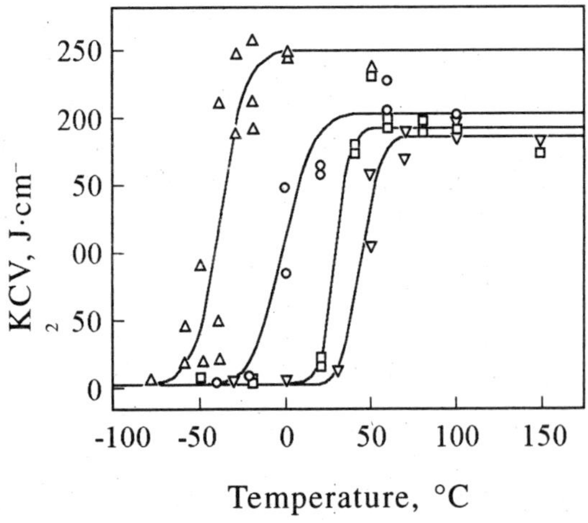

Fig. 6.20. Temperature dependences of impact toughness of non-irradiated specimens (Δ) and specimens of the base metal of 15KhMFA steel irradiated with different neutron fluences. ○ – 9.6×10^{19} cm^{-2}; □ – 1.75×10^{20} cm^{-2}; ▽ – 2.5×10^{20} cm^{-2} [55].

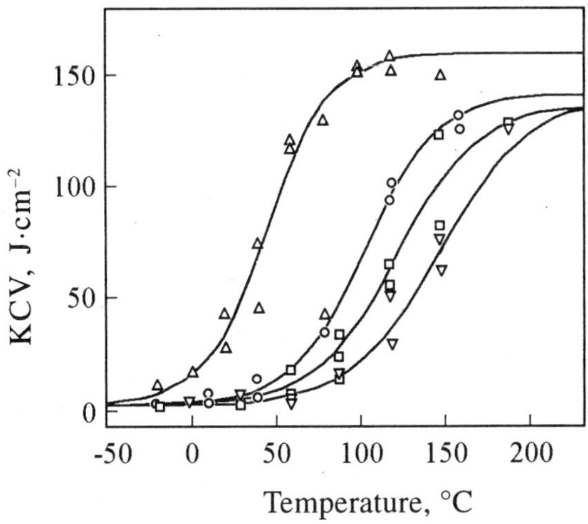

Fig. 6.21. Temperature dependences of impact toughness of non-irradiated specimens (Δ) and specimens of the weld metal of 10CrMoVMn steel irradiated with: ○ – 9.6×10^{19} cm^{-2}; □ – 1.75×10^{20} cm^{-2}; ∇ – 2.5×10^{20} cm^{-2} [55].

only by the type and concentration of the alloying and impurity elements but also by their mutual concentration relationships. The results of these investigations have also been published in the previously cited monographs, reviews and conference proceedings [47–50,52,56]. The concentration relationships represent different elementary function of the content in the given steels, primarily, of the elements such as copper, nickel, phosphorus, silicon, and others. These relationships are usually included in coefficient A of the empirical expressions, approximating the dependence of the variation of the temperature of the ductile–brittle transition T_{DB} of the vessel steels on the fluence by different exponential functions [47,48,53–55]:

$$\Delta T_{DB} = A\Phi^n, \tag{6.1}$$

where n is usually in the range ~0.2–0.5.

In a general case, an increase in the total concentration of elements such as copper, nickel and phosphorus, increases the coefficient of radiation embrittlement A in equation (6.1) and, on the whole, the ductile–brittle transition temperature.

As indicated by a large number of experimental investigations, copper is the most unfavourable impurity in the vessel steels. The negative effect of this element is intensified in the presence of specific concentrations of nickel and phosphors [53,54,56,59,60]. A brief interpretation of the effect of the dissolved elements, in particular, copper and nickel, on the phase instability and radiation hardening and embrittlement of ferritic steels was presented in section 4.6.1, chapter 4. On the basis of experimental experience in the development of vessel steels with improved radiation resistance, in addition to decreasing the content of copper and phosphorus, it is also recommended to decrease to the minimum possible level of the content of S, As, Sb and Sn [47,48,56]. The radiation-stimulated diffusion processes with the participation of this element also support the intensification of embrittlement of the vessel steels under irradiation.

The empirical relationships of the type (6.21), describing the experimental data, are a decreasing function of the flux (n<1), i.e. they reflect a tendency for saturation. The interpretation of the saturation of ΔT_{DB} in vessel steels on the basis of simple modelling representations, taking into account the precipitation of copper from the solid solution, has been published in particular in [61].

On the basis of analysis of experimental data in [62] it was shown that the value of $\Delta\sigma_{02}$ in the radiation of vessel steels, like the value of ΔT_{DB}, is governed by the exponential function of the type 6.1, and the following relationships were determined in [63,64]:

$$\Delta T_{DB} = c_1 \Delta\sigma_y, \quad \Delta T_{DB} = c_2 \Delta H_\mu, \tag{6.2}$$

i.e. the tendency for saturation is also characteristic of radiation hardening. In expressions (6.2), c_1 and $c_2 < 1$.

The saturation of the mechanical properties at specific fluences reflects some quasi-equilibrium structural condition. With further irradiation, as already reported in the section 6.2.1.1, a transition is possible to another quasiequilibrium (or non-equilibrium) structural condition with a corresponding variation of the properties. These kinetic special features are, in our view, efficiently illustrated in particular by the experimental data in [53] (Fig. 6.22). For inspection reference specimens of the base metal (steel 15Kh2MFA), irradiated at a temperature of ~260 °C in a VVER-440 reactor, examination showed distinctive saturation of ΔT_{DB} and $\Delta\sigma_y$ at fluencies of ~(3–4) × 10^{20} cm^{-2}, with a subsequent rapid hardening and embrittlement under further irradiation. It should be mentioned that the saturation of the mechanical properties or the establishment of a quasi-equilibrium structural condition was observed here in the presence of fluences of the order of or higher than the accepted relatively high limiting fluence Φ_c ($\leq 3 \times 10^{20}$ cm^{-2}), corresponding to the service

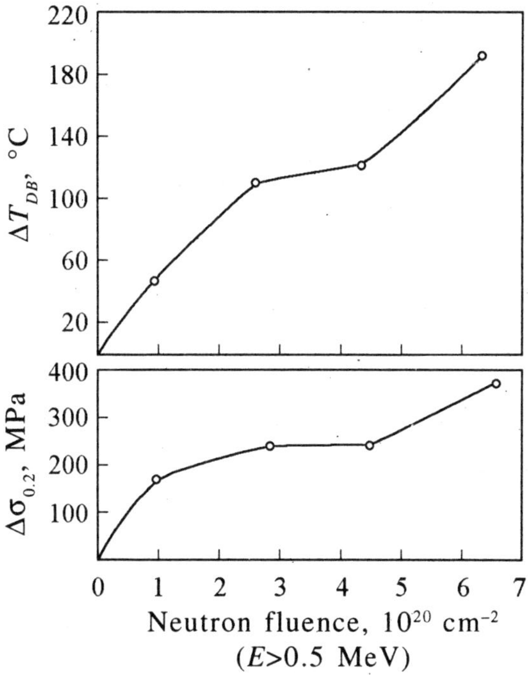

Fig. 6.22. Variation of the ductile–brittle transition temperature and the yield stress of the base metal of 15Kh2MFA steel in relation to neutron fluence [53].

life of the base metal of 15Kh2MFA steal for the given type of reactor. In comparison with the Russian 15Kh2MFA steel with very high radiation resistance, in several foreign low-alloy vessel steels, subjected to irradiation and temperatures of 150–290 °C with no chromium and vanadium or with a very low content of these elements (steels A212B, A302, A533B), the saturation of radiation embrittlement is achieved at considerably lower fluences of ~(2–5) × 10^{19} cm^{-2} [47].

In conclusion, it should again be stressed that the relationships of the variation of the mechanical properties of the vessel steels for the VVER reactors are basically identical with the relationships typical of pure metals (section 6.2.1.1). At the same time, the saturation of the mechanical properties or the establishment of structural quasi-equilibrium in vessel steels (in particular, steels with a higher degree of alloying) is observed at higher fluences. In the case of pure metals, irradiated at temperatures of < 0.4 T_m and at relatively low intensities, saturation was recorded at fluences or radiation doses of ~<10^{19} cm^{-2} (Fig. 6.4,6.6,6.8,6.9). One of the possible reasons for the shift of the value of the fluence in the vessel steels may be associated with the characteristic intensification of the mutual recombination of point defects, typical of alloys. In solid solutions, the process of mutual recombination may be intensified as a result of a decrease of the mobility of the interstitials during their capture by the alloying elements or by the formation low-mobility mixed dumbbells. A similar effect, in particular, was observed by the authors of the present book in electron irradiation of aluminium and its alloys with magnesium and zinc (Fig. 3.13). As indicated by Fig. 3.13, 3.14 and Table 3.2, an increase of the concentration of the alloying elements increases the degree of recombination, the rate of buildup of radiation defects decreases and saturation is obtained at higher radiation doses. In steels, representing multiphase systems, additional recombination of point defects may also take place at different interface boundaries.

As in the case of pure metals, irradiated at temperatures of $<0.4T_m$, radiation embrittlement of the vessel ferritic steels is controlled mainly by the barrier mechanism of hardening. Its physical fundamentals have been examined in section 6.3.2.1.

The Russian and foreign ferritic steels with increased chromium content (7–13%) of the martensitic or ferritic–martensitic grade DI82, EP450, FV448, CRM-12, MANET-I, HT-9, etc. are used or regarded as promising structural materials for the active zones of different atomic reactors [47,65–67]. The steels of this grade and also their reduced-activation Fe–Cr–W compositions, investigated in the section 4.6 and chapter 5, including pearlitic steels with a chromium content of 2–3 %, have also been investigated within the framework of different programmes for fusion reactors [65–71]. As shown in a number of investigations, whose results have been published in, for example, the proceedings of international conferences on materials for fusion reactors [69–71], the reduced-activation compositions of the ferritic steels do not have inferior technological and service properties (including radiation resistance) and in many cases exceed those of Fe–CrMoVNb and Fe–CrMoV prototypes, i.e. there is a good chance of replacing these materials not only in thermonuclear but also nuclear power engineering.

It should be mentioned that the main bulk of the experimental investigations of the mechanical properties in active tensile loading and impact loading has been carried out on ferritic steels irradiated at temperatures of ~300–600 °C. This temperature range corresponds to the realistic or planned working temperatures for the given type of material within the framework of the expected industrial application in atomic power engineering. The large amount of the experimental material obtained in the investigations of the mechanical properties of irradiated reduced-activation pearlitic, martensitic and ferritic-martensitic steels has been systematised in a review in [72]. The results of investigations of the mechanical properties in tensile and impact loading of irradiated ferritic steels of the Fe–9Cr1MoVNb and Fe–12Cr1MoV type, including comparison with the identical data for reduced-activation compositions, has been published in [73–76].

In the investigated temperature range, an increase of irradiation temperature decreases the degree of radiation hardening of ferritic steels and the plasticity of the steels gradually increases. This tendency is also observed up to an irradiation temperature of ~420–450 °C, at which the process of radiation softening becomes dominant and the strength properties may decrease below the values corresponding to the initial, non-irradiated condition [72,73,77]. A characteristic example of the variation of the strength and plasticity of the commercial Fe–9Cr1MoVNb ferritic steel in relation to the irradiation temperature is presented in Fig.6.23 [73]. Identical relationships are also typical of the steels with 12%Cr (Fe–12Cr1MoVW, CRM-12, FV [73,77] and also of reduced-activation compositions with a chromium content of 2 to 12 % [72].

In the softening stage, the concentration of radiation defects of the dislocation type greatly decreases and, on the whole, the initial microstructure changes only slightly. After irradiation at 600 °C, the nature of variation of the microstructure and mechanical properties of the martensitic steels is very similar to the corresponding changes, observed after long-term thermal ageing at the same temperature. In this case, softening like ageing is associated with the rearrangement of martensitic lath boundaries and the increase of the size of the subgrains and precipitates [72,78] as a result of the processes of radiation-enhanced diffusion.

As in the irradiation of pure metals and vessel steels, the ferritic chromium steels at elevated temperatures show saturation of radiation hardening which, however, takes place at considerably higher temperatures (~10 dpa (~1.8×10^{22} cm^{-2})) and 390–400 °C [73,78]). In pure metals and also in the previously investigated vessel ferritic steels, the saturation of radiation hardening and embrittlement is usually detected at fluences corresponding to ~≤0.15 dpa. At temperatures of 390–430 °C, the quasi-equilibrium (or transition) condition may be retained in ferritic steels with 9 and 12 % Cr up to high fluences, corresponding to ~70 [78] or even 100 dpa [79]. At equal fluences, the radiation hardening at elevated temperatures in irradiation is usually maximum for ferritic steels with 12 % Cr [72,74,78,80].

In low-alloy vessel ferritic steels, operating at relatively low temperatures, radiation hardening and embrittlement, like in the case

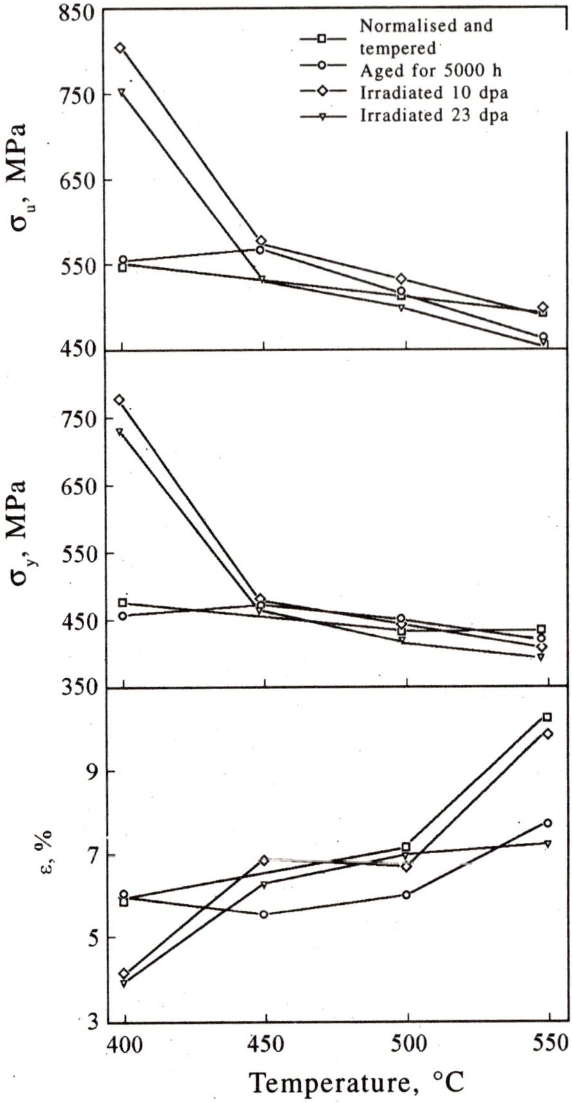

Fig. 6.23. Variation of the ultimate tensile strength, yield strength and total elongation of normalised and tempered Fe–9Cr1MoVNb steel in relation to the temperature of irradiation with neutron fluences corresponding to 10 dpa (○) and 23 dpa (▽). The differences (□) and (○) correspond to the initially non-irradiated state and the state after ageing at irradiation temperatures for 5000 h [73].

of pure metals, is controlled mainly by the structure of radiation defects. In pearlitic, martensitic and ferritic–martensitic steels, irradiated at elevated temperatures, in addition to the radiation defects, the strength and ductility properties are greatly affected by radiation-stimulated phase changes. The phase composition of the pearlitic, martensitic and ferriticomartensitic steels in both the initial condition and after thermal ageing and irradiation was presented in section 4.6.2.

It should be mentioned that the conventional ferritic steels with Nb and Mo have lower stability in comparison with their reduced-activation compositions, especially the optimum composition based on the Fe–9CrWV and Fe–9CrWVTa with a restricted content of manganese ($\leq$0.5%) and vanadium ($\leq$0.2%). Segregations and the radiation-stimulated phases in the irradiated Fe–CrMoV and Fe–CrMoVNb steels are greatly localised at the phase boundaries of different types: prior austenite grains, subgrains, interphase boundaries, and lath martensite boundaries.

The reduced-activation martensitic steels Fe–(7–9)Cr2WVTa with the restricted content of manganese and vanadium are characterised by high phase stability under irradiation. In contrast to their Fe–CrMoV and Fe–9CrWVTa prototypes, phases such as χ, G and α' do not form in them during irradiation. In [81,82] it was shown that their initial phase composition does not change greatly with the precipitation of $M_{23}C_6$, MC and M_6C carbides during irradiation in the temperature range of 390–520 °C, with a fluence of up to 8×10^{22} cm^{-2} (35 dpa). However, in [80,82,84], after irradiation of the steels with the same fluence, examination also showed the formation of the Laves phase at temperatures of 400–425 °C [80,83] and 750 °C [84].

The higher phase stability of reduced-activation ferritic steels with optimum composition in comparison with their Fe–CrMoVW and Fe–CrMoVNb prototypes is evidently one of the main reasons for their higher resistance to radiation hardening and embrittlement at elevated temperatures. Of the reduced-activation ferritic steels of different grades, radiation hardening is less marked in martensitic steels with optimum composition and with the chromium content in the range (7–9)%. In Fig. 6.24 [80], this is shown on the basis of the results of comparative examination of the variation of yield stress σ_y after irradiation of Fe–Cr2WVTa steels with 2.25, 7, 9 and 12 % Cr at a temperature of 425 °C, with a fluence of 36 dpa. In this figure, the value of $\Delta\sigma_y$ is plotted in relation to distance l within the particles of dominant radiation-stimulated precipitates in each of these steels. In the framework of the available barrier models, radiation hardening is usually inversely proportional to the distance between the hardening particles, i.e. $\Delta\sigma_y \sim 1/l$ (section 6.3.2.1). The minimum radiation hardening of Fe–(7–9)Cr2WVTa martensitic steels is associated here with the formation of M_6C carbides [80]. In the

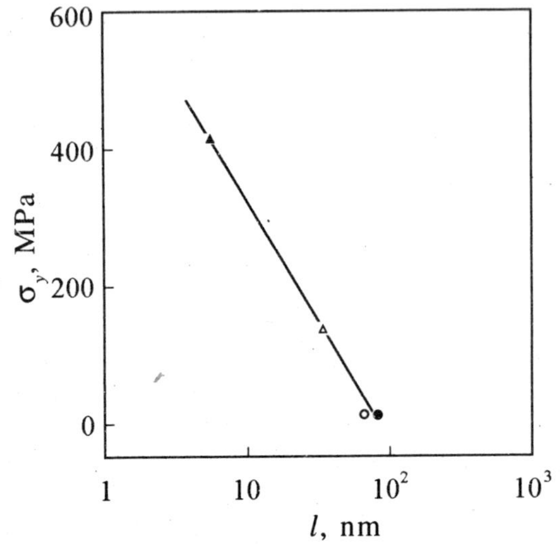

Fig. 6.24. Relationship between the variation of the yield stress and mean spacing between precipitates of phases in reduced-activation Fe–Cr2WVTa steels with 2.25 (Δ), 7 (○), 9 (●) and 12 % Cr (▲), irradiated at 425 °C with a neutron fluence corresponding to 36 dpa [80].

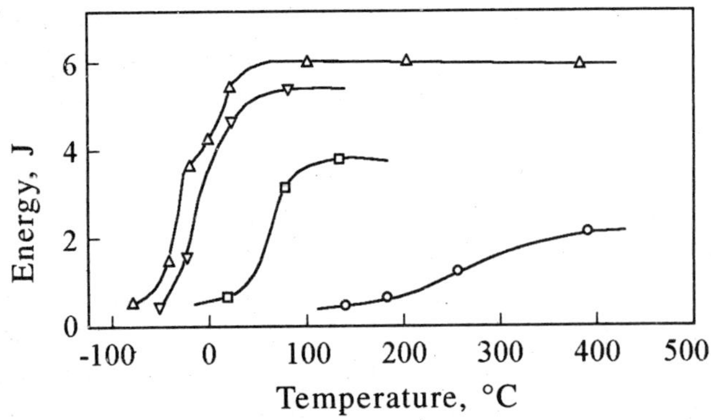

Fig. 6.25. Effect of irradiation temperature on impact toughness of MANET-1 steel. Δ) Non-irradiated condition and condition after irradiation with a neutron fluence corresponding to 5 dpa at temperatures of 300 °C (○), 400°C (□) and 475 °C (▽) [85].

pearlitic steels, the minimum radiation hardening is higher and is determined by radiation-stimulated precipitates of the M_2C carbides. The highest hardening is found in the Fe–12Cr2WVTa ferritic-martensitic steel as a result of the formation of the α'-phase resulting in extensive embrittlement.

The common problem of low-temperature radiation embrittlement in the BCC metals is also characteristic of ferritic chromium steels, irradiated at higher temperatures. The typical dependences of the variation of the temperature of the ductile–brittle transition are shown in Fig. 6.25 on the example of the results of the examination of Fe–Cr–MoVNb MANET-1 martensitic steel, irradiated with a flux of 5 dpa, at temperatures of 300, 400 and 475 °C [85]. As indicated by Fig. 6.25 and the results of other investigations (for example, [75,86]), the shift of the temperature of the ductile–brittle transition decreases with increasing radiation temperature. This dependence also correlates with the variation of the mechanical properties of irradiated ferritic steels at elevated temperatures.

In a number of investigations it was shown [72,75,87,88] that the smallest shift of the temperature of the ductile–brittle transition is characteristic of reduced-activation Fe–9Cr2WV steels and, in particular, steels modified with ~0.070% Ta (Fe–9Cr2WVTa). For example, after irradiation of the Fe–9Cr2WVTa steel with a fluence of ~7 dpa at a temperature of 365 °C, the shift of the temperature of the ductile–brittle transition ΔT_{DB} is only 4 °C [75,87] and after irradiation with a fluence of 36 dpa at 420 °C it is ~40 °C [88]. In the case of ferritic steels with Mo and Nb, the minimum shift of the temperature of the ductile-brittle transition is also recorded for compositions with the content 9 %Cr (Fe–9CrMoVNb) [75,86,89].

The saturation of the shift of the temperature of the ductile-brittle transition in both reduced-activation ferritic steels and in prototypes of these steels with Mo and Nb is usually recorded at fluences in the range from ~10 to 20 dpa [75,70,86,89]. This result is in good correlation with the saturation of radiation hardening and embrittlement at elevated temperatures of irradiated ferritic steels (~10 dpa).

The investigated mechanical properties of the ferritic steels irradiated at elevated temperatures and, in particular, steels with 9% Cr are in complete agreement with the requirements on their application as structural materials for fast reactors. In addition to this, they are also regarded as promising materials for the first wall of fusion reactors. A characteristic special feature of the application of structural materials in the fusion reactors is the higher rate of formation of transmutation of gas products of nuclear reactions and, most of all, helium and hydrogen. The most important problems here are associated with the formation of helium.

The effect of helium on radiation hardening and embrittlement of ferritic steels at elevated temperatures in contrast to austenitic steels is not so strong and is not a limiting factor with respect to the possible application in thermonuclear power engineering [75,90].

At the same time, the high concentrations result in the intensive low-temperature embrittlement of the irradiated ferritic steels. The corresponding results for the 9Cr–1MoVNb and 12Cr–1MoVW steels are presented in a review in [75]. The irradiation of steels in HFIR reactor with a mixed neutron spectrum at temperatures of 300–400 °C to the level of atomic displacements and concentrations of helium 40 and (0.35–1.0) × 10^{-2} at.% correspondingly increases their ductile–brittle transition temperature by 200–240 °C [75]. A further increase of the helium concentration to 2 × 10^{-2} at.% as a result of special introduction of 2% Ni (this simulates the He/dpa ratio for the fusion reactors [90]) displaces after irradiation their transition temperature by ~350 °C [75]. For comparison: after irradiation of the same materials in a harder neutron spectrum of the fast EBR-II reactor at a temperature of 390 °C, the shift of the transition temperature is saturated and at a fluence of 26 dpa the shift for the steels with 12 and 9 % Cr is 144 and 54 °C, respectively [86].

It is evident that this large shift of the transition temperature in relation to the helium concentration cast doubts on the prospect of the application of ferritic steels in thermonuclear power engineering [66] if, of course, efficient methods of suppressing its negative effect are not found.

6.2.1.4. Austenitic steels

The Russian and foreign austenitic chromium–nickel steels of the Kh18N9, Kh18N10T, 0Kh16N15M3B, AISI 304 and 316, PCA and analogues and modifications of these steels are used widely or are regarded as promising structural materials for pipelines and heat exchangers of water–water reactors, vessels, wrappers, and assemblies of fuel elements of fast neutron reactors with liquid metal heat carriers [47–66,91]. These materials, together with reduced-activation austenitic chromium–manganese steels of the type AMCR-0033, Fe–12Cr–20Mn–C, etc. are also regarded as possible structural materials for fusion reactors [65,66,68–71,92]. For hexagonal wrapper tubes of fuel elements and wrappers of fuel assemblies, after austenitising it is recommended to use also cold plastic deformation (usually 20%) in order to decrease the degree of swelling. The feasibility of such an operation may also be important in the case of possible application of austenitic steels as structural materials for fusion reactors.

A number of the relationships governing the variation of the mechanical properties of austenitic chromium–nickel steels in active tensile loading and impact loading at neutron fluxes ~< 30–50 dpa

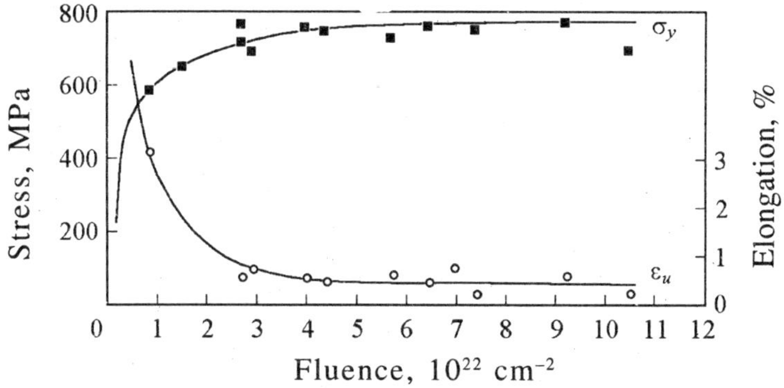

Fig. 6.26. Variation of the yield stress and uniform elongation of annealed 304 steel after irradiation in EBR-II reactor at 370 °C and testing at the same temperature [93].

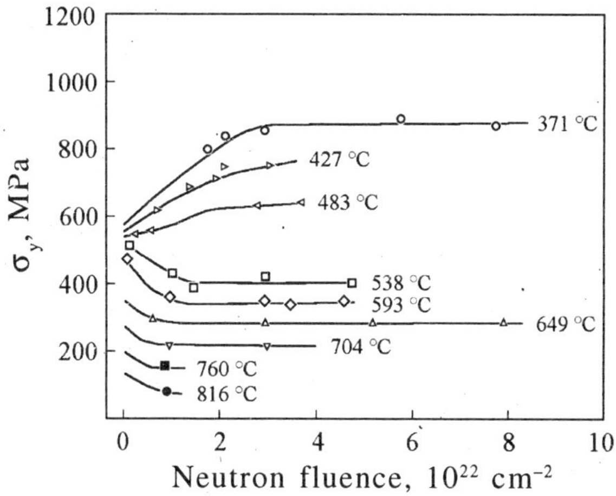

Fig. 6.27. Variation of the yield stress of AISI316 in relation to temperature and neutron fluence in irradiation in EBR-II reactor [91].

(~5 × 10^{22}–1 × 10^{23} cm^{-2}) and at temperatures of > 370–600 °C, and also at higher irradiation temperatures in 'hard' neutron spectra are identical with the previously examined special features of their evolution in the ferritic steels. Characteristic example of the variation of the yield stress and uniform elongation for the given irradiation conditions and also a temperature of 250 °C are presented in Fig. 6.26–6.28 [91, 93,94].

As in the case of ferritic steels, specific temperatures (~<500 °C) result in radiation hardening and embrittlement, and at higher temperatures in radiation softening. Depending on composition, heat treat-

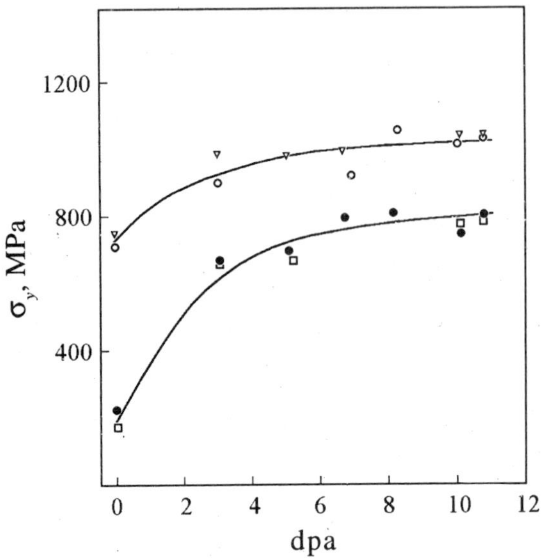

Fig. 6.28. Hardening in annealed (• – steel 316L, □ – steel PCA) and cold-worked (▽ – 20 % c/w steel 316L, ○ – 25 % c/w steel PCA) steel 316L and PCA irradiated at 250 °C, in relation to neutron fluence [94].

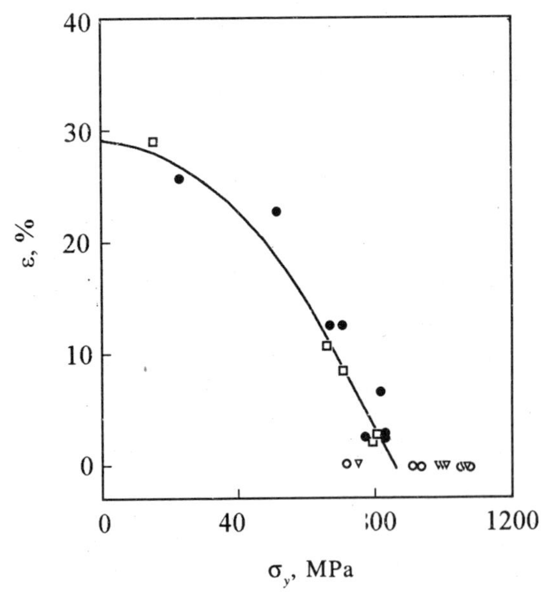

Fig. 6.29. Correlation between the level of radiation hardening and embrittlement in annealed (• – 316L, □ – steel PCA) and cold-worked (▽ – 20 % c/w steel 316L, ○ – 25 % c/w steel PCA) steels 316L and PCA irradiated at 250 °C [94].

ment, deformation and irradiation parameters, the saturation of radiation hardening is observed in different experimental investigations within the limits of the neutron fluences, corresponding to ~5–25 dpa (see, for example [91,93–98]. At relatively low irradiation temperatures, the level of radiation hardening is usually higher in the case of cold-worked steels (Fig. 6.28 [94]) and, usually, correlates with the loss of plasticity in both cold-worked and annealed austenitic steels (Fig. 6.29 [94]).

The intensive embrittlement of austenitic steels ($\varepsilon_u < 1\%$) at the given irradiation temperatures usually reflects the tendency for a reduction in the difference between the ultimate tensile strength and yield stress with increasing fluence [91,96]. A characteristic example of a similar relationship is shown in Fig. 6.30 [98]. The experimental dependences of the type of 6.30 correspond to the semi-empirical relationship [95]:

$$\varepsilon_u = n\left(\frac{1-\sigma_y}{\sigma_u}\right), \text{ where } n \sim 0.5. \qquad (6.3)$$

The investigated relationships are also typical of ferritic steels. At the centre, in relation to the preliminary treatment, the irradiation temperature, the fluence and helium content, the austenitic chro-

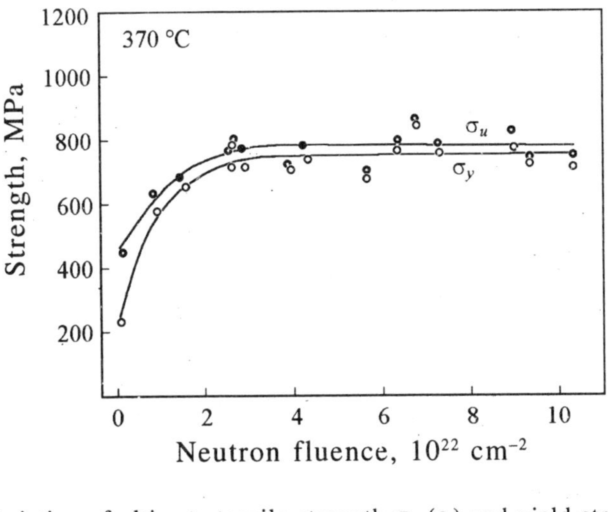

Fig. 6.30. Variation of ultimate tensile strength σ_μ (•) and yield strength σ_y (○) in relation to neutron fluence in annealed 304L steel irradiated at 370 °C in EBR-II reactor and tested at the same temperature [98].

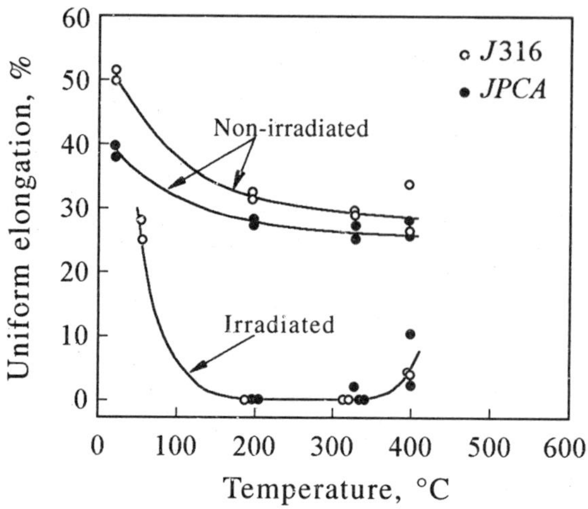

Fig. 6.31. Temperature dependences of the uniform elongation of annealed austenitic steels irradiated with a neutron fluence corresponding to 7–8 dpa [99,102].

mium–nickel steels show special features not typical of ferritic steels. As shown in a number of investigations (for example, [96,99–103]), the temperature dependences of low-temperature radiation embrittlement and hardening of austenitic steels are non-monotonic and are characterised by experimental values (maximum σ_y and minimum σ_u) in the temperature range ~200–350 °C and, according to the results of some investigations [99,102], also in a wider temperature range. Figure 6.31 [99,102] shows this on an example of the steels JPCA and J316 irradiated in the annealed condition.

The maximum values of radiation hardening of the annealed and cold-worked steels in saturation in the given temperature range may reach ~820 and ~1040 MPa, respectively, with a uniform elongation of < 1%. The strongest low-temperature radiation embrittlement is recorded in cold-worked steels [94,99,102]. It should be mentioned that the temperature range of low-temperature radiation embrittlement of the austenitic chromium–nickel steels is almost completely identical with the range of currently planned working temperatures of the ITER reactor [101,105]. It is planned that the first wall of this reactor, regardless of the very low values of ε_u, should be made of a type 316 steel [101,104,105].

The estimate of fracture toughness K_j for mode *I* of loading according to the *I–R* curves for the series of the steels of type 316 and JPCA, irradiated to a fluence of ~3 dpa at a helium content of ~5 × 10^{-3} % at temperatures of 60–125 °C and 20–250 °C and tested at temperatures of 20–250 °C, are presented in [106]. Of course, they show that with the exception of cold deformed 316 steel, the K_j values remain on a relatively high level (from 150 to 400 MPa·m$^{1/2}$) [106]. At the same time it should be taken into account that the maximum low-temperature embrittlement of austenitic steels was recorded at fluences of ~> 6 dpa, whereas the maximum fluence for the ITER reactor is planned for ~9 dpa [103]. In addition to this, the measurements of fracture toughness for mixed loading mode $I_{I/III}$, i.e. taking into account the shear component (evidently, this is more realistic for the operating conditions of the materials of the first wall) may also lead to over-estimating the fracture toughness by reducing it even further.

Another important special feature of radiation damage of a number of austenitic chromium–nickel steels is the excessive embrittlement of these steels at temperatures ~> 400 °C and high fluences (~>30 dpa), correlating with intensive formation of voids and swelling [91, 107–110]. With an increase of the concentration and size of the voids and with an increase in the degree of swelling as a whole, the nature of failure may gradually change from normal transcrystalline ductile dimpled mechanisms to fracture by the mechanism of dislocation channelling [91,100,110].

At specific levels of swelling ~>10 %, both the cold-worked and annealed steels are characterised by almost zero plasticity [107, 108,111,112]. In Fig 6.32 [112] this is shown on the example of ex-

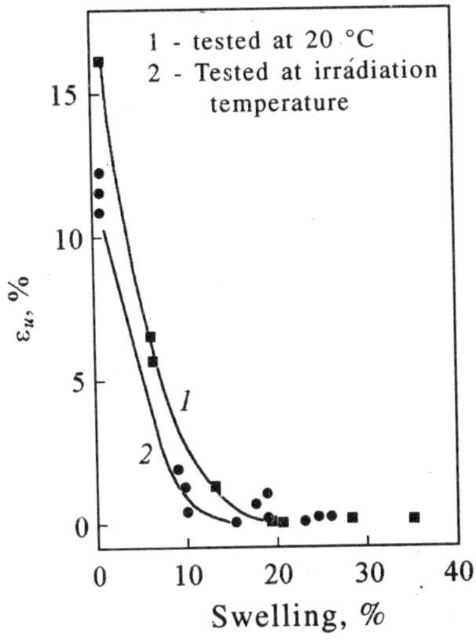

Fig. 6.32. Effect of swelling on uniform elongation of Fe–18Cr–10Ni–Ti steel irradiated at 400–500 °C in a BOR-60 reactor [112].

amination of Fe–18Cr–10Ni–Ti steel subjected to preliminary annealing after its irradiation in a BOR-60 reactor at a temperature of 400–500 °C. According to the results in [107,108,111,112] when reaching some critical levels of swelling, accompanied by a loss of ductility, the steels subjected to preliminary hardening by radiation are also characterised by a rapid softening.

At irradiation temperatures of ~>0.5 T_m, chromium–nickel austenitic steels show a susceptibility to high-temperature radiation embrittlement (HTRE). This type of embrittlement, as already mentioned in section 6.2.1.1, is characterised by a decrease of uniform elongation and by intercrystalline failure. A significant role in the process of HTRE is played by segregation processes of impurity, alloying and transmutation elements, including helium, at the grain boundaries.

The temperature dependence of the variation of uniform elongation, reflecting the tendency for manifestation of HTRE of austenitic steels, is shown in Fig. 6.33 [110]. It is based on the analysis and processing a series of experimental data, including the data cited previously in [101,102]. A large rapid decrease of the values of uniform elongation at temperatures of ~>773 K (500 °C) (Fig. 6.33), detected in the experiments with a relatively high content of transmutation or implanted helium, is usually attributed to the mechanism of high-temperature helium embrittlement [see, for example, Ref. 32, 101, 102, 110, 113–115]. With an increase of the helium content, as shown in [114], for the type 316 steel, there is a tendency for a decrease of the temperature of transition from transcrystalline to integranular failure. In the actual conditions, the mechanism of helium HTRE is more characteristic of operation of austenitic chromium-nickel steels in the reactor was with the mixed neutron spectral (type CM-2 and HFIR) and, in perspective, in fusion reactors, where in comparison

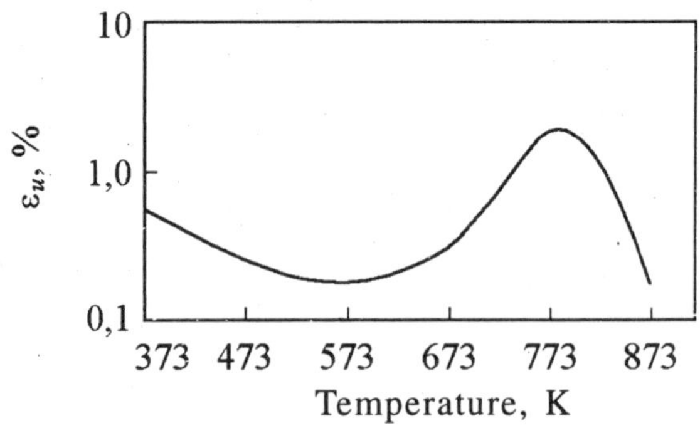

Fig. 6.33. Variation of uniform elongation in annealed austenitic steel in relation to irradiation temperature [110].

with fast reactors (BOR-60 and TFTR), considerably higher concentrations of transmutation helium are formed.

At the same time, it should be mentioned that the important role of helium in the interpretation of HTRE of austenitic chromium–nickel steels is rejected in a number of investigations (for example [91, 116]). It is assumed that the HTRE effect of these materials in neutron radiation is determined to a greater extent by the melting technology, chemical composition, preliminary treatment and also/or by the processes of radiation-stimulating segregation of dissolved elements at the grain boundaries. Nevertheless, special experiments with the testing of helium-implanted austenitic chromium–nickel steels to determine creep, long-term strength and fatigue have clearly confirmed the important role of helium in their high-temperature radiation embrittlement (see, for example [102,117–119]).

The phase changes in chromium–nickel austenitic steels in irradiation have been examined in great detail in section 4.6.3. The main effect on the evolution of the microstructure and mechanical properties of irradiated steels is exerted by radiation-stimulated and radiation-induced processes of segregation, decomposition of their matrix solution and coalescence.

It should be mentioned that a certain role, especially in low-temperature radiation hardening and embrittlement of austenitic steels, may be played by the mechanisms of spinodal decomposition and formation of spatially oriented structures of the type of modulated structures. Similar structural changes have been observed in, for example [120–122] in irradiation of type 316 [120,121] and EP-838 steels [122] with electrons with an energy of 1 and 22 MeV at temperatures of ~40 and 450 °C, respectively. The possible influence of these effects on the mechanical properties of irradiated steels has been discussed in section 6.3.2 in analysis of the mechanisms of radiation hardening and embrittlement.

One of the important relationships governing the phase transformations in chromium–nickel austenitic steels in irradiation is represented, as mentioned in section 4.6.3, by the stabilisation of precipitates of the phases and formation of new radiation-stimulated phrases, enriched with nickel and silicon. These are in particular the η-phase and the phases γ' and G. This process leads to destabilisation of austenite and stimulates the $\gamma \rightarrow \alpha$ transformation, leading to the formation of ferrite and martensite phases characterised by excessive embrittlement. At the same time, nickel depletion of austenite creates favourable conditions for the nucleation of Laves phases, σ and χ, enriched with iron and chromium.

The formation of segregations and precipitates at relatively low irradiation temperatures results in additional radiation hardening and a decrease of plasticity in transcrystalline failure. At elevated and high temperatures, the localisation of segregations and phases at the grain boundaries, and also coalescence of the precipitates in their matrix, stimulates the processes of softening and mixed or intergranular failure.

Attention will now be given to the properties of irradiated austenitic chromium–manganese steels with accelerated decrease of induced radioactivity.

Initial comparative investigations of neutron irradiation at 50 and 125 °C of chromium–manganese steels with a low nickel content of ~5% (EP-838 steel) and 316-type steel [123–125] shows that the mechanical properties of irradiated EP-838 steel in tensile loading in the temperature range 20–750 °C are not lower in comparison with the 316 steel. The authors of [126, 127] carried out mechanical tensile tests of specimens of hexagonal wrapper tubes of EP-838 steel, together with a number of chromium–nickel steels, after irradiation in CM-2 and BOR-60 reactors at higher temperatures (340–800 °C) and fluences up to 60 dpa. The experimental results show that at irradiation temperatures not higher than 400 °C and at the level of radiation damage of up to 30 dpa, the mechanical properties of the steel are not inferior to the mechanical properties of the chromium–nickel steel. At the same time, at high fluences and also irradiation temperatures higher than 440 °C, the plasticity of the steel decreases to almost zero. As noted previously, the identical effect is also characteristic of austenitic chromium–nickel steels (Fig. 6.32).

In the framework of the programme of investigations of reduced-activation materials for the first wall of fusion reactors, a number of compositions have been proposed for nickel-free austenitic steels with an accelerated decrease of induced radioactivity (section 4.6.3.2). The most promising compositions include, in our view, steels of the Fe–12Cr–20Mn–W-C type (The Institute of Materials Science and Metallurgy, Russian Academy of Sciences, the Oak Ridge National Laboratory [66,128–130], and the nickel-free steels of the type AMCR (Euroatom [131]). The chemical composition of AMCR-0033 steel is presented in Table 4.4.

The authors of [132] presented the results of investigations of mechanical properties in tensile loading of Fe–0.1C–12Cr–20Mn–1W and Fe–0.25C–12Cr–20Mn–2W–Ti reduced-activation steels after irradiation up to 10 dpa in a BOR-60 reactor, at temperatures of 350-400 °C. The steels were developed and produced in Russia and the

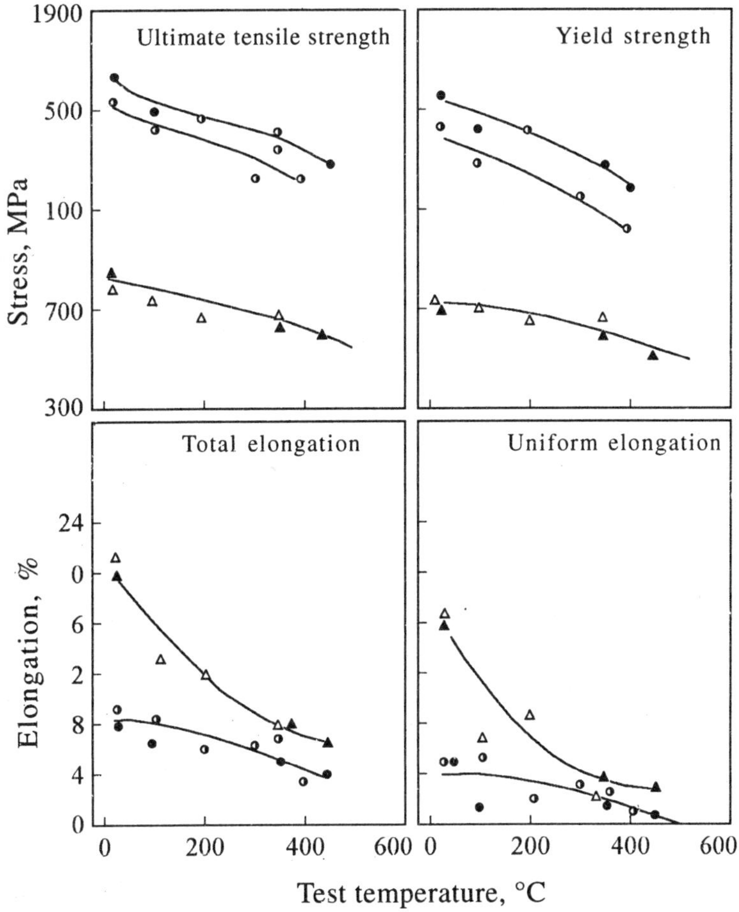

Fig. 6.34 Temperature dependences of the variation of mechanical properties of steels Fe–0.1C–12Cr–20Mn–1W (▲, △) and Fe–0.25C–12Cr–20Mn–2W–Ti (●, ○) after irradiation in BOR-60 reactor at temperatures of 345 °C (○,△) and 400 °C (●,▲) with neutron fluences of 1.6×10^{22} cm^{-2} (○,△) and 2.1×10^{22} cm^{-2} (●,▲) [132].

USA, respectively. The Russian steel was irradiated in the annealed condition, the American steel after austenitising and 20% cold deformation. In tests of the irradiated 12Cr–20Mn–1W steel in the temperature range 20–450 °C it was established that the mechanical properties of the steel are satisfactory and are not inferior to the mechanical properties of the Fe–0.04C–16Cr–11Ni–3Mo–Ti chromium-nickel steel for identical irradiation conditions. The irradiation hardening of cold-worked Fe–0.25C–12Cr–20Mn–2W–Ti steel in the entire temperature range is considerably higher and its total and uniform elongation is lower, but its temperature dependence is less pronounced (Fig. 6.34).

Austenite in Fe–Cr–Mn–C compositions and, in particular, in Fe–12Cr–20Mn–C composition is less stable than in the Fe–Cr–Ni-C steels in both thermal ageing and in irradiation (section 4.6.3.2). Nevertheless, as reported previously in section 4.6.3.2, on the basis of the results of investigations in [129,130,133], the phase stability can be increased and the properties of the Fe–12Cr–20Mn–C steel improved in additional alloying with W, Ti, V, B and P. Actually, as shown recently in [134], this optimum complex alloying increases the irradiation resistance of the Fe–12Cr–20Mn–0.25C basic composition.

Measurements of the mechanical properties in tensile loading a series of chromium–manganese Fe–12Cr–20Mn–0.25C steels, alloyed with different combinations of 1W, 0.1Ti, 0.1V, 0.005B and 0.03P, was carried out after irradiation in a FFTF reactor at temperatures of 420, 520 and 600 °C, with a fluence up to 44 dpa. Tests were carried out at irradiation temperatures and the results were compared with identical data for the 316 steel [134].

The irradiation of all compositions in the pre-annealed condition resulted in irradiation hardening. The lowest degree of irradiation hardening and the loss of ductility were recorded in the steel alloyed with all the five previously mentioned alloying elements. The general and uniform elongation in the steel after irradiation at 420 °C was more than a factor of 4 higher than in other chromium–manganese steels and 316 SS steel. A considerably smaller difference was recorded after irradiation at higher temperatures, resulting in a considerably lower degree of hardening and, consequently, a loss of plasticity. After irradiation at 420 °C of the cold-worked specimens, the plasticity in the majority of chromium–manganese steels was higher than in the 316 SS steel. However, the highest uniform elongation was again recorded for the steel alloyed with all the five elements.

6.2.1.5. Vanadium-based alloys

Initially, the vanadium-based alloys were regarded as promising materials for the active zones of fast reactors with a sodium heat carrier. However, this variant was rejected because liquid sodium intensively transfers oxygen and other interstitial impurities from steels to structural elements made of vanadium, causing rapid corrosion and embrittlement, [135].

At the present time, a number of compositions based on vanadium (mainly the V–Cr–Ti system) are regarded as promising low-activation materials for the first wall of fusion reactors with a lithium heat carrier [66,136,137]. In liquid lithium, the transport of oxygen takes place almost from any metal into lithium [135] and the processes of corrosion and embrittlement, in particular of vanadium, are less intensive.

The high heat conductivity of vanadium-based alloys, the low thermal expansion of these alloys and satisfactory mechanical properties ensure low thermal stresses, formed as a result of the presence of the temperature gradient in the cross-section of the first wall. As regards this parameter, the vanadium-based alloys (V–Ti, V–Cr–Ti) are superior to identical properties of austenitic and ferritic steels, in particular steels 316SS and HT-9 [136,138] which at the present time, as already mentioned, are regarded as promising materials for ITER and DEMO reactors.

The vanadium-based alloys are not inferior or superior to the alternative competing materials (austenitic and ferritic steels) also in a number of other properties, in particular, the decrease of inducing radioactivity, the short-term and long-term mechanical properties and elevated temperatures, and also corrosion resistance in lithium. Nevertheless, as in the case of other BCC alloys, one of the main problems of practical application of these materials in thermonuclear energetics is the problem of low-temperature radiation embrittlement.

On the basis of systematic investigations, the optimum reduced-activation radiation-resistant compositions on the basis of vanadium for the first wall of fusion reactors include the alloys V–(4–5)Ti–(4–5)Cr [137,139–143], including the alloys with different modification elements [137,139,141].

Figure 6.35–6.38 show the results of examination of the strength and ductility properties in tensile and impact loading of V–4Cr–4Ti alloy and other compositions based on vanadium after irradiation in a FFTF-MOTA reactor at temperatures of 420, 520 and 600 °C, up to fluences corresponding to 114 dpa [140,142]. Figure 6.39 shows the dependence of the effect of chromium concentration in V–Cr–(4–5)Ti alloys on the ductile–brittle transition temperature in irradiated alloys and also in non-irradiated materials, containing $<3 \times 10^{-3}$ % H and 6×10^{-2}–1.2×10^{-1} % H [142].

The experimental results show that as regards the set of the properties, the V–4Cr–4Ti alloy is superior to other compositions based on vanadium. At present, this alloy is the optimum and promising vanadium alloy for the first wall and the blanket of the fusion reactors [142,144].

However, it should be mentioned that these highly optimistic results, including those obtained for the radiation resistance of the V–4Cr–

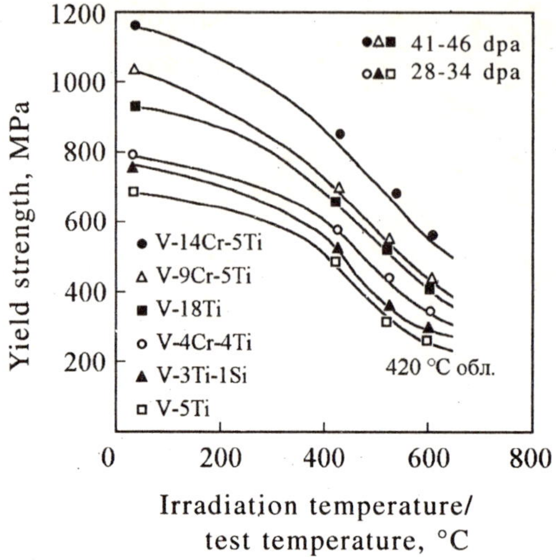

Fig. 6.35. Variation of the yield stress of a series of alloys based on vanadium after neutron irradiation in tensile testing at 20 °C (for an irradiation temperature of 420 °C) and temperatures of 420, 520 and 600 °C, corresponding to irradiation temperatures [140,142].

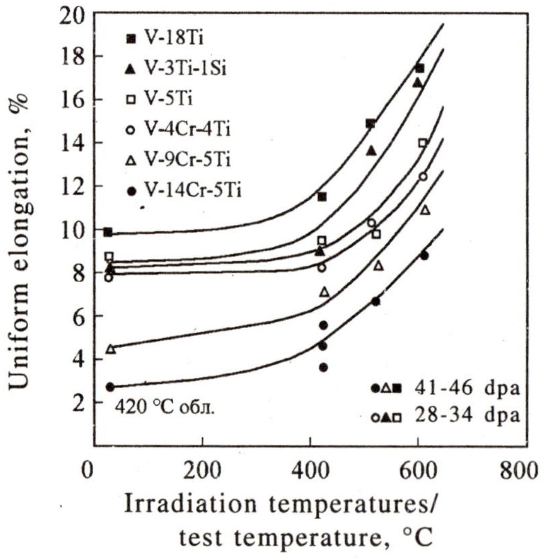

Fig. 6.36. Variation of uniform elongation of a series of alloys based on vanadium after neutron irradiation in tensile testing at 20 °C (irradiation temperature 420 °C) and 420, 520 and 600 °C, corresponding to irradiation temperature [140, 142].

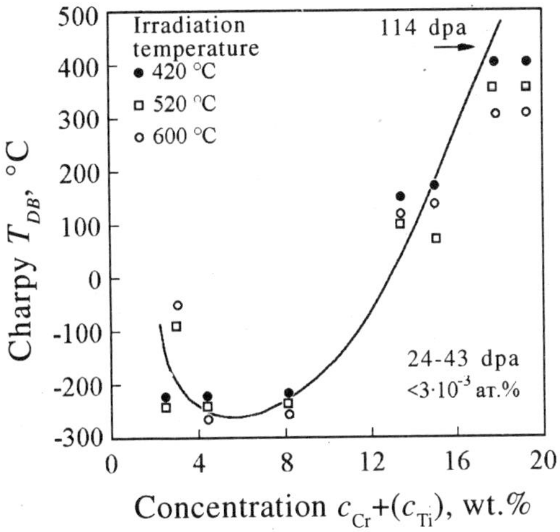

Fig. 6.37. Ductile–brittle transition temperature for a series of alloys based on vanadium irradiated at 420, 520 and 600 °C with neutron fluences corresponding to 24–43 dpa and 114 dpa [142].

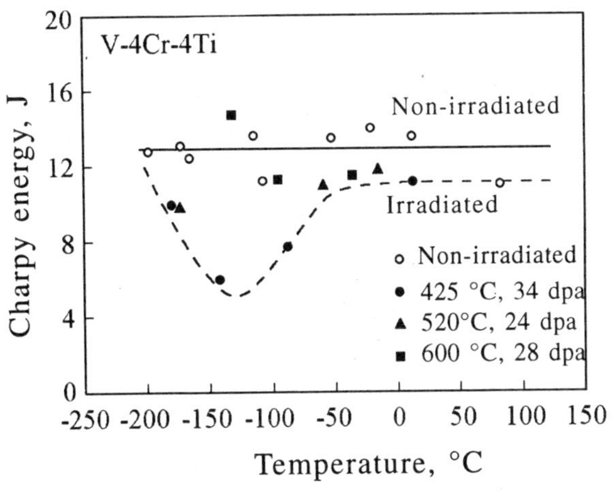

Fig. 6.38. Charpy energy in impact tests of V–4Cr–4Ti alloy in non-irradiated condition and after neutron irradiation at temperatures of 425, 520 and 600 °C up to a fluence corresponding to 34 dpa [142].

4Ti alloy, and a number of other compositions based on vanadium, have been obtained for irradiation temperatures of 420–600 °C. At the same time, recent investigations show [145] that irradiation at lower temperatures (100–275 °C), even with a weak fluence (~0.5 dpa) results in rapid embrittlement of the V–4Cr–4Ti alloy.

This alloy, melted into a 500 kg ingot especially for investiga-

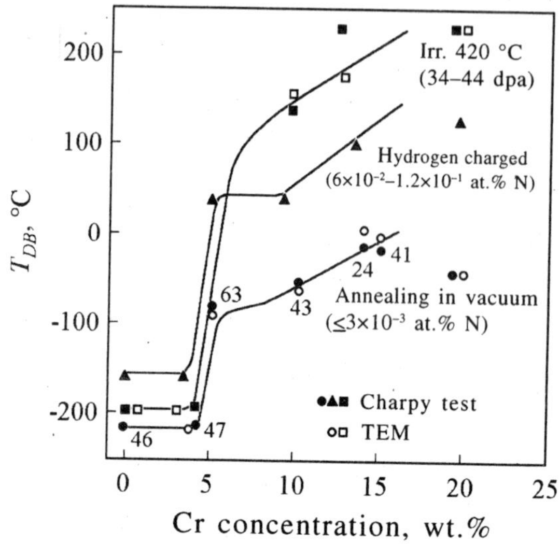

Fig. 6.39. Temperature of the ductile–brittle transition in a series of alloys V–Cr–(4–5)Ti containing hydrogen, or irradiated at 420 °C with neutron fluences corresponding to 34–44 dpa. Alloys: V–5Ti (1), V–4Cr–4Ti (2), V–5Cr–5Ti (3), V–9Cr–5Ti (4), V–14Cr–5Ti (5), V–15Cr–5Ti (6) [142].

tions in the framework of the thermonuclear programme of the USA, contained 3×10^{-2} % O; 8.5×10^{-3} % N and 8×10^{-3} % C. The initial plate for the preparation of specimens from this material was subjected to the optimum recommended annealing (1050 °C) in a vacuum higher than 10^{-5} torr. After preparation, the specimens were annealed at 1000°C in a vacuum better than 10^{-7} torr.

The result of the tensile tests of irradiated specimens at room temperature show a complete loss of work hardening and zero uniform plasticity. The ductile–brittle transition temperature rapidly increased and the Charpy energy in the dynamic test of notched specimens (Fig 6.40*a*) and with an initial crack (Fig. 6.40*b*) decreased. The degree of radiation embrittlement of the alloy increased with increasing radiation temperature.

The results obtained in [145] together with a number of technological problems of production of industrial vanadium alloys, their treatment and welding [66] resulted in additional doubts regarding the application of V–4Cr–4Ti alloy as the structural material of the first wall of fusion reactors. The range of radiation temperatures in [145] completely corresponds to the working temperatures (20–300 °C) accepted at the present time for ITER reactors when using 316 SS steel as the first wall [105]. On the basis of the results published

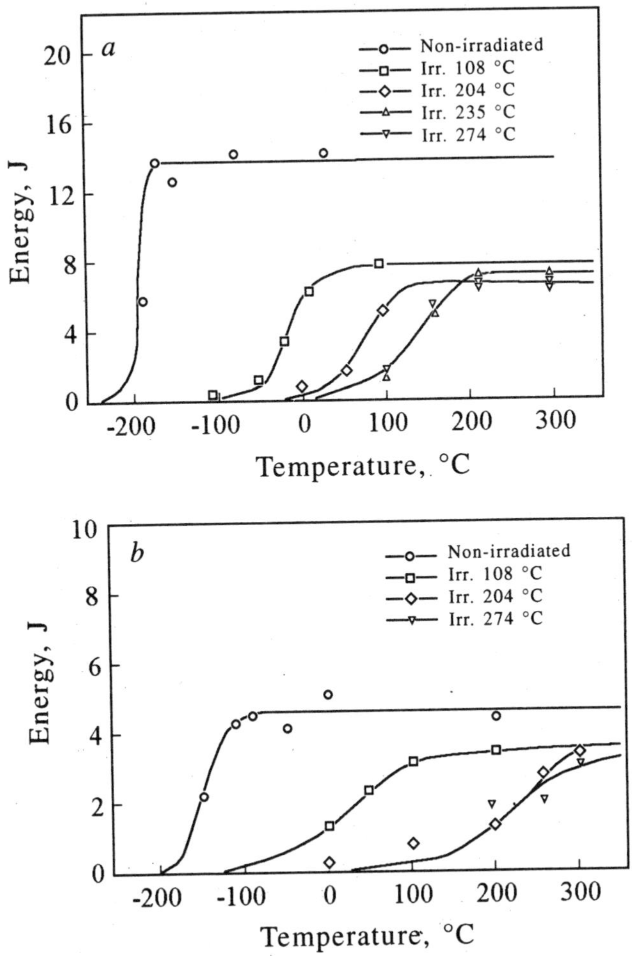

Fig. 6.40. Properties under impact loading of non-irradiated and irradiated specimens of V–4Cr–4Ti alloy with a Charpy notch (*a*) and an initial crack (*b*) [145].

in [145], these temperatures are almost completely unacceptable for the V–4Cr–4Ti alloy. However, in the process of operation of the first wall made of the V–4Cr–4Ti alloy and at higher temperatures it is also essential to consider different non-standard situations which may result in rapid embrittlement and failure.

One of the main advantages of vanadium as a basis for the development of promising structural materials for the first wall of fusion reactors is the accelerated decrease of induced radioactivity (Chapter 5) on the condition of the content in the alloy of the minimum concentration of undesirable (with respect to activation) impurities (Chapter 5, [146]). In the V–4Cr–4Ti alloy, the parameters of the decrease of induced radioactivity in Cr were almost identical with the identical properties of vanadium (Fig. 5.2 and 6.41). The slower decrease of induced radioactivity in titanium (Fig. 5.2 and 6.41) is controlled by the decay of the long life radionuclide ^{39}Ar (Chapter 5 [28,148,149]). Evidently, in the V–4Cr–4Ti alloys, this process will determine the duration of decrease of their induced radioactivity, increasing the period to establishment of the biologically safe level and, at the same time, complicating the problem of utilisation and processing of radioactive waste.

This problem was one of the main reasons for the development of new reduced-activation vanadium alloys, not containing Ti [147, 150]. Titanium was replaced with gallium, characterised by an even higher rate of decrease of induced radioactivity in comparison with V and Cr (Fig. 6.41) [147,150]. The investigations of V–Ga binary compositions, ternary compositions V–Ga–Cr, V–Ga–Ce and V–Ga–Cr alloys with a low Ce concentration (up to 0.07%) show that a number of the selected alloys is not inferior in the initial mechanical properties, heat conductivity and corrosion resistance in lithium to the alloys of the V–Cr–Ti system, including V–5Cr–5Ti alloy [147,150]. Alloying of vanadium with gallium also increases the properties of the alloys in both hot and cold treatment [147,150].

The dilation volume of gallium in vanadium, like that of titanium, is greater than 0. Consequently, taking into account dilation con-

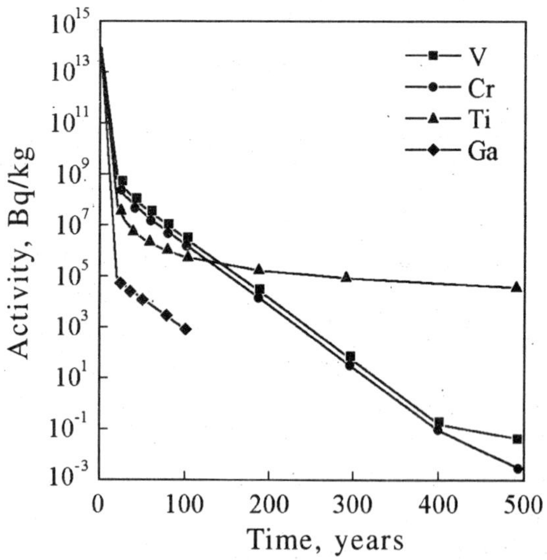

Fig. 6.41. Calculated kinetic dependences of the decrease of induced radioactivity in vanadium, chromium, titanium and gallium [147,150].

siderations of the interaction of radiation defects with the atoms of dissolved elements (Chapter 1), it may bc assumed that the effect of Ga on the radiation resistance in alloys of the V–Ga–Cr system will be identical to the effect of titanium in the alloys of the V–Cr–Ti system. In particular, as in the case of titanium, it is possible to expect the positive effect of Ga on suppressing swelling in the V–Cr alloys.

The results of preliminary investigations of the V–Ga and V–Ga–Cr systems as materials competing with alloys of the V–Cr–Ti system, indicate the promising nature of this direction [147, 150] which in principle should stimulate further investigations of the properties of these materials in the initial condition and after irradiation

6.2.2. The mechanisms of radiation hardening and embrittlement

Previous sections of this chapter were concerned with the main experimental relationships of the variation of the mechanical properties of irradiated metallic materials in active tensile and impact loading and their relationships with the special features of evolution of the microstructure were investigated. In a number of cases, the qualitative or empirical interpretation of these effects was carried out. In this section, attention will be given, in particular, to microscopic considerations, describing the degradation of these properties and also possible methods of decreasing the negative effect of radiation.

6.2.2.1. Radiation hardening

As already mentioned, the interpretation of the microscopic mechanisms of radiation hardening is based on the presentations of the interaction of mobile dislocations with barriers of radiation origin of different type. They include point defects, their complexes with each other and atoms of dissolved elements, clusters of radiation defects, dislocations loos, voids, radiation-stimulated and radiation-induced precipitation of phases and also a dislocation network, formed during irradiation.

Depending on the distance dependence of the interaction potential of obstacles with dislocations in the slip plane, the barriers are subdivided into long-range and short-range [6,16,151]. In the case of the short-range barriers, the characteristic fields of interaction with dislocations do not exceed several atomic spacings. These barriers usually include Peierls–Nabarro barriers, the atoms of dissolved elements, point defects, jogs on the dislocations. In the case of the long-range barriers, the potential of interaction with dislocations changes only slightly with the distance in the range ≥10–15 Å and more. They usually include the dislocations in parallel slip planes, dislocation clusters, grain boundaries, large phase precipitates, and other defects.

The short-range barriers provide the main contribution to the thermally activated component of stresses τ_T (or yield stress $\sigma_y(T)$) and at temperatures of >0 K they may be overcome by moving dislocations both as a result of the applied stresses and thermal activation. It is evident that in the case of the long-range barriers, interacting with the dislocations over large distances, the amplitude of thermal fluctuations is insufficient for overcoming these barriers. In this case, the stresses required for overcoming the barriers, depend only slightly on temperature (through the shear modulus μ) and provide the main contribution to the athermal component of the stresses τ_μ.

It should be mentioned that the athermal elastic mechanism may also dominate in the interaction of dislocations with short-range hard obstacles (the shear modulus of the barrier μ_b is greater than the shear modulus of the matrix μ_m) at specific dimensions of the obstacles and barrier spacings. Orowan proposed a similar mechanism [152] for hardening with non-coherent precipitate particles at the distances between the particles considerably exceeding their size. These barriers are overcome as a result of the merger of dislocation segments, formed during the bending of the dislocation line between adjacent obstacles. The maximum interaction force, determined by the linear tension of the dislocation during passage through these barriers (Orowan's force) is $F = \mu b^2$ [16,110].

Schematically, the temperature dependence of the total shear stress for the two previously examined types of barriers: short-range thermally activated and long-range athermal τ:

$$\tau = \tau_\mu + \tau_T \tag{6.4}$$

is shown in Fig. 6.42 [151]. Figure 6.42 shows that at temperatures higher than some critical temperature T_0, the deformation process may be completely controlled by the athermal mechanism.

The initial modelling considerations of the temperature dependences of radiation hardening, taking into account the short-range thermally activated barriers, were developed by Zeeger [153] and Fleischer [154, 155].

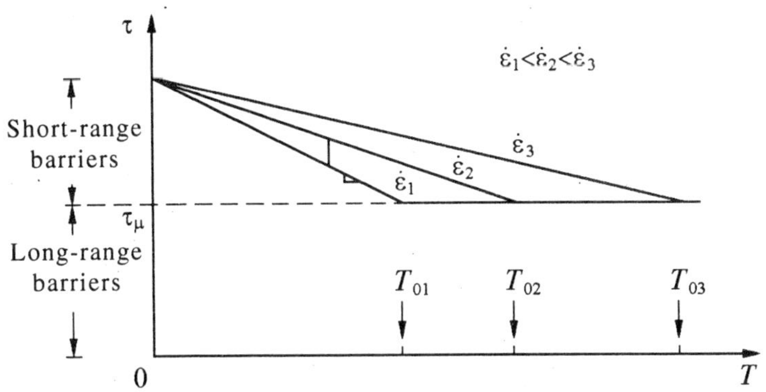

Fig. 6.42. Schematic dependences of flow stress on temperatures [151].

The Zeeger model is based on the mechanism of interaction of dislocations with depleted zones (cutting of these zones) and gives the following temperature dependence:

$$\tau^{2/3} = A - BT^{2/3}. \tag{6.5}$$

In the Fleischer model, radiation hardening is regarded on the basis of the mechanism of interaction of dislocations with barriers, generating tetragonal distortions (interstitials, divacancies, dislocation loops, etc). The temperature dependence of the shear stresses is determined by their relationship:

$$\tau^{2/3} = C - DT^{1/2}. \tag{6.6}$$

In equations (6.5) and (6.6), the parameters *A, B, C* and *D* depend mainly on the strength (hardness) of the barriers and the strain rate.

The results of analysis of the temperature dependences of radiation hardening within the framework of Zeeger and Fleischer modelling considerations, mainly for copper and aluminium, and also for a number of other metals, have been described in considerable detail in a number of articles (for example, [3–6, 14–16] and [6, 156–158] respectively).

It should be mentioned that the interpretation of the experimental data in these and other similar investigations is highly ambiguous: in a number of investigations, preference has been given to the Zeeger or Fleischer model, or it has been reported that none of the theories is fully adequate to the experiment. The latter conclusion is fully justified, especially for neutron radiation.

In irradiated metallic materials, there is usually a spectrum of various obstacles, and the type, concentration and ratio between these obstacles may greatly change depending on temperature and radiation dose. These factors are not taken into account in the Zeeger and Fleischer models. In addition, as indicated by previous examination, the subdivision of barriers into thermally activated short-range and athermal long-range is highly conventional and may also depend on the evolution of the defects–phase structure during irradiation. In every case, the radiation hardening mechanism may be determined on the basis of analysis of the kinetics of the process and accurate information on the structure of radiation damage.

As an example, attention will be given to experimental data presented in Fig. 6.3 for the variation of the yield stress of copper during irradiation with electrons with an energy of 2 MeV at a temperature of <31 K. It is shown [5] that the radiation hardening of copper is proportional to the square root of the concentration of radiation defects ($c^{1/2}$). A similar relationship follows from the Fleischer model [154,155]:

$$\Delta\tau = A\mu c^{1/2}\Delta\varepsilon, \tag{6.7}$$

where A is a constant <1, and $\Delta\varepsilon$ is the tetragonal distortion.

In the conditions of irradiation of copper, corresponding to the given experimental conditions, only the isolated Frenkel pairs appear, and the free migration of point defects, as shown in [159,160] does not take place (the temperature of the start of free migration of interstitials in the substage I_E in copper $\geq$~50 K). No dislocation loops form as a result of the clustering of radiation defects and the process of radiation hardening is controlled by the buildup of isolated interstitials, forming tetragonal distortions. Thus, in this case all the examined facts fully correspond to Fleischer modelling considerations.

The correct analysis of experimental kinetic and temperature dependences of radiation hardening, in particular, copper and aluminium, especially at all temperatures, including those presented in Fig. 6.2, 6.3, 6.5, 6.6 is not only of theoretical but also practical interest. The low-alloyed alloys based on these elements are used or examined as promising stabilising materials of superconducting magnetic systems of fusion reactors (alloys of copper and aluminium [161,162]) and structural materials of space systems (aluminium alloys). In the latter case, the degradation of the mechanical properties may be caused by the effect of electrons and protons of MeV energies of artificial and natural radiation belts of the Earth in the temperature range ~120–420 K [163,164].

In addition to the classification of the barriers, examined previously – short-range and long-range, thermally activated and athermal – they are also subdivided with respect to strength or hardness. The strength or hardness of the barriers is determined by the maximum value of the force acting on a dislocation during the passage of obstacles of a specific type. An increase in the shear stress and the yield stress, determined by barrier hardening, is equal to respectively:

$$\Delta\tau = \frac{F_{\max}}{bl} = \frac{\alpha\mu b}{l} \quad (6.8)$$

$$\Delta\sigma_y = M\Delta\tau = \frac{M\alpha\mu b}{l} \quad (6.9)$$

In equations (6.8) and (6.9), l is the distance between the obstacles, $\alpha \leq 1$ is the coefficient of the strength of the barriers of the given type in relation to the Orowan force $F = \mu b^2$, M is the corrected Taylor factor, linking the shear stress of single crystals and the yield stress of polycrystals under tensile loading. For the FCC and BCC metals, its value is 3 and 2.75 [110,165].

For discrete obstacles (clusters of defects, dislocations, voids, particles of phase precipitates, etc), the distance between the barriers l is usually assumed to be equal to $l = (d\rho)^{-1/2}$ [16,110] and in the case of dislocations it is inversely proportional to the square root of dislocation density $l = \rho_d^{-1/2}$. Consequently, equation (6.9) for discrete and dislocation barriers has the following appropriate form:

$$\Delta\sigma_y = M\alpha\mu b(d\rho)^{1/2}, \quad (6.10)$$

$$\Delta\sigma_y = M\alpha\mu b\rho_d^{1/2}, \quad (6.11)$$

In equation (6.10), d and ρ is the diameter and density of the discrete barriers, respectively.

For dispersed precipitates of phases, the authors of a number of publications (for example, [165–167]) presented modified equations (6.10) taking into account the degree of coherence (coherent, semi-coherent and non-coherent) and the shape of the particles, and also the relationships between the size of the particles and the distance in the slip plane. Identical data for hardening resulting from the presence of voids taking into account the size distribution and the ratio of the diameters and barrier spacings have been presented in [168].

The types of barrier in relation to the Orowan's force ($F = \alpha\mu b^2$, α=1) are subdivided into strong (α>0.5), weak (α<0.25) and barriers of intermediate or medium strength (0.25<α< 0.5) [16,110]. The strongest barrier usually include voids and large phase precipitates, intermediate strength barriers large dislocation loops and fine-dispersion phases. The clusters of defects, small dislocation loops and also small gas bubbles and dislocations are regarded as weak barriers.

In particular, it should be mentioned that the validity of specific theoretical assumptions regarding the mechanisms of radiation hardening (and the equations expressing this hardening) and the strength of the barriers may be determined only on the basis of experiments in which the mechanical properties are periodically and simultaneously measured during irradiation and the microstructure for the dominant evolution of a specific single type of defect is analysed. For example, for the interpretation, evaluation or modelling of radiation hardening, determined by dislocation loops, a number of authors (for example, [169–172]) used the equation derived in [173] instead of equation (6.10)

$$\Delta\sigma_y = \beta\mu b d\rho^{2/3}. \qquad (6.12)$$

where β is a numerical parameter.

In this case, the authors know of no literature experimental data in which the validity of equation (6.12) is confirmed on the basis of the previously-mentioned kinetic analysis method. In contrast to equation (6.12), the validity of equation (6.10) within the framework of such analysis has been established in a large number of experimental studies (see, for example [11,174,175]). As examples of the validity of equation (6.10), Fig. 6.43 [174] and Fig. 6.44 [11] show the kinetics of radiation of σ_y and H_μ in relation to the parameter $(d\rho)^{1/2}$ copper and vanadium in the process of evolution of clusters and dislocation loops of the interstitial type in neutron and electron radiation, respectively. The dependence in Fig. 6.44 was obtained in comparison of the experimental data shown in Fig. 6.8 and 6.12.

Table 6.2 shows the results of evaluation of the relative strength of barriers α on the basis of the experimental data for a number of FCC and BCC metals and alloys. The data for Ni and a number of austenitic Fe–Cr–Ni steels represented our generalised results, presented

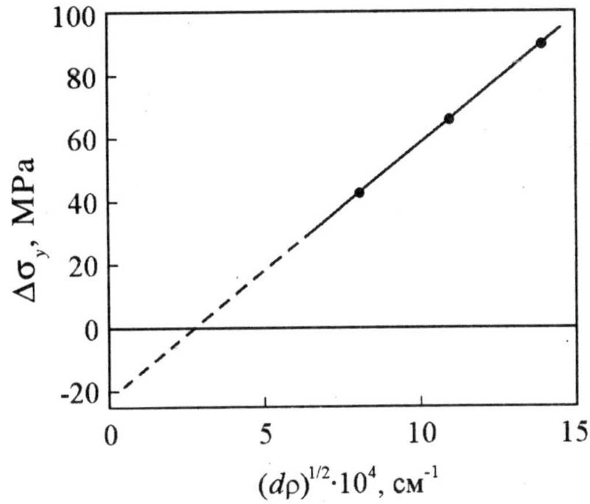

Fig. 6.43. Kinetics of hardening of copper in irradiation with neutrons with an energy of 14 MeV at 90 °C [174].

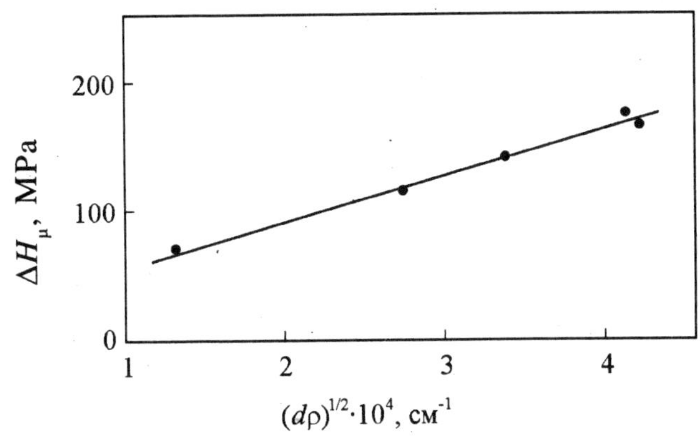

Fig. 6.44. Kinetics of hardening of vanadium in irradiation with electrons with an energy of 2.2 MeV at a temperature of 100 °C [11].

in [110]. The values of α, presented in Table 6.2 with references to the sources are indicated by the asterisk, are the results of our evaluation of the values of α on the basis of analysis of the experimental data presènted in the cited studies. Our calculations of the values of α for the voids and loops were carried out using equation (6.10) and Taylor's factor M, equal to respectively 3 and 2.75 for the FCC and BCC metals. In cases in which in the cited investigations the radiation hardening was measured by the Vickers microhardness method [177, 181,11]), the following relationship between H and σ_y was used

Table 6.2. Strength of barriers in irradiated metals and alloys

Strength of barriers	Type of barrier	Material	α	Literature
	Orowan barriers			
	FCC metals and alloys			
Strong barriers	Voids	Ni	~1	[110]
	Voids	Fe–Cr–Ni steels	~1	[110]
	Voids: $\bar{d}$ = 52 nm, ρ = 8.7×10^{19} m^{-3}	Al	0.76	[177]*
	Large phase precipitates	Fe–Cr–Ni steels	~1	[110]
Medium strength barriers	Frank loops	Fe–Cr–Ni steels	0.33–0.45	[110]
	Frank loops: d = 9–29 nm, ρ ~ 10^{14}–10^{15} m^{-3}	Fe–Cr–Ni steels	0.45	[178]
	Fine MC carbides	Fe–Cr–Ni steels	0.33–0.45	[110]
Weak barriers	Clusters of defects: d = 2.4–2.7 nm, ρ = 2.5×10^{22}–8.1×10^{22} m^{-3}	Cu	0.23	[174]
	Fine loops and clusters of defects	Fe–Cr–Ni steels	~0.2	[110]
	Clusters of defects	Fe–Cr–Mn steels	0.11–0.24	[175]
	Small bubbles	Fe–Cr–Ni steels	0.2	[110]
	Dislocations	Fe–Cr–Ni steels	–-0.11–0.2	[110]
	Dislocation network	Fe–Cr–Ni steels	0.16	[178]
	BCC metals and alloys			
Strong barriers	Voids: $\bar{d}$ = 6.5 nm, ρ = 1.9×10^{21} m^{-3}	V	1.0	[19]*
	Voids: $\bar{d}$ = 4 nm, ρ = 5×10^{22} m^{-3}	Mo	0.77	[179]*
	Loops: $\bar{d}$ = 6.3–3.4 nm, ρ = 3.8×10^{22}–m^{-3}	Nb	0.64	[20]
	MC_2 carbides	Steel Fe–2Cr2WVTa	0.71	[80]*
Medium strength barriers	Loops: $\bar{d}$ = 7 nm, ρ = 1.4×10^{22} m^{-3}	V	0.39	[19]*
	Loops: $\bar{d}$ = 7.4–83 nm, ρ = 2.1×10^{20} m^{-3}	V	0.35	[179]*
	Loops a<100>: $\bar{d}$ = (21–15) nm, ρ = 2.1×10^{21}–8×10^{21} m^{-3}	V	0.38	[180]
	Loops: $\bar{d}$ = 6.6 nm, ρ = 2.3×10^{22} m^{-3}	Mo	0.38	[179]*
	α–phase	Steel Fe–12Cr2WVTa	0.40	[80]*
	M_6C carbides	Steel Fe–9Cr2WVTa	0.32	[80]*
Weak barriers	Clusters of vacancies (microvoids): $\bar{d}$ = 3 nm, ρ = 1.5×10^{22} m^{-3}	V	0.22	[11]*
	M_6C carbides	Steel Fe–7Cr2WVTa	0.22	[80]*
	Loops: $\bar{d}$ = 4.5 nm, ρ = 2×10^{23} m^{-3}	V	0.20	[19]*
	Loops $a/2$ <111> $\bar{d}$ = 3.5–4.5 nm, ρ = 6×10^{21}–3.9×10^{21} m^{-3}	V	0.16	[180]
	Loops: $\bar{d}$ = 2.5–2.2 nm, ρ = 1.3×10^{23}–1.6×10^{23} m^{-3}	Nb	0.125	[20]

(see, for example [64, 176]):

$$\Delta H = 2.8\Delta\sigma_y. \tag{6.13}$$

As indicated by Table 6.2, the relative strength of the barriers basically corresponds to their classification presented previously. At the same time, the numerical values of τ cannot be regarded as completely reliable. Firstly, in different investigations concerned with the evaluation of the hardening parameters, the authors used different relationships between the shear stress τ and the yield stress σ_y, determined in the experiments. For example, in the evaluation of α on the basis of the experimental data in [177,174,179,19,11], the correction factor between τ and σ_y was represented by Taylor's factor, in study [178] it was the Mises factor equal to $\sqrt{3}$ [176] $\left(\Delta\sigma_y = \sqrt{3}\tau\right)$, and in [20] and [180] the Tresca factor equal to 2 [176] ($\sigma_y = 2\tau$). In [175], the correction factor between τ and $\sigma_{0.2}$ was not considered at all, i.e. it was assumed to be equal to unity.

As already mentioned, the strength of the barriers of even the same type may also greatly depend on their dimensions and density. In the case of small obstacles dislocations may overcome these obstacles also as a result of different thermally activated dislocation reactions in direct contact interaction with the barriers [181]. For the thermally activated process of cutting of the loops, the increase of the flow stress is determined by the equation [182]:

$$\Delta\tau = \tau_\mu + \frac{U_0 - mkT}{V} \tag{6.14}$$

where τ_μ is the athermal component of the stresses, determined by the bending of the dislocations on the loops (equation (6.8), and $(U_0 - mkT)/V$ is the component of effective stresses for the process of cutting of the loops.

In equation (6.14), the activation volume $V = \bar{l}b^2$, $\bar{l}$ is the mean distance between the loops, U_0 is the activation energy for passing through an obstacle. The parameter $m = \ln(NAb\nu_0/\varepsilon')$, where N is the number of moving dislocations, A is the area, travelled by the dislocation after overcoming an obstacle, ν_0 is the frequency of oscillations of the dislocation line, ε' is the strain rate.

For small barriers, thermal fluctuations may support overcoming these barriers and, at the same time, decrease their effective strength, determined by equation (6.10). With an increase in the size and/or the hardness of these obstacles, thermal fluctuations have a smaller and smaller effect on their effective strength. The experimental confirmation of this role (in particular, the size factor) follows, in our view, from analysis of certain results, presented in Table 6.2.

In [19], vanadium was irradiated at temperatures of 70 and 200 °C. The size of the interstitial dislocation loops for the given temperatures was $\bar{d} = 4.5$ nm and $\bar{d} = 7$ nm. The estimate of the strength of the barriers were respectively $\alpha = 0.2$ and $\alpha = 0.39$. An additional confirmation of this dimensional relationship is also represented by the results obtained for the radiation hardening of vanadium in the process of electron radiation at 100 °C in [11]. They are presented in Fig. 6.4.4 and in Table 6.2. These data indicate that in the range of variation of the size of the loops from 7.4 to 83 nm, whilst retaining their constant density during irradiation, the hardening kinetics does not change at constant $\alpha = 0.35$, which in fact is very close to the value $\alpha = 0.39$, estimated for the loops with $\bar{d} = 7$ nm using the results in [19]. A large increase in the value of the effective coefficient of the strength of the barriers in relation to the value of $\alpha = 0.2$ to $\alpha = 0.35$–0.39 with an increase of the diameter of the loops from $\bar{d} = 4.5$ nm to $\bar{d} = 7$ nm and its constancy in subsequent increases of the dimensions is evidently associated with the transition from the thermally activated mechanism of intersection of the loops to the athermal elastic mechanism of overcoming the barriers of the Orowan type.

The identical conclusion regarding the effect of the size factor may also be made on the basis of the results published in [180]. For the loops with a diameter of $\bar{d} = 3.5$–4.5 nm (loops $a/2$ <111>) and $\bar{d} = 15$–21 nm (loops a<111>) in irradiated vanadium, the estimate of α gave values of 0.16 and 0.38, respectively. In this case, it should be mentioned that the authors of [180] have explained the results not on the basis of the dimensional factor but on the basis of different crystallographic orientation of the loops. However, this interpretation is not justified because in the previously examined results of [11] the value of $\alpha = 0.35$ was obtained in particular for the loops $a/2$<111> with a size of 7.4 nm.

On the basis of the energy approach, the maximum size of the obstacles in cutting may be evaluated using the equation of the type of (6.15) [183,184]:

$$d_c = \frac{C\mu b^2}{\gamma} \tag{6.15}$$

where γ is the energy of the interface, formed as a result of overcoming a barrier. Constant C in equation (6.15) is equal to ~(1.2–2.2). For $\gamma \cong 1000$–2000 erg/cm^2 the value of d_c does not exceed 5 nm.

The critical size of the obstacles overcome by the thermally activated mechanism also depends on the ratio between the shear modulus and the Burgers vectors of the barrier and the matrix. In particular, for the process of cutting obstacles, it is determined by the equation

[165,167]:

$$d_c = \frac{42\mu_m b_m^2}{\mu_b b_b} \tag{6.16}$$

The indexes m and b in equation (6.16) belong to the parameters of the matrix and the barrier, respectively.

At $d > d_c$ the mechanism of overcoming the barrier changes – the dislocation may bypass the barrier by the Orowan-type mechanism.

The strength or hardness of the barriers may also change as a result of high-intensity radiation-stimulated non-equilibrium or equilibrium segregation of the impurity and alloying elements. The process of non-equilibrium segregation during irradiation results in a significant change in the composition of the matrix solid solution in the vicinity of the sinks, including at the loops and voids. The dislocation loops in this case may be represented by the Guinier–Preston type formations or loops with the stoichiometry of the phases. The variation of the chemical composition and formation of the phases are also detected in the vicinity of the voids. The basic theoretical presentations regarding the mechanism of non-equilibrium segregation processes during irradiation and the results of appropriate experimental investigations have been examined this sufficient detail in the sections 2.2, 2.3, 3.7.2, 3.8.1, 3.8.2 and 4.4.1.

The formation of equilibrium segregations at the sinks is evidently possible in the case of interstitial impurities at not too high concentrations of these impurities in the matrix solid solution. In thermodynamic equilibrium, i.e. without irradiation, the interstitial impurities migrate by the interstitial diffusion mechanism. On the basis of considerations regarding the size factor, it may be assumed that mixed dumbbell configurations, which also include interstitials, do not form during radiation (see section 1.3). Therefore, it is not justified to assume that the diffusion mechanism of the interstitials changes in the radiation conditions. This means that their diffusion flows on the sinks in the radiation process are not associated with diffusion flows of interstitials and vacancies which are also the main reasons for non-equilibrium segregation. At the same time, at relatively high initial concentrations of interstitials in the matrix solid solution concentration supersaturation of sinks with these elements may take place as a result of the formation of impurity atmospheres [185] or chemical compounds (phases).

Regardless of the obvious importance of the problem, the theoretical explanation of the effect of segregation processes on radiation hardening is not yet available, and the estimation of the parameters of hardness of discrete barriers (loops, voids, clusters of defects) and dislocations is usually carried out using the following equations:

$$\alpha = \frac{\Delta\sigma_y}{M\mu b(d\rho)^{1/2}} \tag{6.17}$$

$$\alpha = \frac{\Delta\sigma_y}{M\mu b(\rho_d)^{1/2}} \tag{6.18}$$

We shall examine possible aspects of the effect of radiation-stimulated segregation processes on hardening and the coefficient of hardness of the barriers on the basis of simple modelling representations. In this case, we shall restrict analysis to the case in which the matrix and the vicinity of the sinks are characterised only by changes of the chemical composition of solid solutions and the radiation-stimulated the phases do not form.

1. As a result of segregation, the shear modulus μ_s and the Burgers vector b_s (b_s~= the lattice parameter a) of the solid solution in the vicinity of a sink will differ from the appropriate parameters of the matrix solid solution μ_m and b_m. The elementary estimates using equations (6.10) and (6.17) for the effective coefficient of hardness of the barrier with segregations α_s in this case give:

$$\alpha_s = \alpha\frac{\mu_s b_s}{\mu_m b_m}. \tag{6.19}$$

where α is the coefficient of hardness of the barrier without segregation. However, it should be mentioned that the concentration dependences of the sheer modulus and the lattice parameter are very weak [186, 187] and their changes during solid-solution hardening can hardly have any significant effect on the coefficient of hardness of the barriers.

2. The voids and relatively large dislocation loops are very rigid barriers for the dislocations and are overcome by them by an Orowan-type mechanism [110]. In solid-solution hardening of the area around these sinks, as a result of segregation processes there are two possibilities for the modification of the given mechanism of hardening and the

variation of the coefficient of hardness of the barriers.

If the coordinates of bypassing an obstacle by dislocations do not change, overcoming of the barrier by the dislocations should in this case be controlled also by the stresses required for movement of the dislocations in the hardened solid solution with the concentration of the dissolved elements $c_b > c_m$, where c_m is the content of the dissolved elements in the matrix solid solution. In accordance with the considerations regarding solid-solution hardening [185], an additional increase of the yielding limit as a result of the formation of segregations in the vicinity of the barrier is:

$$\Delta\sigma = M\mu\varepsilon^n (c_b - c_m). \tag{6.20}$$

In this equation, the quantities M and n according to the Mott–Nabarro and Cottrell theories are M = 2 and 2.5 and n = 1 and 4/3, respectively, ε is the factor of the linear dimensional correspondence.

Taking into account (6.20), the general equations for the variation of the yield stress $\Delta\sigma_s$ and the effective coefficient of hardness of the barrier with segregations α_s have the form:

$$\Delta\sigma_s = \alpha_s M\mu b(d\rho)^{1/2} = \alpha M\mu b(d\rho)^{1/2} + M\mu\varepsilon^n (c_b - c_m) \tag{6.21}$$

$$\alpha_s = \alpha + \frac{\varepsilon^n (c_b - c_m)}{b(d\rho)^{1/2}} \tag{6.22}$$

In equations (6.21) and (6.22), α is the coefficient of hardness of the barrier without segregations.

If the coordinates of bypassing a barrier by the dislocations in the vicinity of a sink with segregations change, i.e., the dislocations will bypass the segregation fully or partially (depending on the profile degree of hardening), the increase of the yield stress will be determined mainly by the decrease of the barrier spacing $l=(d\rho)^{-1/2}$ as a result of an increase of the size of the obstacle d. In this case, the coefficient α_s is:

$$\alpha_s = \frac{\alpha l}{l_s} \tag{6.22}$$

Here l and l_s are the appropriate barrier spacings.

In addition to the radiation hardening as a result of an increase in the hardness of the obstacles, the process of radiation-stimulated segregation must also be accompanied by softening as a result of a decrease in the concentration of the impurity and alloying elements in the matrix solid solution during their movement to the sinks. In this case, as shown in the section 6.2.1.1 on the example of the experimental data for neutron-irradiated vanadium (Fig. 6.14), the effect of radiation softening as a result of the depletion of the matrix solid solution during the departure of dissolved elements to the voids (Table 6.1) may also dominate in comparison with the radiation hardening determined by the barrier mechanism. A decrease in the yield stress as a result of a decrease of the concentration of the atoms of dissolved elements in the solid solution during their segregation on sinks by analogy with equation (6.20) may be represented in the form:

$$\Delta\sigma = M\mu\varepsilon^n \left(c_b - c_m\right). \tag{6.24}$$

where c_s is the total concentration of the atoms of dissolved elements that departed to the sinks during irradiation.

The general equation for the decrease of the yield stress as a result of segregation at discrete barriers and solid-solution softening is written in the form:

$$\Delta\sigma_s = \alpha_s M\mu b(d\rho)^{1/2} - M\mu\varepsilon^n c_s. \tag{6.25}$$

With an increase of temperature, the intensity of radiation hardening, determined by the variation of the hardness of the barriers during segregation, should decrease. This is associated with a general tendency for a decrease in the susceptibility to radiation-stimulated segregation with increasing temperature (sections 2.2, 2.3, 3.7.2, 3.8.1, 3.8.2). The total temperature effect of radiation hardening will, however, be determined also by the effect of the second member of equation (6.25) which shows a tendency for a decrease with increasing temperature. This is indicated more convincingly on the basis of identical analysis for the process of radiation-stimulated equilibrium segregation. As already mentioned, during irradiation this segregation mechanism may operate in the case of interstitial impurities.

For the case of equilibrium segregation of the atoms of dissolved elements on the dislocation loops, equation (6.25) may be converted taking into account equation (6.26) [185]:

$$c_d = c_0 \exp(U / kT), \tag{6.26}$$

where c_d and c_0 is the concentration of dissolved elements on the dislocation and in the solid solution, U is the energy of binding of these elements with the dislocation.

Using equation (6.26), it may be shown that the total concentration of the atoms of dissolved elements on the dislocation loops with density ρ and mean diameter d is:

$$c_s = c_0 \frac{\pi d \rho \exp(U / kT)}{Na} \tag{6.27}$$

here N is the number of atoms in the unit volume, a is the lattice spacing.

Consequently, equation (6.25) has the following form:

$$\Delta\sigma_s = \alpha_s M \mu b (d\rho)^{1/2} - M \mu \varepsilon^n c_0 \frac{\pi d \rho \exp(U / kT)}{Na} \tag{6.28}$$

Equation (6.28) shows that a decrease of $\Delta\sigma_s$ as a result of weakening of the segregation effect of hardening with increasing temperature (decrease in α_s) will be compensated to a certain extent by the hardening of the matrix solid solution as a result of transition of the atoms from segregations on the loops into the solution. It is also possible that the total effect in this case will also be hardening.

The experimental results obtained in the evaluation of the effect of segregation processes on radiation hardening and the hardness of the barriers are very scarce. We know only of several investigations in which this analysis was carried out for neutron-irradiated niobium [20,188] and vanadium [21] with interstitial impurities: oxygen and carbon. In these investigations, the results show an increase of the degree of radiation hardening and the hardness of the barriers – dislocation loops with increasing radiation temperature from 27 to 202 °C [20] (Table 6.2) or annealing temperature after irradiation [21,188]. These effects are linked with the intensification of the processes of formation of segregation of the interstitials on the loops with increasing radiation temperature or subsequent annealing temperature. Unfortunately, as shown below, the reliability of these results is very doubtful.

The interpretation by the author of the experimental data in [20] cannot be regarded as sufficiently justified because of the following reasons. Firstly, when evaluating the hardness of barriers using equation (6.17) the value of $\Delta\sigma$ was accepted for the irradiated specimens with an oxygen concentration of 0.048 at.%, and the electron microscope data for the parameters of the loops were averaged out for the specimens, containing 0.04 and 0.07 at.% of oxygen. Secondly, as indicated by Table 6.2, the given average data for the loop parameters from [20] (the size and density of the loops) change with increasing radiation temperature. Taking into account the previously made considerations, this fact makes it also possible to assume the possibility of changes of the hardness of loops as a result of changes of the mechanism by which the loops are overcome.

The author's interpretation of the experimental data in [21,188] is also ambiguous; in these studies, annealing of V and Nb at a temperature of up ~400 °C after irradiation at 150 °C resulted in further hardening of these materials. Within in the framework of the modelling assumptions made in this book, an increase in the annealing temperature increases the tendency for the weakening of segregations on the sinks and, consequently, hardening of the matrix of the solid solution. For the equilibrium segregation of interstitial impurities, this is directly indicated by equation (6.28). If the second member of the right-hand side of the equation decreases more rapidly than the first number with increasing temperature, hardening will take place. Consequently, since the hardness of the barriers in [21] and [188] was estimated using equations (6.9) and (6.17), the apparent increase of the strength of the barriers could have been detected.

It should be mentioned that the increase of the concentration of impurity and alloying elements together with the intensification of the segregation processes leads to additional radiation hardening as a result of a decrease of the distance between the barriers l (increase of $(d\rho)^{1/2}$). In section 6.2.1.1, we referred to the appropriate experimental data [9,16,17,19–22] and discussed briefly the results obtained for neutron-irradiated aluminium–magnesium alloys [22] (Fig. 6.13).

The main effect of the decrease of the barriers spacing in alloying is associated with an increase in the density of the barriers – dislocation loops, voids, etc – as a result of their heterogeneous nucleation on the atoms of the dissolved elements. It is characteristic that the total power of the barriers, proportional to $(d\rho)^{1/2}$ increases, regardless of the simultaneous increase of the degree of mutual recombination, decreasing the concentration of introduced point defects. A similar conclusion follows obviously from the results of analysis of the experimental data for the electron radiation of aluminium-based alloys, presented in Table 3.2 and in Fig. 3.14.

In steels and alloys, especially at irradiation temperatures of ~$\geq 0.4T_m$, radiation-stimulated precipitates of phases may contribute significantly to or even control the radiation hardening process. The results of analysis of the radiation-stimulated phase transformations and the data on the hardening phases in industrial and advanced metallic materials for nuclear power engineering are presented in Chapter 4 and in section 6.2.1.2–6.2.0.5 for alloys based on aluminium and vanadium, and austenitic and ferritic steels. A characteristic example of the controlling effect of radiation-stimulated precipitation of phases on the hardening of irradiated reduced-activation steels of the Fe–CrWVTa type is presented in Fig. 6.24 [80]. Our estimates of the coefficients of hardness of the hardening phases α in the given steels within the framework of the barrier model in accordance with equation (6.17) are presented in Table 6.2.

In irradiation of materials, radiation hardening is not neccessarily controlled by the evolution of some single type of defects or phase precipitates. At specific temperatures, in particular, in neutron irradiation, radiation hardening is determined by the superimposition contribution of the defects. Appropriate estimates are usually obtained using three types of approximations: additive, root mean square and mixed (equations (6.29)–(6.31) respectively):

$$\Delta\sigma_y = \sum_i \Delta\sigma_i \tag{6.29}$$

$$\Delta\sigma_y = \left(\sum_i \Delta\sigma_i^2\right)^{1/2} \tag{6.30}$$

$$\Delta\sigma_y = \left(\sum_i \Delta\sigma_i\right)^{1/2} + \Delta\sigma_d \tag{6.31}$$

In the additive and root-mean-square (rms) approximations, the summation of the contribution to radiation hardening from different obstacles is carried out independently of the type of these obstacles. In equation (6.31), the hardening resulting from the discrete obstacles is added up within the framework of the rms approximation taking into account the linear contribution from dislocation long-range acting barriers.

Equations (6.29)–(6.31) are used for both the analysis of experimental data and for modelling the variation of the mechanical properties of irradiated materials in order to predict their degradation in re-

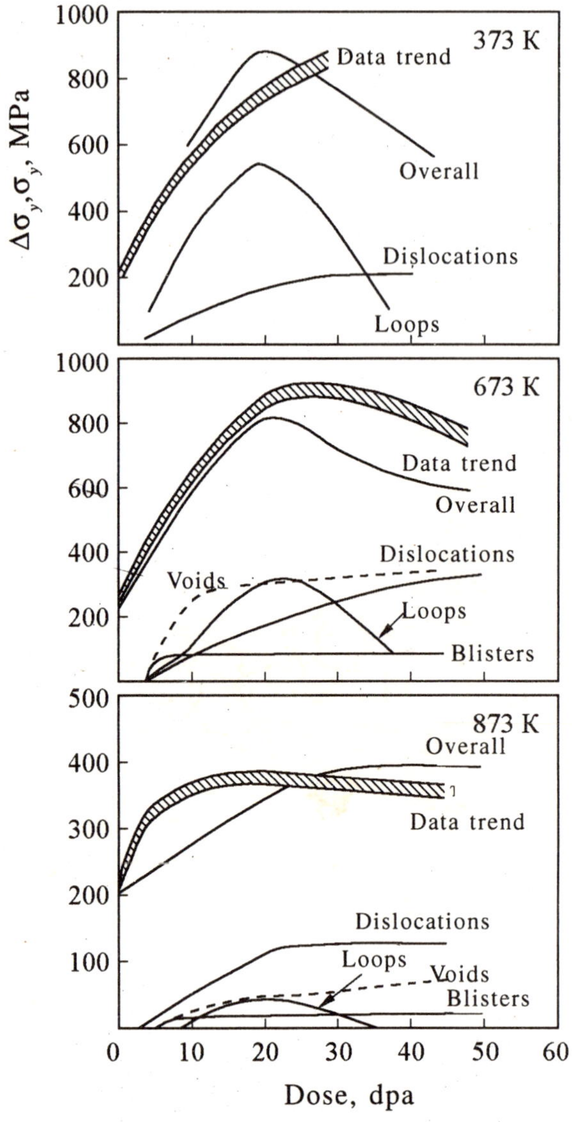

Fig. 6.45. Calculated dependences of variation $\Delta\sigma_y$ for individual defect components of radiation hardening of type 316 steel at temperatures ~100, 400 and 600 °C [110].

lation to temperature and radiation dose (see, for example [95,110]). Figure 6.45 shows the results of calculations of the dose dependence of radiation hardening of austenitic chromium–nickel steels for temperatures of 100, 400 and 600 °C [110]. They were carried out on the basis of analysis of experimental data and appropriate modelling considerations regarding the mechanisms of the evolution of the structure of radiation damage. Numerical estimates were obtained using equations

(6.10), (6.11) and (6.31), and also the values of the strength of barriers for chromium–nickel steels, presented in Table 6.2 with references to [110].

The dependences of $\Delta\sigma_y$ for individual components of hardening and total changes $\Delta\sigma_y$, calculated from equations (6.10), (6.11) and (6.31), correspond to the thin lines in Fig. 6.45. The total changes $\Delta\sigma_y$ are in a satisfactory agreement with the dependences Δs_y, reflecting the tendency for variation of radiation hardening (thick lines in Fig. 6.45), estimated in [110] on the basis of the averaged-out data for different austenitic steels. It should be mentioned that the results of calculations carried out using equations (6.10), (6.11) and (6.31) were obtained without taking radiation-enhanced phase changes into account. Evidently, this is one of the reasons for the different between these values and the predictions obtained on the basis of averaged-out data for radiation hardening.

6.2.2.2. Radiation embrittlement

The main parameters of radiation embrittlement are:

1. A decrease in uniform elongation
2. An increase in the ductile-brittle transition temperature
3. A decrease in fracture toughness.

Low-temperature radiation embrittlement is a direct consequence of radiation hardening. Uniform elongation decreases with increasing radiation dose as a result of a more marked increase of yield strength σ_y in comparison with ultimate strength σ_u (equation (6.3)). The difference $\sigma_u - \sigma_y$, is:

$$\sigma_u - \sigma_y = \int_{\varepsilon_f}^{\varepsilon_r} \frac{d\sigma}{d\varepsilon} d\varepsilon \qquad (6.32)$$

where ε_r and ε_f are the values of the relative elongations, corresponding to the ultimate tensile and yield strength values.

Equation (6.32) shows that a tendency for the decrease in the difference between the ultimate and yield strength with increasing radiation dose is determined by a decrease of work hardening $d\sigma/d\varepsilon$ and uniform elongation $\varepsilon_u = \varepsilon_r - \varepsilon_f$. This conclusion is in complete agreement with the available experimental data (see, for example, Fig. 6.1).

Using equation (6.32), equation (6.3) may be transformed to the following form:

$$\varepsilon_u = \frac{n \int_{\varepsilon_f}^{\varepsilon_r} \frac{d\sigma}{d\varepsilon} d\varepsilon}{\sigma_y} \qquad (6.33)$$

Equation (6.33) shows directly that the decrease of uniform elongation during irradiation is caused by the simultaneous increase of the yield stress and the decrease of work hardening. In the case of zero uniform elongation, there is a complete loss of work hardening [145] which is in complete agreement with equation (6.33).

The mechanism of radiation hardening – the effect of radiation on the yield stress σ_y in equation (6.33) – has been examined sufficiently in the previous section. As regards the effect of radiation on work hardening $d\sigma/d\varepsilon$, at present, it is only possible to provide a qualitative explanation of this effect, taking into account the considerations of work hardening of non-irradiated metals.

The hardening of non-irradiated metallic materials in plastic deformation is caused by the increase of the density of dislocations. The different stages of work hardening are characterised by the multiplication of dislocations, the interaction of dislocations with each other and with other obstacles, and also by the formation of dislocation barriers of different types. Hardening is non-monotonic, and with an increase of the degree of deformation and, correspondingly, the total power of dislocation barriers, the degree of hardening decreases.

The identical tendency is also recorded in the irradiation of metallic materials with increasing dose: concentration and/or the size of radiation defects – dislocation barriers increases, and work hardening decreases. Thus, in both cases, the effect of plastic deformation on the increase of the total power of the dislocation barriers decreases monotonically (i.e. the relative contribution of work hardening to total hardening constantly decreases with increasing radiation dose or deformation). In this case, it is also essential to consider the probability of a situation in which an increase of the degree of defect supersaturation on the decrease of the work hardening is affected by the processes of recovery which decreases the concentration or strength of dislocation barriers [110]. The complete loss of work hardening and, consequently, the zero uniform elongation (equation (6.33)) is associated in all likelihood with the strong barrier pinning of the dislocations and with complete suppression of dislocation slip.

The increase of the temperature of the ductile–brittle transition T_{DB} in irradiation is also the direct consequence of radiation hardening. In the simplified form, this is shown schematically in Fig. 6.15. Within the framework of the analytical approach, the effect of radiation on the ductile–brittle transition temperature may be investigated on the basis of the well-known Cottrell–Petch relationships [189]:

$$\left(\sigma_1 d^{1/2} + K_y\right) K_y = \beta\mu\gamma \tag{6.34}$$

or

$$\sigma_y K_y d^{1/2} = \beta\mu\gamma. \tag{6.35}$$

The equations (6.34) and (6.35) are completely equivalent. In these equations, σ_1 is the resistance to the movement of the dislocations in the slip plane, d is the grain diameter, K_y is the parameter characterising the strength of blocking of the dislocation sources, β is a coefficient which depends on the type of deformation, γ is the effective surface energy of the crack.

The Cottrell–Petch relationships (6.34)–(6.35) were obtained on the basis of the well-known Griffith equation which links the critical size of the crack nucleus $2l$ with the lower stresses σ_f for crack propagation:

$$\sigma_f = \left[\frac{2E\gamma}{\pi l\left(1-\nu^2\right)}\right]^{1/2} \tag{6.36}$$

In equation (6.36), E is Young's modulus, ν is the Poisson coefficient.

In the course of subsequent analysis, we shall restrict ourselves to an example in which the temperature dependence of radiation hardening is fully controlled by the effective stresses of cutting of dislocation loops by mobile dislocations (equation (6.14)). In fact, it should be mentioned that the identical temperature dependence of the flow stresses is also observed in intersection of 'forest' dislocations [190] which, for example, is also characteristic of the deformation of non-irradiated metallic materials.

Taking into account the radiation hardening caused by the dislocation loops (equation (6.14)):

$$\Delta\sigma_y = M\Delta\tau = \sigma_\mu + \frac{M(U_0 - mkT)}{V} \tag{6.37}$$

equation (6.35) assumes the following form:

$$\left[\sigma_y^0 + \sigma_\mu + \frac{M(U_0 - mkT)}{V}\right] K_y d^{1/2} = \beta\mu\gamma \tag{6.38}$$

In equation (6.38), σ_y^0 is the yield stress of the non-irradiated material.

As a result of further transformations, the following equation is obtained for the the ductile–brittle transition temperature T_{DB}:

$$T_{DB} = \frac{\left[\left(\sigma_y^0 + \sigma_\mu\right)V + MU_0\right] - \dfrac{V\beta\mu\gamma}{K_y d^{1/2}}}{KmM} \tag{6.39}$$

Analysis of equation (6.39) shows that the increase in the ductile–brittle transition temperature is proportional to the contribution of the athermal component of the stresses σ_μ and the value of the activation barrier U_0 for the process of cutting of the loops. The additional increase of T_{DB} may be caused by the increase of the parameter of the strength of blocking of dislocation sources K_y as a result of radiation-stimulated segregation. As in the case of non-irradiated metallic materials, a decrease of plasticity in the case of irradiation may be partially compensated by a decrease of the grain size d as a result of the appropriate heat treatment.

We shall investigate the difference compiled from a number of the members of equation (6.39):

$$V\sigma_\mu - V\left[\left(\frac{\beta\mu\gamma}{K_y d^{1/2}}\right) - \sigma_y^0\right] \tag{6.40}$$

In this equation, the activation volume is $V = \bar{l}b^2$, where $\bar{l}$ is the mean distance between the loops, being $\bar{l} = (d\rho)^{-1/2}$.

Quantity $V\sigma_\mu = b^2\bar{l}\sigma_\mu = b^2(d\rho)^{-1/2}\sigma_\mu$ is indepeindent of $\bar{l} = (d\rho)^{-1/2}$, because of equations (6.9) and (6.10). From this equation, because in radiation hardening always $\left[\left(\beta\mu\gamma/K_y d^{1/2}\right)-\sigma_y^0\right]>0$, (Fig. 6.15), the difference (6.40) is maximum at the minimum value $\bar{l}=(d\rho)^{-1/2}$. This leads to a very important conclusion: the value T_{DB} in equation (6.39) is maximum for the radiation temperature which results in the maximum density of dislocation loops ρ or, more accurately, quantity $(d\rho)^{1/2}$. Because of equations (6.9) and (6.33), the same radiation temperature will also be characterised by the maximum radiation hardening and the minimum uniform elongation, respectively.

These relationships are in an excellent agreement with the experimental data. For example, in the neutron-irradiated V–4Ti–4Cr alloy, the shift of T_{DB} increased with an increase of the radiation temperature (Fig. 6.40). This was accompanied by a corresponding increase of radiation hardening (σ_y) and by an increase of the density of clusters of radiation defects [145].

In the austenitic Cr–Ni steels, irradiated in the temperature range of 200–350 °C, examination showed the maximum radiation hardening and the anomalously low uniform elongation $\varepsilon_u < 1\%$ (Fig. 6.35). The maximum radiation hardening corresponds to the temperature of ~330 °C, at which the density of dislocation loops of the interstitial type and the value $(d\rho)^{1/2}$ are also maximum [110,178,191].

An additional contribution to radiation hardening and, consequently, to the increase of T_{DB} and the decrease of ε_u may also be provided (as shown in the previous section) by the increase of the barrier strength of the loops α as a result of the process of radiation-stimulated segregation.

In austenitic steels, the anomalously low temperature embrittlement is evidently not only the consequence of radiation hardening, caused by the interaction of moving dislocations with the dislocation loops of the interstitial type. In section 4.6.3.1, we presented the results of a series of experimental investigations of phase changes in irradiated Fe–Cr–Ni (steel 316 SS) and Fe–Cr–Ni–Mn (steel EP-838) austenitic steels, including the formation in them of modulated structures during the radiation-stimulated spinodal decomposition [120–122,192]. Its mechanism is presented in section 4.3.2, on the basis of the results published in [193].

The theoretical fundamentals of hardening during the spinodal decomposition have been investigated in [194,195]. It was shown [195] that the maximum stresses, acting on the dislocations in the slip plane, are:

$$\sigma_m = 0.8 A \eta Y. \quad (6.41)$$

In equation (6.41), A is the amplitude of the wave of concentration modulation, η is the factor of linear dimensional mismatch, Y is the function of elastic constants.

The authors of [195] also showed the direct correlation of radiation hardening with the formation of the modulated structure in the neutron-irradiated Ni–12 at.% Ti alloy.

The formation of the modulated structures in the 316 SS and EP-838 steels was also reported in [120–122,192] in a relatively wide temperature range (20–450 °C) and a wide range of doses (8×10^{18}–6×10^{21} cm^{-2}) of irradiation with electrons with energies of 1, 2.3 and 21 MeV. The experimental data, presented in [192], show that the radiation hardening of EP-838 steel after irradiation with electrons with an energy of 21 MeV at 450 °C, as in the case of the irradiated Ni–12 at.% Ti alloy [195], is associated with the formation of a modulated structure. Therefore, it may be possible that the process of radiation-stimulated spinodal decomposition also contributes to the anomalous embrittlement of neutron-irradiated austenitic steels (Fig. 6.31).

As in the model of the ductile–brittle transition, assumptions on the propagation of cracks represent the basis of estimates of important service characteristics of structural metallic materials, i.e. their fracture toughness. In linear fracture mechanics, the work required for the separation of the material at the tip of a crack (per unit of formation of the new surface) is [151]:

$$G = \frac{1-\nu^2}{E} K_I^2 \quad (6.42)$$

where K_I is the stress intensity factor. For an isolated crack with length $2l$ at the X axis:

$$K_I = \sigma_0 (\pi l)^{1/2}, \quad (6.43)$$

where σ_0 is the tensile stress, acting on the X axis.

The given crack will propagate only when the values of G_C and K_I in equations (6.42)–(6.43) will exceed, under the effect of the applied stresses, some critical values G_C and K_{IC}:

$$K_{IC}^2 = \frac{E}{1-\nu^2} G_C \tag{6.44}$$

In equation (6.44) K_{IC} is the fracture toughness for the plane-strain state.

In the case of the isolated crack, the fracture stresses are:

$$\sigma_f = K_{IC}(\pi l)^{1/2.} \tag{6.45}$$

In an ideal case, i.e. an absolutely elastic solid, and assuming that $G_C = 2\gamma$, the relationship (6.45) taking into account equation (6.44), is transformed to the Griffith equation (6.36).

Fracture toughness K_{IC} degrades very rapidly during irradiation (see, for example [51,52,96,101,110,196,197]) and for materials of a number of structures and structural elements of atomic reactors it is one of the standard criteria in the evaluation of their efficiency. It should be mentioned that, regardless of the rapid embrittlement as a result of a significant loss of uniform elongation, the steels, irradiated to very high fluences may retain the dimpled mechanism of transcrystalline structure [91,96]. This indicates the ductile mechanism of crack growth as a result of the formation, merger and coalescence of microvoids and the ductile nature of fracture as a whole.

In addition, we shall examine the possible reasons for the decrease of fracture toughness in irradiation. The authors of [96] obtained a relationship between the fracture toughness of the initial material K^0_{IC} and irradiated material (K_{IC}):

$$K_{IC} = K_{IC}^0 \left[\frac{(\sigma_u + \sigma_y)\varepsilon_u}{(\sigma_u^0 + \sigma_y^0)\varepsilon_u^0} \right]^{1/2} \tag{6.46}$$

In this equation, the values of the ultimate strength σ_u and σ_u^0, yield strength σ_y and σ_y^0, and uniform elongation ε_u and ε_u^0 relate to the irradiated and non-irradiated conditions, respectively.

The equation (6.46) does not make it possible to determine, on the basis of analytical examination, the tendency for the variation of fracture toughness (without quantitative estimates) and the factors controlling its evolution during irradiation. At the same time, using equations (6.32) and (6.33) it may be presented in the following form suitable for analysis:

$$K_{IC} = AK_{IC}^{0}\left[n\int_{\varepsilon_f}^{\varepsilon_r}\frac{d\sigma}{d\varepsilon}d\varepsilon\left(2+\frac{\int_{\varepsilon_f}^{\varepsilon_r}\frac{d\sigma}{d\varepsilon}d\varepsilon}{\sigma_y}\right)\right]^{1/2} \tag{6.47}$$

In equation (6.47) $A=\left[\left(\sigma_u^0+\sigma_y^0\right)\varepsilon_u^0\right]^{-1/2}$.

Analysis of equation (6.47) shows that the decrease of fracture toughness in irradiation and also of uniform elongation (equation (6.33)) is caused by the simultaneous increase of the yield strength and the decrease of work hardening. In this case, the functional dependence of the degradation of fracture toughness (primarily on σ_y) is weaker than that in the case of uniform elongation. This result is also in agreement with experiments and appropriate estimates of fracture toughness. The dependence makes is also possible to explain the retention, during irradiation, of the permissible service level of fracture toughness (and plasticity as a whole), even at relatively low values of uniform elongation.

The main structural components, controlling the processes of low-temperature radiation hardening and embrittlement in both pure metals and in alloys are the clusters of defects and dislocation loops of the interstitial type. With increasing irradiation temperature their concentration passes through a maximum and then gradually decreases. The diffusibility of the point defects (especially vacancies) increases and the intensity of the processes of radiation-stimulated phase changes becomes higher. The radiation hardening and embrittlement are controlled on an increasing scale by the process of interaction of dislocations with voids and phase precipitates in both the matrix and at the interfaces of different types. There is a stronger tendency for embrittlement and mixed (transcrystalline and intercrystalline) fracture during the deceleration of groups of dislocations in large carbide and intermetallic phase precipitates in the matrix and at the grain boundaries. Fractured by cleavage, these particles are potential nuclei of cracks in accordance with the Griffith equation (6.36). The probability of occurrence of these processes increases with an increase of the size of the particles $2l$, and this is supported by radiation-enhanced coalescence (section 4.5).

An important factor of the decrease of the temperature of transition from transcrystalline to mixed mechanisms of fracture in irradiated metallic materials in comparison with the non-irradiated materials is the process of radiation-stimulated segregation of impurity, alloying and transmutation elements at the grain boundaries. In addition to phase embrittlement of the boundaries, this process also supports a decrease of their cohesion strength. At the same time, the segregation processes can stimulate a decrease of the cohesion strength at the interface boundaries inside the grains. A similar effect is recorded in, for example, irradiated austenitic chromium–nickel steels, modified with tantalum [97]. The titanium carbide, formed in alloying with titanium, efficiently traps the transmutation helium at the matrix-particle interface boundaries and, consequently, suppress the processes of swelling and high-temperature helium embrittlement. At the same time, concentrating inside the grains at the matrix–TiC boundaries, helium causes the nucleation of cracks and, consequently, stimulates transcrystalline failure.

In section 6.2.1.4 we examined the experimental data on the anomalous embrittlement of austenitic steels, irradiated with relatively high fluences (≥30 dpa). In this case, embrittlement correlates with intensive formation of voids and swelling and is accompanied by softening. At swelling lelevs of ≥10%, the plasticity decreases to almost 0 in both cold-worked and annealed steels (Fig. 6.32). The process of deformation of the materials is accompanied by the formation of local deformation channels (dislocation channelling). The coalescence of the voids with the formation of microcracks in these channels is evidently the most probable mechanism of embrittlement and fracture of these irradiated materials [109].

As mentioned in section 6.2.1.1, one of the main factors of the loss of high-temperature plasticity and intensification of the process of intergranular failure of neutron-irradiated metallic materials is the formation of bubbles of transmutation helium at the grain boundaries. These effects are especially evident in austenitic chromium–nickel steels (section 6.2.1.4) because of the large sections of formation of helium, mainly on nickel, by the reaction (n,α) on both thermal neutrons and on fusion neutrons. In this case, it is characteristic that the plasticity in both short-term tensile and long-term creep and fatigue tests decreases with increasing test temperature [117,198,199].

High-temperature helium embrittlement is detected in austenitic steels at temperatures ≥500 °C, when the diffusion flow of helium from the matrix to the grain boundaries is relatively intensive. On the basis of the results obtained in [198,199], the kinetic stages of the development of helium embrittlement are controlled by the following mechanisms:

1. The formation of helium bubbles at the grain boundaries

2. The stationary growth of the bubbles as a result of the inflow of the atoms of helium and vacancies

3. The unstable growth of the bubbles with the preferential absorption of vacancies followed by the transformation of vacancies into voids.

4. The nucleation and growth of cracks as a result of void coalescence.

As mentioned previously, the alloying of austenitic chromium–nickel steels with titanium is one of the possible methods of suppressing their high-temperature helium embrittlement.

6.2.3. Irradiation creep

In section 6.2.1, we systematised the main experimental relationships of the variation of the mechanical properties in active tensile loading and impact loading of pure metals, solid solutions and a number of industrial steels and alloys.

Unfortunately, the experimental data on the effect of irradiation on creep cannot be processed by similar efficient systematisation: in many cases, they are highly ambiguous or even contradicting. For example, in [200,201] and [202], the authors detected respectively the intensification and deceleration of irradiation creep in aluminium. Other similar examples are also available.

Naturally, it is more difficult to obtain reliable experimental information on irradiation creep than on the mechanical properties of irradiated metallic materials in tensile and impact loading. This is associated mainly with the following reasons:

1. If in the latter case the radiation defects during application of stresses represent only obstacles to mobile dislocations, then the stresses in creep have a constant direct effect on the formation and evolution of the defect phase structure.

2. Creep is one of the mechanisms of dimensional instability of irradiated materials which also may be determined by the processes of void formation, radiation growth, and radiation-stimulated evolution of phase composition. To a certain degree, all these processes are inter-linked and influence the analysis of experimental data. There are many cases in which it is not possible to separate these processes and determine their mutual effect.

3. In contrast to the static tests of the mechanical properties, in analysis of the creep results, a significant role is played by the kinetic

factor, i.e. it is necessary to obtain complete information on the special features of the time dependence of the process.

Because of these and a number of other reasons, experimental investigations of the irradiation creep of pure metals and modelling systems are very limited and, basically, are concerned with important materials of the active zones of nuclear reactors: austenitic (a large majority of investigations) and ferritic steels. Therefore, in subsequent considerations, attention will be given only to a number of general established relationships and special features and we shall consider only several experimental examples without differentiation, in contrast to section 6.2.1, with respect to the classes of metals and alloys.

In section 6.2.3.2 we shall pay attention to the considerations of microscopic mechanisms of irradiation creep.

6.2.3.1. Experimental data

An important basis for examining irradiation creep were the experiments with the relaxation of elastic stresses under irradiation which is in fact creep in the range of low strains and stresses not exceeding the elasticity limit. Stress relaxation, accelerated by irradiation, was detected both in a number of metals and simulation systems and in structural industrial materials [203,204]. As an example, Fig. 6.46 shows the comparative course of stress relaxation in 99.99% silver at a temperature of 90 °C without irradiation and with irradiation with electrons with an energy of 2.3 MeV [205]. Evaluation of the activation energy of radiation (temperature range 30–90 °C) and thermal stress relaxation (temperature range 90–130 °C) gave the values of 0.2 and 0.48 eV.

Figure 6.47 shows the typical thermal creep curve. In many cases, identical dependences are also recorded in irradiation creep. However, as shown in a number investigations, irradiation may modify the nature of the dependences of the given curve, especially in section I and III of creep stages.

Figure 6.48 shows the results of investigations of the effect of electron radiation on creep of aluminium with a purity of 99.999% and the Al–0.82 at.%Zn alloy at a temperature of 40 °C [201]. In both pure aluminium and in the alloy, irradiation is accompanied by a linear increase of the creep rate to the maximum value with the tendency for a decrease of the rate with further irradiation.

The presented dependences greatly differ from the curves of stage I of thermal creep which are characterised by the monotonically decreasing form with subsequent exit to the stationary creep rate (Fig. 6.47).

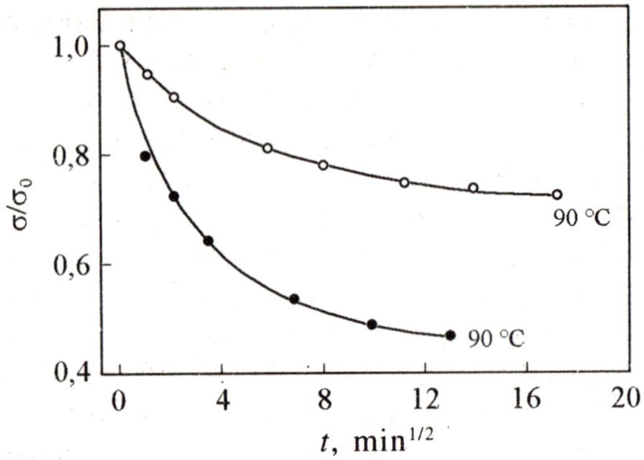

Fig. 6.46. Kinetics of stress relaxation in silver at 90 °C without irradiation (○) and in electron irradiation (●) [205].

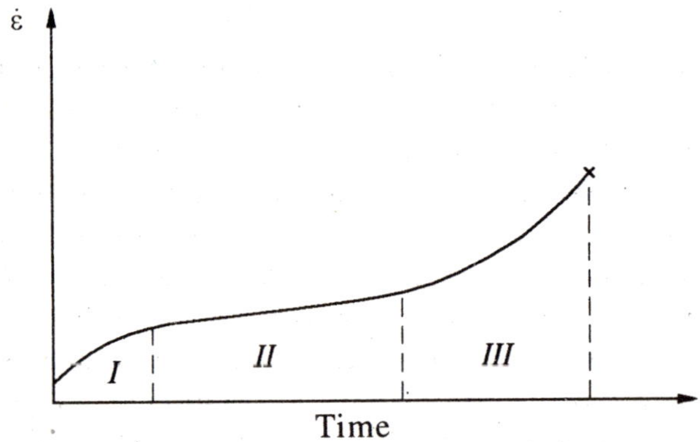

Fig. 6.47. Typical thermal creep curve: I – stage of primary or unsteady (transitional creep); II – stage of secondary or steady-state creep; III – stage of accelerated creep; × – long-term strength.

The dependences shown in Fig 6.48 can be interpreted more efficiently on the basis of the considerations regarding the mechanism of radiation-enhanced diffusion (section 4.2.1) and diffusion parameters of interstitials and vacancies (section 2.2).

The rate of thermal creep at elevated temperatures is controlled by dislocation climb as a result of diffusion flow of the vacancies. In irradiation, both the vacancies and interstitials diffuse to the dislocations. In pure aluminium, the diffusibility of interstitials is considerably higher than the mobility of the vacancies (their migration energies are 0.1 and 0.57 eV [206]). Consequently, up to a specific period of time, dislocation climb and the rate of irradiation creep

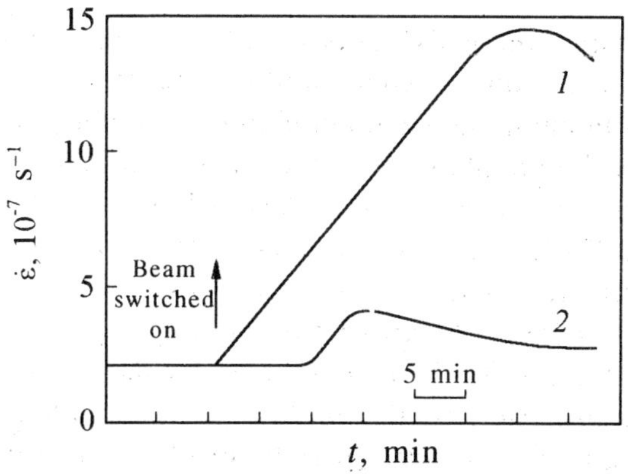

Fig. 6.48. Variation of the creep rate of pure aluminium (1) and Al–0.82 at.% Zn alloy (2) in irradiation with electrons with an energy of 2 MeV at 40 °C [201].

are controlled only by the diffusion flow of interstitials. With time, the diffusion flows of the vacancies and, consequently, the recombination rate of the vacancies with interstitials on the dislocations increases. This decreases the creep rate, as also shown in Fig 6.48.

The exit to the quasi-stationary creep rate within the framework of the given considerations should be determined, in all likelihood, by the time to establishment of the dynamically equilibrium concentrations of vacancies and interstitials for the combined mechanism of radiation-enhanced diffusion in accordance with equation (4.5):

$$\tau = \frac{1}{Z_v^d \rho_d D_v} \tag{6.48}$$

The equation shows directly that the exit to quasi-equilibrium is controlled, in addition to dislocation density ρ_d, by the diffusibility of the vacancies.

As shown in section 2.2, the effective diffusibility of interstitials in alloys may greatly decrease as a result of the formation of low-mobility mixed dumbbell configurations. For the Al–0.82 at.%Zn alloys, this has been directly established in the examination of the formation and growth of the interstitial dislocation loops in the radiation process in a high-voltage electron microscope [18] (section 3.7.2).

The decrease of the diffusibility of the interstitials in the Al–0.82 at.%Zn alloys results, in contrast to pure aluminium, in the formation of an incubation period on the creep curve (Fig. 6.48): in this case, the interstitials require a certain time for arrival on the dislocations in comparison with pure aluminium where process takes place almost instantaneously. This result is in good correlation with the previously mentioned experimental data on the nucleation and growth of loops in aluminium and aluminium–zinc alloys under radiation: in comparison with pure aluminium, the nucleation time of the interstitial dislocation loops in the Al–0.82 at.%Zn alloy greatly increases (Fig. 3.13 and 3.14, Table 3.2). The results of these investigations show that the diffusibility of the vacancies in the Al–0.82 at.%Zn alloy increases (the constant of exit of the vacancies to the loops $K_V \sim D_v$ in Fig. 3.14 and Table 3.2 increases). Similar data were also obtained in [207] in examination of the mobility of quenched vacancies in aluminium and aluminium–zinc alloys by electron–positron annihilation.

The decrease of the diffusibility of the interstitials with a simultaneous increase of the mobility of the vacancies in the Al–0.82 at.%Zn alloy results in the intensification of the mutual recombination of point defects in the matrix. Consequently, this process and the increase of the diffusion coefficient of the vacancies D_v result in a decrease in the intensity of the excess flow of the interstitials to the dislocation is $Z_i^d D_i c_i - Z_v^d D_v c_v$ and this also decreases the maximum on the creep curve and displacement of this maximum to lower radiation doses. The increase of D_v in equation (6.14) also decreases the time of exit in the alloy to quasi-equilibrium. This is also in agreement with the experimental dependences presented in Fig. 6.48.

In an analytical review in [208], the problem of the transition period of creep of intermetallic materials was investigated with a special reference to the conditions of cyclic pulsed radiation, characteristic of the operation of fusion reactors. As in the previously examined example of aluminium, the initial period of irradiation is characterised by a large increase of the creep rate followed by a decrease of the rate. With further cycles of pulsed irradiation, creep jumps with subsequent relaxation are reproduced. This is characterised by a tendency for a monotonic decrease of the rate of thermal creep after completion of each consecutive cycle of irradiation. This is associated with the radiation-induced changes of the defect–phase structure. In [208], the authors proposed methods of suppressing the negative effect of pulsed irradiation on the creep rate, based on the selection of the appropriate material for the first wall and the optimum parameters of pulsed radiation (temperature, duration and frequency of pulses). It should also be mentioned that the examined effects were detected mainly in special experiments with the radiation of materials with charged particles using high-sensitivity testing equipment.

As in thermal creep, the phase transformations may also change the deformation rate and modify the nature of the curve of irradiation creep during continuous irradiation. Depending on the dimensional relationships between the matrix and the phase precipitates, deformation during creep may be accelerated, slow down or may even be negative. This effect was detected in, for example, [209–211] in neutron irradiation of a number of austenitic steel. Similar deformation dependences in irradiation creep of cold-worked steels 316, AMCR 0033 and PCA are shown in Fig. 6.49 [209]. As assumed in [210], the negative dimensional effects in AMCR 0033 steels are associated with the formation of α-ferrite. At the same time, the acceleration of deformation in the conditions of irradiation creep of austenitic steels is usually associated with the precipitation of TiC and/or TiO phases [210,211].

Of the greatest interest for reactor materials science are the investigations of experimental relationships in the second and third stages of irradiation creep. As already mentioned, the experimental data on irradiation creep are highly ambiguous and cannot be efficiently systematised, even for the same class of materials. Nevertheless, a number of a relatively general relationships have been obtained.

The experimental results show that in comparison with thermal creep, the rate of steady irradiation creep depends far less extensively on temperature. In Fig. 6.50–6.51 this is indicated on the basis of experiments with proton and deutron irradiation of austenitic steels 321 and 316L, respectively. In a number of experiments on cold-worked austenitic steels, examination showed the almost complete

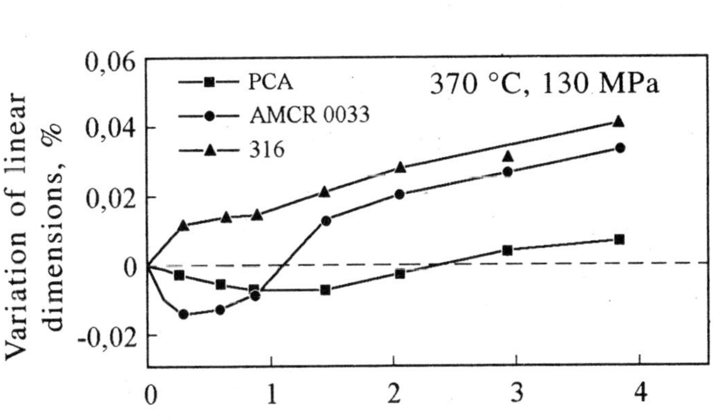

Fig. 6.49. Variation of the linear size of cold-worked 316, AMCR 0033 and PCA steels in creep testing under uniaxial tensile loading in relation to neutron fluence. [209].

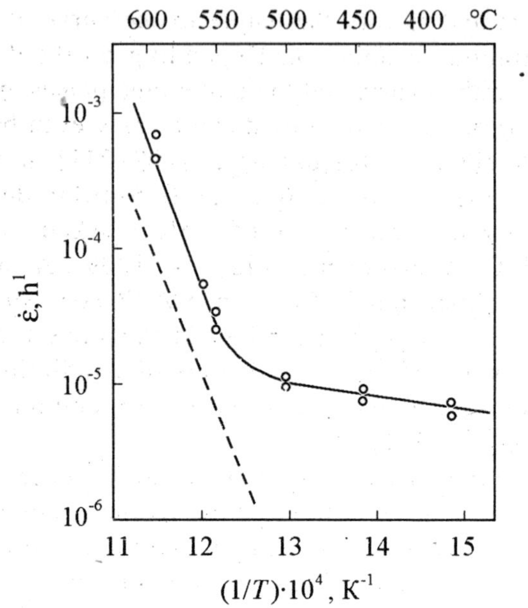

Fig. 6.50. Temperature dependences of the rate of thermal creep (dotted line) and irradiation creep (solid line) in proton irradiation of cold-worked 321 steel [212].

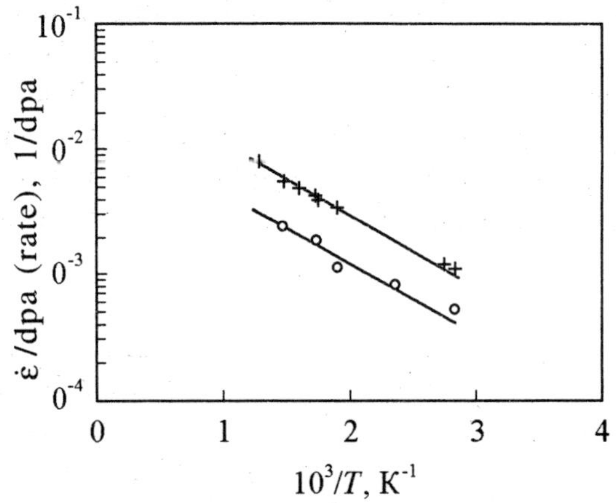

Fig. 6.51. Temperature dependences of steady-state creep in irradiation with deutrons with an energy of 19 MeV of annealed (○) and cold-worked (+) steel 316L in the temperature range 80–400 °C, activation energy of irradiation creep Q = 0.11 eV [213].

independence of the rate of irradiation creep on temperature (Fig. 6.52, 6.53). In this case, the cold working of austenitic steels usually increases the rate of irradiation creep in the steady stage [213,215, 216] (for example, Fig. 6.51).

In analysis of the kinetics of the process of irradiation creep of

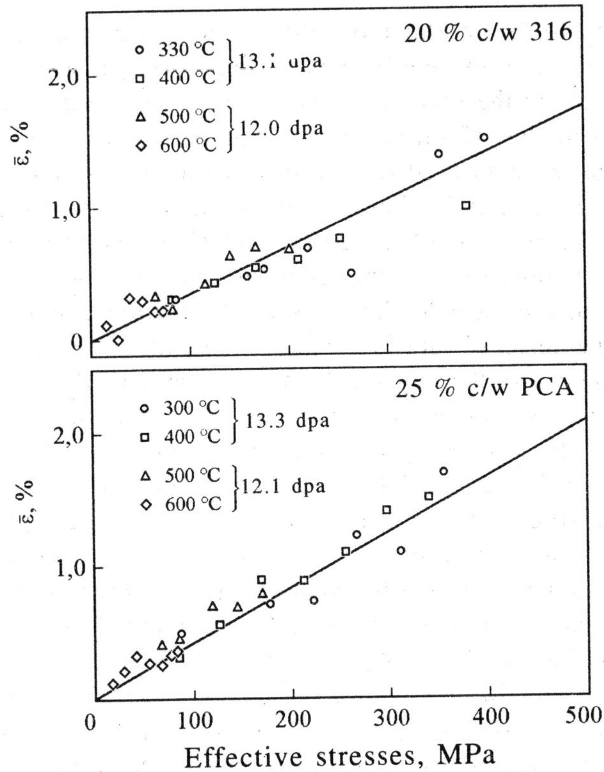

Fig. 6.52. Experimental data for irradiation creep in cold-worked 316 and PCA steels irradiated with neutrons in the temperature range 300–600 °C [214].

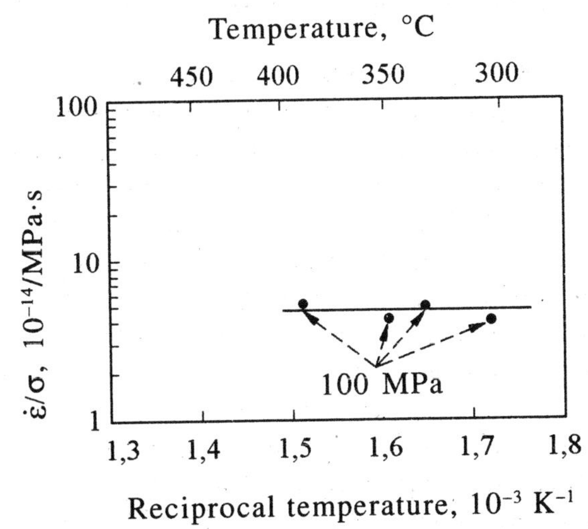

Fig. 6.53. Stress-normalised rate of irradiation creep of cold-worked (20%) steel AMCR-0033 in relation to temperature in neutron irradiation [209].

austenitic steels on the basis of generalisation of the set of the experimental data in neutron irradiation in [91] it was concluded that the creep rate in the second – steady-state stage – may remain constant until the process of radiation swelling starts to take place. In the given stage, especially at relatively low temperatures and/or fluences, the linear dependence of creep rate $\dot{\varepsilon}$ on stress σ [91,211,215] (for example, Fig. 6.52) is observed in a relatively large number of cases. At the same time, in a number investigations [216–218] it was shown that with an increase of the stresses the proportionality between $\dot{\varepsilon}$ and σ in the austenitic steels is not retained and the creep rate increases (the exponent n in the relationships $\dot{\varepsilon} \sim \sigma^n$ becomes $n > 1$). An identical tendency is also observed with approach to temperatures at which the role of thermal creep becomes more important [91,215,218]. In this case, an increase of temperature reduces the stresses of transition from the linear to nonlinear dependence [217].

The nonlinear dependence between $\dot{\varepsilon}$ and σ and the exponent $n > 1$, as shown by a detailed analysis of the experimental data [219], is highly characteristic of the irradiated austenitic steels. Similar data were published in [91,220]. For steel HT-9, in particular, irradiated at a temperature of 400 °C, exponent n in the relationships $\dot{\varepsilon} \sim \sigma^n$ is equal to approximately 2 [220]. At the same time, for this steel, irradiated at a temperature of 330 °C, the linear relationship between $\dot{\varepsilon}$ and σ is retained at stresses up to 500 MPa [211]. It should also be noted that the linear relationship between $\dot{\varepsilon}$ and σ was obtained in the previously cited study [220] for the ferritic Fe–9CrMo steels, although the radiation parameters of the steel are identical to HT-9 steel.

Figure 6.54 shows the dependence of the creep rate in the steady-state on stresses in the V–4Cr–4Ti alloy, irradiated in a BR-10 reactor at a temperature of ~445 °C [221]. In this case, the linear relationship between $\dot{\varepsilon}$ and σ is retained in the entire range of applied stresses, but at stresses of ~110–120 MPa the creep rate rapidly increases.

As already mentioned, the dimensional changes, accompanying phase transformations, may modify the form of the dependence of deformation on both stresses and temperature. An identical role is also played by the stress-accelerated swelling process. Similar examples of the results of [222] for the reduced-activation Fe–12Cr–19MnW steel, irradiated in an MOTA-2B reactor, are presented in Fig 6.55.

With an increase of temperature and fluence, the increasingly important role in the process of deformation of steels irradiated under stress is played by the mechanism of the effect of swelling on creep. The experimental data obtained in reactor experiments on pressu-

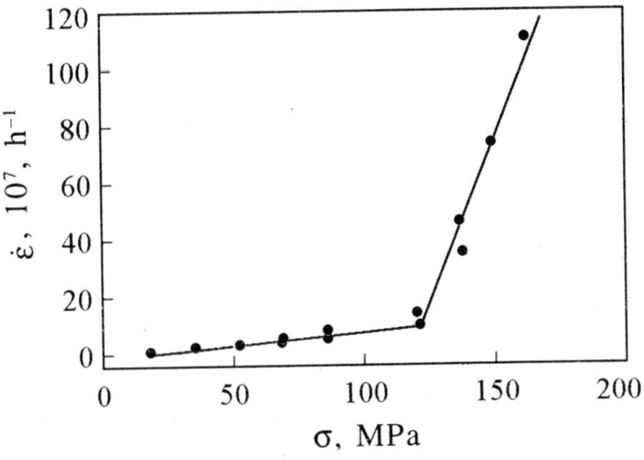

Fig. 6.54. Dependence of the rate of irradiation creep on stresses in V–4Cr–4Ti alloy [221].

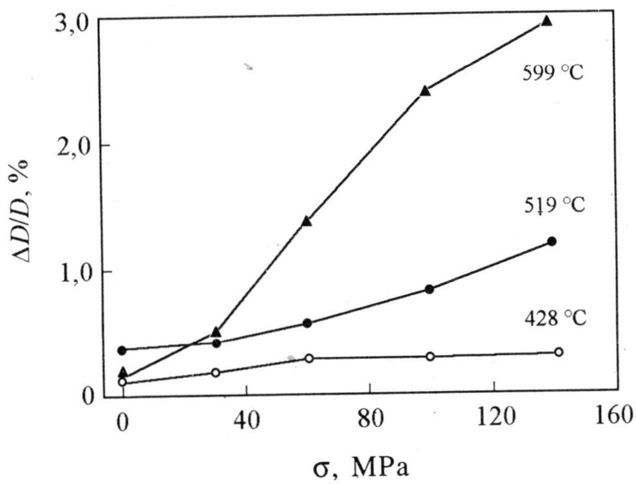

Fig. 6.55. Changes of the diameter of pressurised pipes in steel Fe–12Cr–19MnW after irradiation in MOTA-2B reactor [222].

rised tubes under pressure are usually analysed on the basis of semi-empirical relationships linking the effective creep rate and the swelling rate of the type [91,215]:

$$B = \frac{\bar{\dot{\varepsilon}}}{\bar{\sigma}} = B_0 + D\dot{S} \tag{6.49}$$

$$A = \frac{\bar{\varepsilon}}{\bar{\sigma}} = B_0 + \alpha \dot{S} \quad (6.50)$$

In equations (6.49) and (6.50), parameter B_0 represents the creep rate independent of swelling, and the components $D\dot{S}$ and $\alpha\dot{S}$ determine the fraction of the creep rate determined by the swelling process. Correlation coefficient D reflects the relationship of creep and swelling at any specific moment of time, whereas coefficient α is the quantity D averaged out with respect to fluence.

Figures 6.56 and 6.57 show the most convincing examples of the direct relationship between creep and swelling. In the majority of similar experiments, the correlation between creep and swelling was not so unambiguous but was sufficiently distinctive. This correlation depends on composition, preliminary treatment and irradiation parameters [91]. This is also confirmed by the experiments in which it was shown that creep and swelling depend in the same manner on these parameters [91,224]. The average-out estimates of the parameters B_0 and D for austenitic steels, according to the data published in [91,224–227], give the values of the orders of B_0=(1–2)× 10^{-6} MPa^{-1}× dpa^{-1} and D = ~0.6×10^{-2} MPa^{-1}. Identical estimates for the ferritic steels according to the experimental results obtained in [91, 220, 228] for B_0 and D give values B_0 = 0.5×10^{-6} MPa^{-1}× dpa^{-1} and D = $(0.7–1.0)\times10^{-2}$ MPa^{-1} are very similar to those for austenitic steels.

With an increase of the rate and level of swelling in reactor experiments on pressurised pipes, a reversed relationship is detected

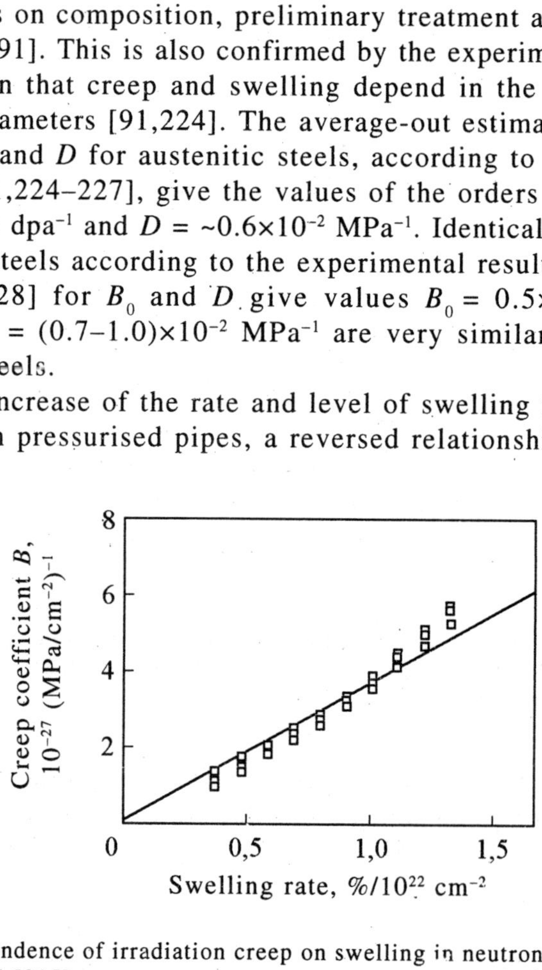

Fig. 6.56. Dependence of irradiation creep on swelling in neutron-irradiated steel 304L at 415 °C [215].

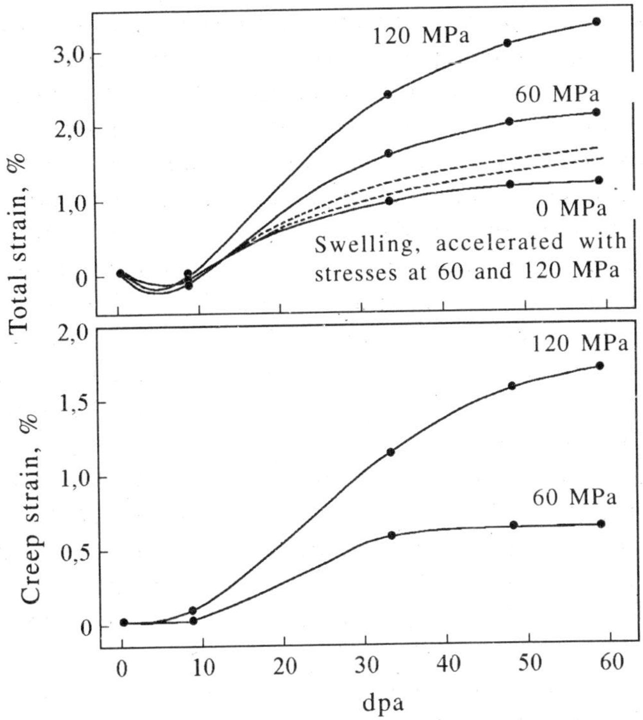

Fig. 6.57. Swelling and creep strain of pipes under pressure made of DIN 1.4864 stainless steel irradiated with neutrons at 420 °C [223].

between creep and swelling starting from a specific moment when the rate of volume swelling approaches the limiting value (~1%/dpa): the creep rate starts to decrease with a tendency to saturation or even complete 'disappearance' [91,229,230]. The interpretation of these effects in the above studies is based on the analysis of evolution of the degree of anisotropy of the structure of radiation defects in relation to applied stresses in the process of neutron irradiation.

In conclusion of the analysis of experiments of this type, following [91], it is necessary to mention that the conditions of examination of the dimensional instability of the materials in reactor experiments on pressurised pipes are not adequate to the actual conditions existing in fuel tubes of heat-generating elements. In modelling experiments on pressurised pipes, the high level of the stresses stimulates the creep process preceding swelling.

In contrast to this, the initial level of stresses in real fuel pins is very low and gradually increases with the release of transmutation gases from fuel. In these conditions, on the other hand, the swelling process precedes the creep process with a constant increase of its role in the development of dimensional instability. In this case, since the pressure and, consequently, the stresses in the materials of the pressurised tubes of the heat-generating elements continuously increase, the development of dimensional instability in this case, in contrast to the experiments on fuel pin cladding, is admittedly of the non-stationary nature. Consequently, in the combination of these factors, both the kinetics of the process and the creep and swelling parameters for real and modelling conditions (and also their dependence on the irradiation parameters) differ. Therefore, it is evident that the development of matters of obtaining reliable information on the dimensional instability of materials of structural elements of reactors in the real conditions on the basis of modelling experiments is a very urgent task.

As established in a large number of experimental investigations (for example [91,211,228,231]) the ferritic steels of the HT-8 type and their reduced-activation analogues are characterised by a considerably higher resistance to irradiation creep in comparison with austenitic steels of the type 316 and PCA up to temperatures of ~500–550 °C. As reported in [231–233], the best set of the properties, such as the resistance to creep and swelling, and also the minimum shift of the temperature of the ductile–brittle transition is recorded in the case of reduced-activation martensitic steels of the type Fe–(8–9)Cr–2WVTa. As regards the set of these properties, the steels are considerably superior to austenitic steels which should result in their use instead of austenitic steels in the active zones of fast reactors [91], and they should also be regarded as promising materials for the first wall and the blanket of the fusion reactor. However, in the latter case, as already mentioned on the basis of investigations [75] (see section 6.2.1.3), the application of these materials depends greatly on solving the problem of suppressing their low-temperature embrittlement at high concentrations of transmutation helium.

6.2.3.2. The mechanism of irradiation creep

To a first approximation, the microscopic mechanisms of irradiation creep, proposed for the interpretation of experimental data can, in our view, be subdivided into four basic groups.

The mechanisms of the first group are based on the assumption that under the effect of external applied non-hydrostatic stresses, the crystal and defective structure of metallic materials becomes anisotropic in relation to the flows of point defects. Similar mechanisms include the mechanism of preferential nucleation of the dislocation loops of the interstitial and vacancy type in planes of the crystal lattice orthogonal and parallel in relation to the direction of applied stresses [234,235] (the mechanism of nucleation of dislocation loops, initiated by stresses). The same group includes the mechanism of preferential absorption of point defects by dislocation loops and edge dislocations with the Burgers vectors in a specific orientation in relation to the direction of applied stresses [236–239]. In this case, the interstitials and vacancies are preferentially absorbed by dislocation loops and edge dislocations with the Burgers vectors parallel and perpendicular, respectively, in relation to the axis of applies stresses, determining the creep deformation as a result of the process of growth of the loops or dislocation climb (the mechanism of stress-initiated preferential absorption of point defects by dislocations arranged in a specific orientation).

The same group also includes a mechanism in which irradiation creep is the result of the process of dislocation slip in the field of obstacles which are overcome by dislocation climb as a result of preferential absorption of point defects by dislocations with the Burgers vectors with a favourable orientation in relation to the axis of applied stresses [239] (as in the previous case). This mechanism may be defined as the slip mechanism, initiated by dislocation climb under the effect of applied stresses.

The mechanisms of the second group include mechanisms based on the assumption that the diffusion flows of point defects on dislocations sinks become anisotropic in relation to the axis of applied stresses [240–242] (the mechanism of anisotropic diffusion of point defects to the dislocations).

The mechanisms of the third group are based on assumptions on the misbalance of the flows of interstitials and vacancies to the dislocations and voids [243]. In this case, the interstitials are preferentially absorbed by dislocations, and the vacancies by the voids. Under the effect of the stresses, the dislocations climb in the field of obstacles and overcome these obstacles as a result of excess flows of interstitials (the creep process), and the vacancies, merging into voids, initiate the swelling process [243] (the slip–climb mechanism, determined by the misbalance of the flows of point defects).

The fourth group includes the combined mechanism of climb + slip–climb, proposed in [243–245]. The creep rate in this case is represented in the form of the total effect of the climb mechanisms, initiated by stresses [236–239] and slip–climb, initiated by the misbalance of the flows of the interstitials and vacancies to the dislocations and voids [243].

A number of the previously mentioned mechanisms of irradiation creep is presented in a series of analytical review studies [246–255]. We shall examine these mechanisms of irradiation creep in greater detail.

I. The stress-initiated nucleation of dislocation loops

This mechanism was proposed for the first time in [234] for the formation of dislocation loops in atomic displacement cascades. In [235], the following equation was derived for the concentration of nuclei of dislocation loops in one of the three orthogonal planes of the lattice, normal to the axis of applied stress σ:

$$\rho_l = \rho_l \frac{\left[\exp\left(\frac{\sigma b^3 n}{kT}\right) - 1\right]}{\left[\exp\left(\frac{\sigma b^3 n}{kT}\right) + 2\right]} \tag{6.51}$$

where ρ_l is the total concentration of the loops, n is the number of interstitials in the nucleus of dislocation loops.

The rate of irradiation creep for the given mechanisms, neglecting the thermodynamic concentration of vacancies (in contrast to [235]) is determined by the equation:

$$\dot{\varepsilon} = f\rho_l \left(Z_i^l D_i c_i - Z_v^l D_v c_v\right) \tag{6.52}$$

where Z_i^l and Z_v^l are the parameters of the efficiency of absorption of interstitials and vacancies by dislocation loops.

As reported in [247], this mechanism may operate in metallic materials with a low initial density of dislocations and only at relatively low radiation doses.

II. The mechanism of stress-initiated preferential absorption of point defects by dislocations oriented in a specific direction

At present, this mechanism is used most frequently for the analysis of experimental data for irradiation creep. Therefore, we shall examine in detail its physical fundamentals, taking into account the fact that a number of these fundamentals are also used in other models of irradiation creep, and also in analysis of swelling processes.

As in the description of the processes of nucleation and growth of dislocation loops and voids in the absence of external applied stresses σ (section 3.7–3.8), the creep mechanisms are based on following the equations for the difference of the flows of the point defects to sinks of different type. At $\sigma = 0$, the resultant flow of the interstitials to the dislocations is:

$$Z_i^d D_i c_i - Z_v^d D_v c_v \tag{6.53}$$

The parameters $Z_{i,v}^d$ in equations (6.52) and (6.53) take into account the force interaction of point defects as centres of dilation distortions in the lattice, with the hydrostatic field of stresses of the dislocations (equation (3.14)) and are equal to [245,256]:

$$Z_{i,v}^d = \frac{2\pi}{\ln\left(\dfrac{2R}{L_{i,v}^d}\right)} \tag{6.54}$$

where

$$L_{i,v}^d = \frac{\mu b(1+\nu)\Delta V_{i,v}}{3(1-\nu)\pi kT} \tag{6.55}$$

In the equations (6.54) and (6.55) $R = (\pi\rho_d)^{-1/2}$, $\Delta V_{i,v}$ are the dilation volumes of the interstitials and vacancies, ν is the Poisson coefficient.

In the conditions of nonuniform applied stresses, the energy of interaction of the point defects with the dislocations $E_{i,v}^{(\beta)}$ has the form [245,246]:

$$E_{i,v}^{(\beta)} = E_{i,v}^0 + \delta E_{i,v}^\beta = \frac{\mu b}{\pi}\left[\frac{(1+\nu)\Delta V_{i,v}}{3(1-\nu)} + \varepsilon V_{i,v} A_{i,v}^{(\beta)}\right]\frac{\sin\theta}{r} \tag{6.56}$$

and the equation for $Z_{i,v}^{(\beta)}$ taking into account

$$\delta L_{i,v}^{(\beta)} = \frac{\mu b \varepsilon V_{i,v} A_{i,v}^{(\beta)}}{\pi kT} \tag{6.57}$$

is expressed in the form [236,245,256]:

$$Z_{i,v}^{(\beta)} = Z_{i,v}^{d}\left\{1+\frac{3(1-\nu)Z_{i,v}^{d}\varepsilon V_{i,v}A_{i,v}^{(\beta)}}{2\pi(1+\nu)\Delta V_{i,v}}\right\} \tag{6.58}$$

where $\varepsilon = \sigma/E$ is the elastic deformation, carried out by applied stresses σ, E is the Young modulus, $V_{i,v}$ are the volumes of appropriate point defects.

As shown in [235–238, 257], the quantities $A_{i,v}^{(\beta)}$ and, consequently, $Z_{i,v}^{(\beta)}$, depend on the orientation of the Burgers effect of the dislocations in relation to the axis of applied stresses. For cubic crystals, it is usually assumed that there are dislocations with the Burgers vector parallel to the three cubic axes (β = 1,2,3) with the density $\rho_d^{(1)} = \rho_d^{(2)} = \rho_d^{(3)} = \frac{1}{3}\rho_d$. Assuming that the dislocations of the type β=1 have the Burgers vector parallel to the axis of the applied stresses, and the dislocations of the type β = 2, 3 have a perpendicular vector, in accordance with the calculations in [257] the following equations are used to estimate the quantities $A_{i,v}^{(1)}$ and $A_{i,v}^{(2,3)}$ in equation (6.58):

$$A_{i,v}^{(1)} = -\left\{\frac{(1-\nu)(1+\nu)\Delta K_{i,v}}{3(1-\nu)K+(1+\nu)\Delta K_{i,v}}+\frac{5(1+\nu)(2-\nu)\Delta\mu_{i,v}}{15(1-\nu)\mu+2(4-5\nu)\Delta\mu_{i,v}}\right\} \tag{6.59}$$

$$A_{i,v}^{(2)} = A_{i,v}^{(3)} = -\left\{\frac{(1-2\nu)(1+\nu)\Delta K_{i,v}}{3(1-\nu)K+(1+\nu)\Delta K_{i,v}}-\frac{5(1+\nu)^2\Delta\mu_{i,v}}{15(1-\nu)\mu+2(4-5\nu)\Delta\mu_{i,v}}\right\} \tag{6.60}$$

In the equations (6.59) and (6.60):

$$\Delta K_{i,v} = \Delta K_{i,v}^{*} - K \tag{6.61}$$

$$\Delta\mu_{i,v} = \Delta\mu_{i,v}^{*} - \mu \tag{6.62}$$

where K and μ are the moduli of bulk and shear elasticity of the matrix, and $K_{i,v}^{*}$ and $\mu_{i,v}^{*}$ are the effective moduli of bulk and shear elasticity of the interstitials and vacancies.

As already mentioned, in accordance with the concept proposed in [236–238], the edge dislocations of the type $\beta=1$ preferentially absorb interstitials, and those of type $\beta=2$, 3 absorb the vacancies thus controlling at the same time the creep process both as a result of the growth of the dislocation loops and as a result of climb of the network dislocations. Within the framework of the analysis of the flows of the point defects on dislocations of the type $\beta=1$ with $Z_{i,v}^{(1)}$ and the dislocations of the type $\beta = 2,3$ with $Z_{i,v}^{(2)} = Z_{i,v}^{(3)}$, the following equation was obtained in [245,257] for the rate of irradiation creep:

$$\dot{\varepsilon} = \frac{2}{9}\rho\left[\Delta Z_i D_i c_i - \Delta Z_v D_v c_v\right] \tag{6.63}$$

where $\rho = \rho_l$ is the total concentration of the dislocation loops of the interstitial type, or $\rho = \rho_d$ is the total density of the network dislocations.

In accordance with the considerations [236,245,256], described previously, the values of $\Delta Z_{i,v}$ in equation (6.63) are:

$$\Delta Z_{i,v} = Z_{i,v}^{(1)} - Z_{i,v}^{(2)} = \frac{3(1-\nu)\left(Z_{i,v}^{d}\right)^2 \varepsilon V_{i,v} a_{i,v}}{2\pi(1+\nu)\Delta V_{i,v}} \tag{6.64}$$

where

$$a_{i,v} = A_{i,v}^{(1)} - A_{i,v}^{(2)} = -\frac{15(1+\nu)\Delta\mu_{i,v}}{15(1-\nu)\mu + 2(4-5\nu)\Delta\mu_{i,v}} \tag{6.65}$$

On the basis of the estimates [237,258], the values of μ_i^* and μ_v^* in equation (6.62) are: $\mu_i^* = 0$ and $\mu_v^* = \mu$. Consequently, in equation (6.65) for $a_{i,v}$ $\Delta\mu_i = -\mu$ and $\Delta\mu_v = 0$. Consequently, $\Delta Z_v = 0$ and equation (6.63) for $\dot{\varepsilon}$ is simplified:

$$\dot{\varepsilon} = \frac{2}{9}\rho \Delta Z_i D_i c_i \tag{6.66}$$

i.e. irradiation creep within the framework of this mechanism is fully controlled by the flows of interstitials to the dislocations. At the same time, it should be stressed that the values of the dynamically equilibrium concentrations of the point defects $c_{i,v}$ also depend on the diffusion parameters of the vacancies.

In sections 3.5 and 4.2.1, on the basis of solving the system of equations (3.3)–(3.4), we presented equations for the concentration of interstitials c_i and vacancies c_v, and also the coefficient of self-diffusion $D_{i,v}c_{i,v}$ included in equations (6.63) and (6.66). The solutions for $D_{i,v}c_{i,v}$ on the basis of the system of equations (3.3)–(3 4) were obtained in section 4.2.1, including for stationary mechanisms of the introduction and annealing of radiation defects: the mechanism of mutual reombination of vacancies and interstitials, the linear mechanism (annealing of point effects on dislocations sinks), and the combined mechanism. The latter mechanism also takes into account the mutual recombination of point defects and the annealing of these defects on the dislocations and, evidently, it is most efficient for describing irradiation creep in a wide range of test temperatures (depending on temperature, the relative contribution of both mentioned mechanisms changes). The expressions for $D_{i,v}c_{i,v}$ in accordance with the equations (3.16)–(3.21) have the following form:

$$D_{i,v}c_{i,v} = \frac{F(\eta)G}{S_{i,v}} \tag{6.67}$$

The function $F(\eta)$ is determined by equation (3.19):

$$F(\eta) = \frac{2}{\eta}\left[(1+\eta)^{1/2} - 1\right] \tag{6.68}$$

in which the recombination parameter η is:

$$\eta = \frac{4RG}{D_v D_i S_v S_i} \tag{6.69}$$

In equations (6.67) and (6.69), G is the rate of introduction of free point defects, $S_{i,v} = Z^d_{i,v}\rho_d$ is the power of dislocations sinks, R is the mutual recombination constant.

As shown in section 3.5, function $F(\eta)$ determines the fraction of point defects, absorbed by the dislocation sinks in relation to the total rate of introduction G. At $\eta \to 0$, the value of $F(\eta)$ tends to unity, i.e. all introduced point defects travel to the sinks and their mutual recombination is not important. If the value of η is high, then $F(\eta) = 2/\eta^{1/2} \to 0$ and mutual recombination plays a controlling role.

On the basis of transformations using equations (6.64), (6.65) and (6.67), equation (6.66) may be represented in the following form:

$$\dot{\varepsilon} = A\frac{Z_i^d \sigma}{\nu_i \mu} F(\eta) G \tag{6.70}$$

In equation (6.70), $\nu_i = \Delta V_i / V_i$ is the relative dilation volume of the interstitial, $\mu = E/2\,(1+\nu)$ is the shear modulus. Our numerical estimates of constant A using equation (6.64) for $\Delta Z_{i,v}$ give the value A=0.15. In [238] and [259], the values of A in equation (6.70) were estimated as equal to 0.12 (randomly distributed dislocations) and 0.31 for an isotropic crystal with the concentration of favourably oriented dislocations ρ_A=1/3, respectively.

It is evident that for the creep process and, in particular, irradiation creep, and the increase dislocation density is a significant feature. Consequently, in a number of theoretical studies (for example, [236, 258]), the product $D_i c_i$ in equation (6.66) is estimated on the basis of the linear mechanism of annealing of point defects, i.e. the process of mutual recombination of point defects is completely ignored. The product $D_i c_i$ for the stationary linear mechanism is:

$$D_i c_i = \frac{G}{Z_i^d \rho_d} \tag{6.71}$$

and the equation for $\dot{\varepsilon}$ has the following form:

$$\dot{\varepsilon} = A\frac{Z_i^d \sigma}{\nu_i \mu} G \tag{6.72}$$

Equation (6.72) shows that, in this case, the rate of irradiation creep does not depend on temperature and dislocation density and is directly proportional to radiation intensity.

As already mentioned in the previous section, in a number of reactor experiments on austenitic chromium–nickel steels, examination showed the almost complete independence of the creep rate on temperature (Fig 6.52 and 6.53). However, the results presented in Fig. 6.50 and 6.51 indicated the completely determined temperature dependence of the rate of irradiation creep.

It should also be mentioned that the linear dependence of the creep rate on radiation intensity in the experimental investigations is not always detected. This is especially characteristic of relatively lower radiation temperatures. Similar results were obtained in, for example [216,217], in which the creep rate of cold-worked austenitic steels at radiation temperatures of 280 °C [216] and 420 °C [217] changed in proportion to the square root of the radiation intensity ($\dot{\varepsilon} \sim G^{1/2}$).

In contrast to equation (6.72), equation (6.70), expressed by means of function $F(\eta)$, makes it possible, as already mentioned, to interpret more efficiently experimental data in a relatively wide temperature range, including the relationship describes previously. Depending on the irradiation parameters, dislocation density and the diffusion characteristics of point defects, the creep rate of $\dot{\varepsilon} \sim \rho_d D_i c_i = F(\eta)G$ is in this case an ambiguous and non-monotonic function of both temperature and the rate of introduction of points defects. In this case, the activation energy of irradiation creep E_c may change from $E_c = E_v^m / 2$ at low temperatures to $E_c = 0$ at high temperatures, and its dependence on the rate of introduction of point defects was varied in the range from $G^{1/2}$ to G (section 3.5 and 4.2.1). The absence of temperature dependence of $\dot{\varepsilon}$ and its proportionality G is characteristic of high dislocation densities and practically corresponds to the linear mechanism of annealing of radiation defects (equation (6.72)).

The relationships described above illustrate clearly the results of numerical calculations of the temperature dependences $F(\eta)\ G = \rho_d D_i c_i \sim \dot{\varepsilon}$ (Fig. 6.58) and $F(\eta)G$ on the rate of introduction of point defects (Fig 6.59) for 316 chromium–nickel austenitic stainless steel. In reality, they represent the stress-normalised rates of irradiation creep ($\dot{\varepsilon}/\sigma = CF(\eta)G$), where the constant C, according to equation (6.70) is $C = A(Z_i^d / \nu_i \mu)$. Figure 6.58 also shows the temperature dependences of $\rho_d D_v c_v^e$ (straight lines) proportional to the rate of thermal creep. Calculations were carried out using the parameters for point defects from [260–262]. Their values are presented in Table 6.3.

On the basis of numerical calculations, presented in Fig. 6.58 and 6.59 it is convenient, in our view, to also mention the following relationships:

1. The rate of irradiation creep increases with increasing dislocation density, which is in complete agreement with the experimental data on the irradiation of cold-worked and annealed steels (Fig 6.50 and 6.51), examined in the previous section. In fact, this conclusion does not follow from equation (6.72). In this case, the value of $\dot{\varepsilon}$ is independent of dislocation density.

2. With an increase of dislocation density, the difference in the rate of irradiation creep at the same irradiation intensity but at different

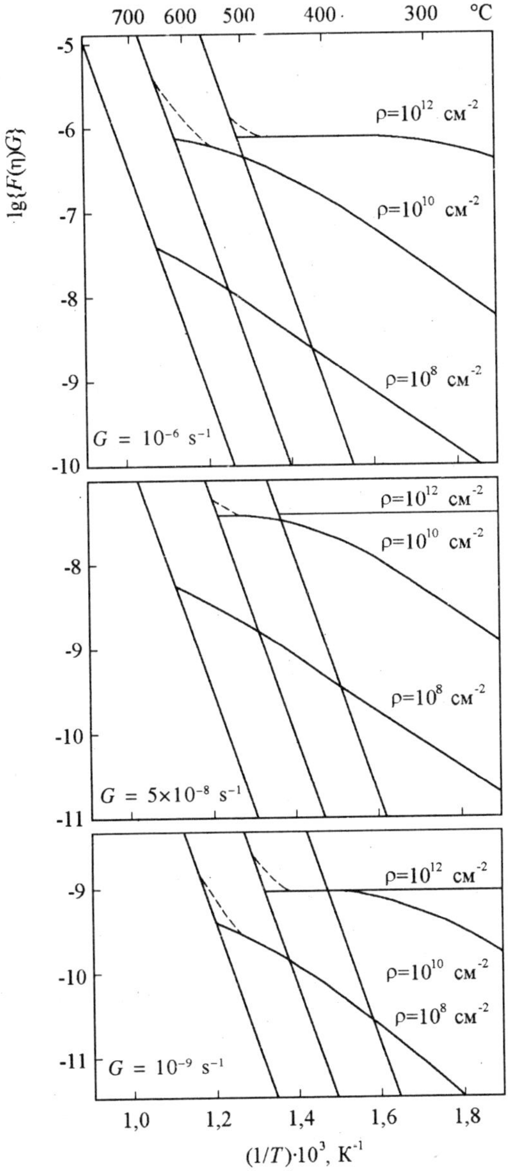

Fig. 6.58. Calculated temperature dependences of parameter $F(\eta)G \sim \dot{\varepsilon}$ for 316-type austenitic steel.

temperatures decreases.

3. With an increase of radiation intensity, the difference in the rate of irradiation creep at the same radiation intensity but at the different temperatures increases.

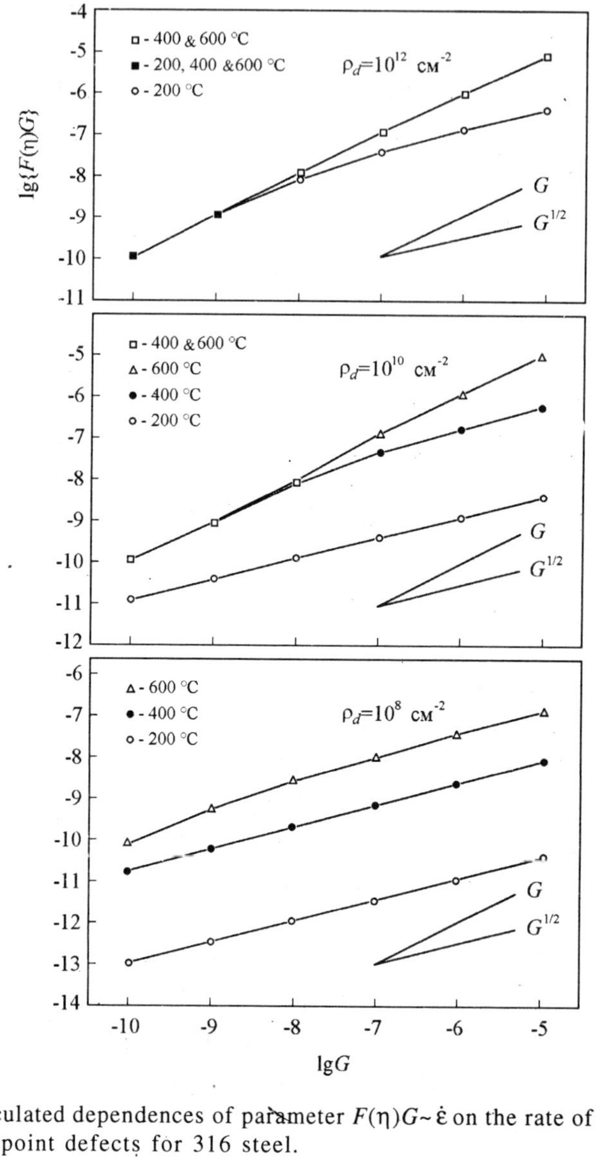

Fig. 6.59. Calculated dependences of parameter $F(\eta)G \sim \dot{\varepsilon}$ on the rate of introduction of irradiation point defects for 316 steel.

We have intentionally paid special attention to the role of diffusion mechanisms, dislocation density and the rate of introduction of point defects in the interpretation of the results obtained for irradiation creep because these relationships are also typical of other mechanism of irradiation creep which will be examined below. To a certain degree, the presented calculated data show that the ambiguity of the interpretation of the results for irradiation creep may be associated,

Table 6.3. Values of parameters of point defects for numerical calculations of the value of $F(\eta)G \sim \dot{\varepsilon}$ in 316 austenitic chromium–nickel steel

Parameter	Symbol	Value
Vacancy formation energy, eV	E_v^F	1.5
Vacancy migration energy, eV	E_v^m	1.4
Interstitials migration energy, eV	E_i^m	0.9
Pre–exponential factor of the coefficient of diffusion of vacancies, m^2s^{-1}	D_v^0	5×10^{-5}
Pre–exponential factor of the coefficient of diffusion of interstitials, m^2s^{-1}	D_i^0	4×10^{-6}

in particular, with the effect of a number of insufficiently reliable or difficult to control parameters, such as the rate of introduction of freely migrating point defects and their diffusion properties (Chapters 1 and 2) and also the dislocation density. These parameters actually determine the diffusion mechanism in irradiation creep.

Returning to the examined creep mechanism, it is important to note the following:

1. When examining the mechanism of growth of the dislocation loops, the role of the mechanism may be regarded only in a very limited time period, especially for annealed metals and FCC alloys. The main reason here is associated with the fact that it applied stresses stimulate the transformation of loops with stacking faults into perfect loops in their interaction with acting dislocations, with subsequent accelerated growth of the loops and the formation of a dislocation network.

2. A number of experimental data obtained for the stage of steady irradiation creep (the climb of the network dislocations) may be interpreted qualitatively with high accuracy of the basis of the given mechanism (see dependence of stresses, radiation intensity and temperature). At the same time, as shown by analysis [241,245,250,251, 263–265], the theoretical estimates within the framework of this mechanism gives lower rates of irradiation creep those detected in the experiments and, consequently, in many cases this mechanism can not be treated as the controlling one. However, as already mentioned, this discrepancy may also be associated with the insufficiently correct selection of certain calculation parameters, such as the rate of introduction of free-migrating point defects and their diffusion properties, and also dislocation density.

III. The slip mechanism initiated by dislocation climb under the effect of applied stresses

This mechanism was proposed in [239]. In fact, it is a modified slip-climb mechanism, proposed in [243], and examined in greater detail in [244,245]. In the model [239], as in [243–245], attention is given to a network of dislocations where the distance between the dislocations is assumed to be $\lambda = (\pi\rho)^{-1/2}$. Under the effect of applied stresses σ, the appropriate components of the network are bent between the obstacles, separated by the distance λ, until their linear tension becomes equal to σ. In this case, creep is limited by the magnitude of deflection of the given segment $\varepsilon=\sigma/E$.

However, if the climb process is possible, the pinned dislocation overcomes barriers and is again pinned and bent to the next climb.

The creep rate in the framework of these model is:

$$\dot{\varepsilon} = \varepsilon(\pi\rho)^{1/2} v_c \tag{6.73}$$

In [243–245], the rate of dislocation climb v_c was calculated on the basis of the excess flow of interstitials on dislocations $Z_i^d D_i c_i$-$Z_v^d D_v c_v$ on the condition that the given flow is compensated by the preferential absorption of vacancies by the voids: $Z^v_v D_v c_v - Z^v_i D_i c_i$. This model, based on the misbalance of the flows of interstitials to the dislocations and voids, will be examined later.

In contrast to this approach, in the calculation of the climb rate of the dislocations v_c in [239], the authors used the same physical principles and equations examined in detail in the description of the previous mechanism of irradiation creep. Here, as previously, the climb of dislocations, overcoming the obstacles, takes place as a result of the preferential absorption of interstitials by the dislocations, with the Burgers vector paralled to the axis of applied stresses.

In the form identical with (6.66), the equation for the rate of irradiation creep within the framework of this mechanism has the following form [239]:

$$\dot{\varepsilon} = \frac{4\varepsilon}{9b}(\pi\rho)^{1/2} \Delta Z_i D_i c_i \tag{6.74}$$

Taking into account equation (6.67) for $D_i c_i$ to the function $F(\eta)$ and the estimate ΔZ_i using equation (6.64), equation (6.74) can be transformed into the following form:

$$\dot{\varepsilon} = A\frac{Z_i^d \sigma^2}{v_i \mu^2 b} F(\eta) G \rho_d^{-1/2} \quad (6.75)$$

where $A = 0.20$.

The main differences between the previous and this mechanism may be summed up as follows:

1. In the first case, deformation during creep takes place as a result of dislocation climb, and in the latter case as a result of dislocation slip.
2. In the first case, the creep rate depends in a linear manner on the stresses, whereas in the latter case, this dependence is quadratic.
3. The dependence of the creep rate on dislocation density in the first case is stronger.

In section (6.2.3.1) we presented a number of experimental data for steels characterised by the non-linear dependence of $\dot{\varepsilon}$ on σ on in particular, the dependence $\dot{\varepsilon} \sim \sigma^2$. The results of these investigations may be qualitatively interpreted in favour of the realistic possibility of operation of the previously examined mechanism of irradiation creep. However, the quantitative estimates of the creep rates in the framework of the given mechanism are almost completely comparable with the appropriate estimates, typical of the mechanism of stress-initiated dislocation climb in irradiated materials, examined previously, i.e. below the values observed in the experiments. However, also in this case, the difference between the experimental data and theoretical estimates may be associated with the incorrect selection of a number of calculation parameters.

IV. The mechanism of anisotropic diffusion of point defects to dislocations

As mentioned previously, the investigated mechanism of irradiation creep, based on the preferential absorption of interstitials by dislocations with the Burgers vector parallel to the axis of applied stresses, give too low values of the creep rate in comparison with the experimental values. The model of anisotropic diffusion of point defects under the effect of applied stresses, proposed in [240–242] makes it possible to improve the quantitative agreement of the theory with experiments because in estimates it yields considerably higher values of the rate of irradiation creep.

The model is based on the assumptions according to which the external force field changes the anisotropy of the energy of point defects in the saddle configuration, i.e. the probability of directions of the elementary jumps in relation to the axis of applied stresses.

The expression for the energy causing the diffusion drift of rate point defects to sinks has the following form [266]:

$$E(r)=\varepsilon_{ij}(r)P_{ij}-\frac{1}{2}\alpha_{ijkl}(r)\varepsilon_{ij}(r)\varepsilon_{kl}(r) \qquad (6.76)$$

In this equation, $\varepsilon_{ij}(r)$ is the total strain in the vicinity of the defect with the coordinate r, including the strain caused by elastic defects and external stress, P_{ij} is the dipole-force tensor, and α_{ij} is the elastic polarizability of the point defects. Tensor P_{ij} characterises the magnitude and the direction of three orthogonal dipole forces, causing elastic atomic displacements in the vicinity of the point defects, examined within the framework of the linear theory of elasticity.

Analysis shows [242] that the influence on creep strain of the effect associated with the anisotropy of the point defect in the saddle configuration (the first term in equation (6.75)) is considerably stronger than the effect of the mechanism of absorption of point effects by favourably oriented dislocations (the second term in (6.75)). It is also characteristic that for the examined mechanism of anisotropic diffusion, the dislocation–defect interaction depends far more markedly on the orientation of the dislocation line in relation to the axis of applied stresses than on the direction of its Burgers vector.

In [242] it was shown that the equation for irradiation creep, controlled by the mechanism of anisotropic diffusion, may be presented in the form identical with equation (6.72). Numerical estimates of coefficient A in equation (6.72), carried out for copper and iron [243, 267] give however values for the mechanism of anisotropic diffusion which are more than an order of magnitude higher than the values calculated for the mechanism of irradiation creep which is determined by the preferential absorption of interstitials by the dislocations with the Burgers vectors favourably oriented in relation to the axis of applied stresses. Identical estimates for zirconium also give similar results [268].

V. The slip–climb mechanism, determined by the misbalance of the flows of point defects

In principle, all the previously examined mechanism of irradiation creep in the pure form may occur separately or together only under the condition that voids do not yet form. The realisation of these mechanisms requires not too high temperatures and relatively low fluences and stresses. With an increase of these parameters, the probability of the nucleation and growth of the voids increases and, consequently, the probability of transition to the mechanism, based on the misbalance of the flows of point defects to the dislocations and voids, increases.

As already mentioned, this creep mechanism was proposed for the first time in [243] and examined in greater detail in [244,245]. Its main principles have already been explained in this book when examining the slip mechanism, induced by the climb of dislocations under the effect of applied stresses. The creep rate within the framework of this mechanism in the general form is described by equation (6.73) in which the mean rate of climb of the dislocation v_c is determined by the expression:

$$v_c = \frac{1}{b}\left(Z_i^d D_i c_i - Z_v^d D_v c_v\right) \tag{6.77}$$

In this case, the concentration of point defects $c_{i,v}$, included in equation (6.76), must be determined on the basis of solving a system of equations, which takes into account the growth of voids during radiation:

$$\frac{dc_i}{dt} = G - Rc_v c_i - Z_i^d \rho_d D_i c_i - 4\pi\rho_v r_v Z_i^v D_i c_i \tag{6.78}$$

$$\frac{dc_v}{dt} = G - Rc_v c_i - Z_v^d \rho_d D_v c_v - 4\pi\rho_v r_v Z_v^v D_v c_v \tag{6.79}$$

$$\frac{dr_v}{dt} = \frac{1}{r_v}\left(Z_v^v D_v c_v - Z_i^v D_i c_i\right) \tag{6.80}$$

In the equation (6.78)–(6.80), r_v is the mean radius of the voids, $Z_{i,v}^v$ is the efficiency of absorption of point defects by voids with mean radius r_v, ρ_v is the concentration of the voids.

On the basis of the highly realistic assumption that the following equality is fulfilled in the steady state creep stage:

$$\frac{dc_i}{dt} = \frac{dc_v}{dt} \neq 0 \quad (6.81) \tag{6.81}$$

or

$$\frac{dc_i}{dt} = \frac{dc_v}{dt} = 0 \tag{6.82}$$

after algebraic transformations of the system (6.78)–(6.79):

$$\rho_d \left(Z_i^d D_i c_i - Z_v^d D_v c_v\right) = 4\pi r_v \rho_v \left(Z_v^v D_v c_v - Z_i^v D_i c_i\right) \tag{6.83}$$

The left-hand part of equality (6.83) is the total excess flow of interstitials to all dislocations ρ_d, and the right-hand part is the identical flow of vacancies to all voids ρ_v, equal to the swelling rate $\dot{S}$:

$$\dot{S} = 4\pi r_v \frac{dr_v}{dt} \rho_v = 4\pi r_v \rho_v \left(Z_v^v D_v c_v - Z_i^v D_i c_i\right) \tag{6.84}$$

Comparison of equations (6.77), (6.83) and (6.84) gives

$$v_c = \frac{1}{b}\left(Z_i^d D_i c_i - Z_v D_v c_v\right) = \frac{\dot{S}}{b\rho_d} \tag{6.85}$$

Substitution of (6.85) into equation (6.73) gives directly the relationship between the creep and swelling rates:

$$\dot{\varepsilon} = \varepsilon \left(\frac{\pi}{\rho_d b^2}\right)^{1/2} \dot{S} \tag{6.86}$$

Taking into account that $\varepsilon = \sigma/E$ and $E = 2(1+\nu)\mu$, from equation (6.86)

$$\dot{\varepsilon} = A \frac{\sigma}{\mu b} \dot{S} \rho_d^{-1/2} \tag{6.87}$$

where $A = 0.7$.

VI. The combined mechanism of irradiation creep

This mechanism has been proposed in [244,245]. The creep rate is presented as the result of the total effect of the mechanisms of climb, induced by stresses [226–239] and slip–climb, determined by the misbalance of the flows of interstitials to dislocations and voids [243–245].

The proposed partial variant of the combined mechanism [244, 245] may be presented in a more general form, adding the creep rate for any previously examined mechanisms, associated and not associated with swelling. On the basis of this approach, the summation of, for example, equations (6.76) and (6.87) for the appropriate slip–climb mechanisms gives:

$$\dot{\varepsilon}_\Sigma = \frac{\sigma}{\mu b \rho_d^{1/2}} \left[A_1 \frac{Z_i^d \sigma}{v_i \mu} F(\eta) G + A_2 \dot{S} \right] \qquad (6.88)$$

The constants A_1 and A_2 from equations (6.75) and (6.87) are equal to respectively 0.2 and 0.7.

The same procedure can be used for presenting other combinations of the rates of irradiation creep not associated $\dot{\varepsilon}(\sigma)$ and associated $\dot{\varepsilon}(\sigma, \dot{S})$ with swelling, i.e.:

$$\dot{\varepsilon}_\Sigma = \dot{\varepsilon}(\sigma) + \dot{\varepsilon}(\sigma, \dot{S}) \qquad (6.89)$$

The mechanisms not associated with void formation and swelling may evidently determine the rate of irradiation creep at relatively low temperatures. With increasing temperature and radiation dose, the probability of the processes of void formation and swelling increases. In this case, depending on temperature and fluence, these mechanisms may act together to different degrees simultaneously with a tendency according to which the mechanism, associated with swelling at large fluences, becomes dominant.

In conclusion, it should be mentioned that the examined combined mechanism is in fact a microscopic analogue of semi-empirical relationships (6.49) and (6.50), linking the effective rate of creep and swelling. As showed in section (6.2.3.1), these relationships are used usually in the analysis of data for irradiation creep in reactor experiments on pressurised tubes. A number of results of similar analysis for irradiated austenitic and ferritic steels have also been presented in section 6.2.3.1.

6.3 SWELLING

In section 3.8, we examined in considerable detail the current general microscopic considerations regarding the processes of voids formation leading to an increase of the volume of irradiated metallic materials, i.e. swelling:

$$S = \Delta V(\%) = \left(\frac{4}{3}\right)\pi \sum_i \rho_i r_i 100\% \qquad (6.90)$$

In this equation, ρ_i and r_i are the concentration and the radius of the i-population of the voids.

We analysed the possible mechanisms of the effect of dissolved elements on the nucleation and growth of voids in pure metals and solid solutions. In particular, special attention was given to investigating modelling representations regarding the effect of the diffusibility of point defects during their interaction with dissolved elements on the process of mutual recombination of point defects, vacancy supersaturation and the rate of nucleation of voids. Analysis was also carried out of the effect of segregations of dissolved elements on the dislocations and voids on the rate of nucleation and growth. However, in this case, the swelling resistance of specific steels and alloys was not investigated in relation to the chemical and phase composition.

In the present section, special attention is given to the analysis of the swelling of metallic materials for which the process is important in practice. They include, primarily, the materials of active zones of nuclear reactors, and also promising materials of the first wall and the blanket of fusion reactors, investigated at the present time. Altogether, these are austenitic and ferritic steels and also vanadium-based alloys. Special attention is given to the analysis of the swelling of austenitic chromium–nickel steels which are used widely in nuclear power engineering and for which this problem is very important in comparison with other grades of these materials.

Prior to examining specific materials, attention will be given to the most general macroscopic relationships of the process of swelling of metallic materials.

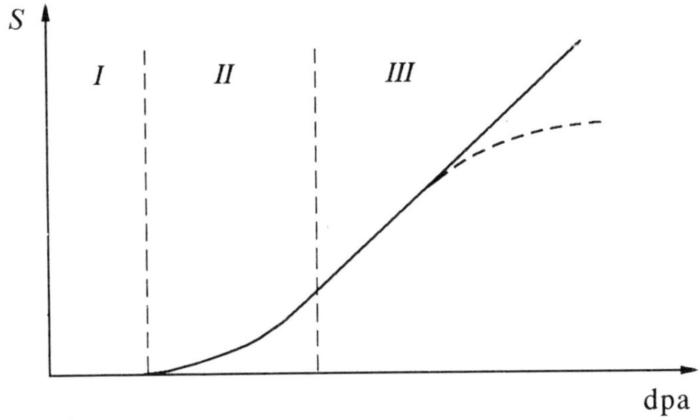

Fig. 6.60. Schematic dependence of swelling of metals and alloys under isothermal irradiation [269].

Figure 6.60 shows the schematic dependence of the swelling of metals and alloys during isothermal irradiation [269]. Usually, the swelling curve contains the following characteristic time periods: I – incubation period, II – transition period, III – the period of stationary swelling. The incubation period is usually associated with the formation of a network of dislocations [270, 271] and of void nuclei of the critical size [271]. The transition period is characterised by the rapid growth of avoids with further transition to the stationery swelling rate. The result of a large number of investigations showed that both the incubation and transition periods are highly sensitive to the chemical and phase composition, cold deformation, temperature and other irradiation parameters. The dose at which the swelling is ~0.1% is referred to as the conventional swelling threshold [269]. In the steady stage III, the swelling rate in relation to the material and other parameters may vary in the range (0.1–1)%/dpa [91,269].

In some cases, after the period of stationary swelling there is a tendency for saturation (the dotted linc in Fig. 6.60) [91,269] (see section 3.82) and even to dissipation of voids [272]. However, as reported in [91,269], these effects are not typical of irradiated metals and alloys.

The temperature dependence of swelling is usually almost the bell-shaped (Fig. 6.61). In section 3.5, attention was already paid the characteristic temperature ranges of radiation damage during the introduction and thermally activated annealing of non-correlated point defects. Therefore, the temperature range I corresponds to the temperatures in which the process of mutual recombination is dominant. With increasing temperature, the processes of departure of defects to the

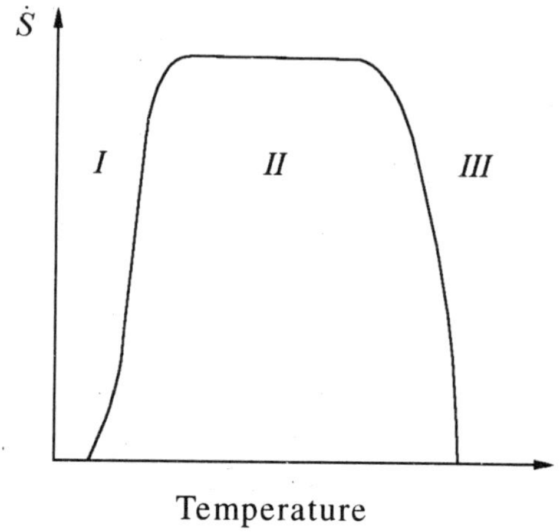

Fig. 6.61. Schematic temperature dependence of swelling rate [91,269].

sinks and, in particular, voids are intensified. The rate of growth of the voids increases, and the swelling rate approaches a stationary value (regime II). In the temperature range corresponding to regime III, the swelling rate decreases as the result of the intensification of the process of emission of vacancies from the voids.

The processes of swelling in the above-mentioned metallic materials will now be investigated.

6.3.1. Austenitic chromium–nickel steels

One of the main directions of development of austenitic chromium-nickel steels with increasing resistance to swelling and creep is based on the modification of the composition of the standard industrial steel AISI 316 (its chemical composition is presented in Table 4.4 in section 4.6.3.1).

In appropriate programmes in the USA and Japan, the modified steel 316 and a number of other steels are regarded mainly as the material of active zones for fast reactors with liquid metal heat carriers [273,274]. The approximate service parameters of the materials being developed should be on the level: the working temperature range 370-700 °C, fluence ~3×10^{22} cm^{-2} ($E > 0.1$ MeV) [270, 274]. This fluence corresponds to ~150 dpa. The 316 steel used currently in reactor construction does not satisfy these requirements (neither with respect to swelling nor long-term strength).

The development of steels, corresponding to the above-mentioned parameters, is essential for increasing the degree of burn-out of nuclear fuel and the service period of structures and structural elements of the active zone and, consequently, increasing the reliability and economic efficiency of reactors.

The modification of the composition of AISI 316 steel aimed at increasing its swelling resistance and long-term strength is based at present on the development of optimum compositions by means of combined alloying with elements such as titanium, carbon, phosphorus, silicon and boron with a slight increase of the nickel concentration and a decrease of the chromium concentration combined with cold deformation. One of the approximate compositions of the modified steels in the USA (USPCA) and Japan (JPCA) is presented in Table 4.4 (section 4.6.3).

In previous investigations of the compositions of the typical AISI 316 steel modified either with phosphorus and titanium [273] or phosphorus [275] it was shown that the phosphorus in concentrations of ≥0.02%, both individually [275] and combined with titanium [273], greatly suppresses swelling in both cold-worked (10%) [275] and in annealed materials [273]. For example, in annealed 316 steel, modified with 0.2% Ti, after irradiation at a temperature of 540 °C with a fluence corresponding to ~76 dpa, swelling decreased several times with an increase of the phosphorus concentration from 0.035% to 0.08%, and at a phosphorus concentration 0.08% it was ~6% [273]. In cold-worked 316 steel with 0.02% P, not containing titanium, after irradiation at a temperature of 550 °C with a neutron fluence of 9.4× 10^{22} cm^{-2} ($E > 0.1$ MeV) the swelling was <5% [275].

A very suitable example of the strong effect on swelling of complex alloying of austenitic steels of the type D9 (Fe–0.04C–13.7Ni–16.2Cr-2.5Mo–2.0Mn–1.5Si–0.1Zr) with titanium and phosphorus in combination with cold deformation (20%) is presented in Fig. 6.62 [276].

Further analysis of this problem shows that advances have been made recently in the development of chromium–nickel steels with increased resistance to swelling both on the basis of development of different modifications of PCA-type steels [91,274,277] and on the basis of Fe–16Cr–17Ni [278,279] and Fe–15Cr–(20–25)Ni [91, 274,280] systems with different content of titanium, silicon and phosphorus, especially in combination with cold deformation. As shown in [278, 279], in the alloys of the Fe–16Cr–17Ni systems the strongest effect in suppressing swelling is shown by alloys either with 0.1%P or 0.1% P+0.25% Ti.

Figures 6.63 and 6.64 show respectively the temperature and dose dependences of the swelling of a series of cold-worked steels whose compositions are presented in Table 6.4 [274]. The graphs indicate that steels with increased silicon content (0.8%) are characterised

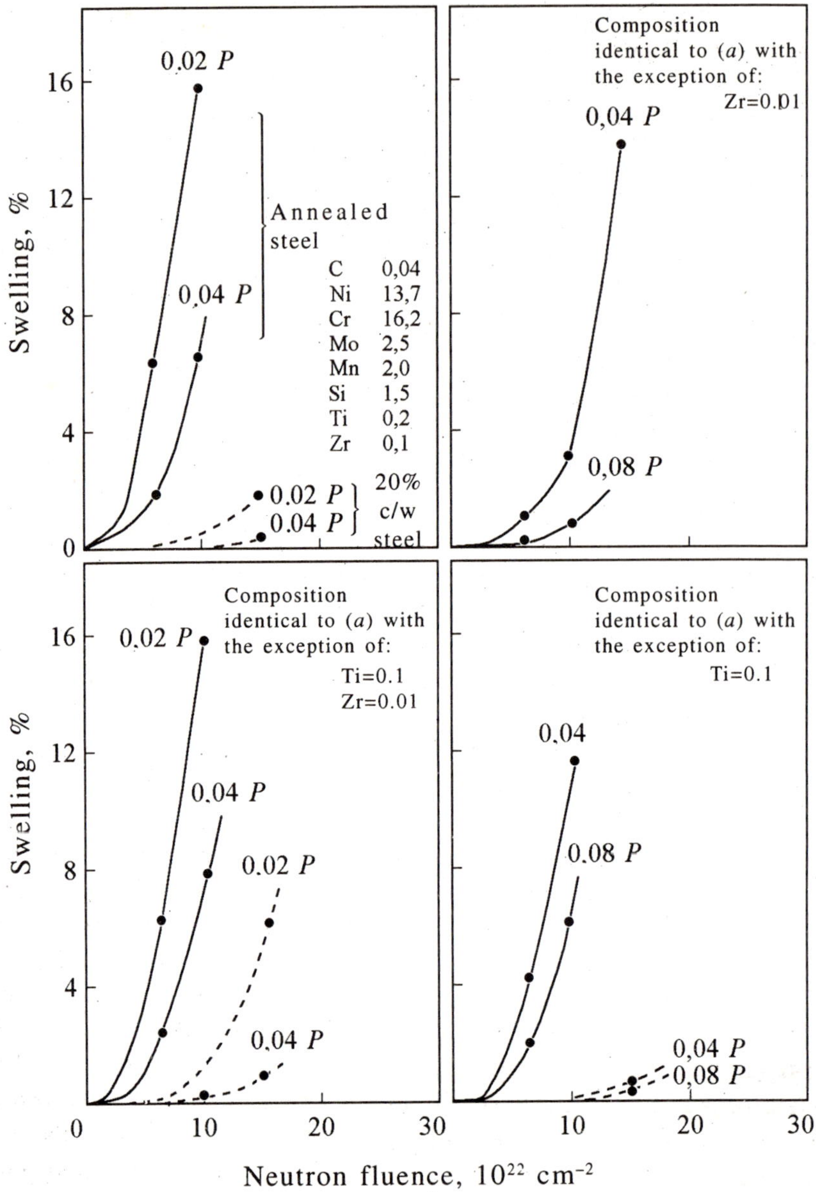

Fig. 6.62. Effect of alloying with titanium and phosphorus and also cold plastic deformation, on swelling of D9 austenitic steel, irradiation temperature 540 °C [276].

by relatively high swelling resistance at 120 dpa in the entire investigated temperature range. Especially high swelling resistance in the vicinity of the low temperature peaks of swelling at ~400 °C (here the degree of swelling of the materials is the highest) was recorded

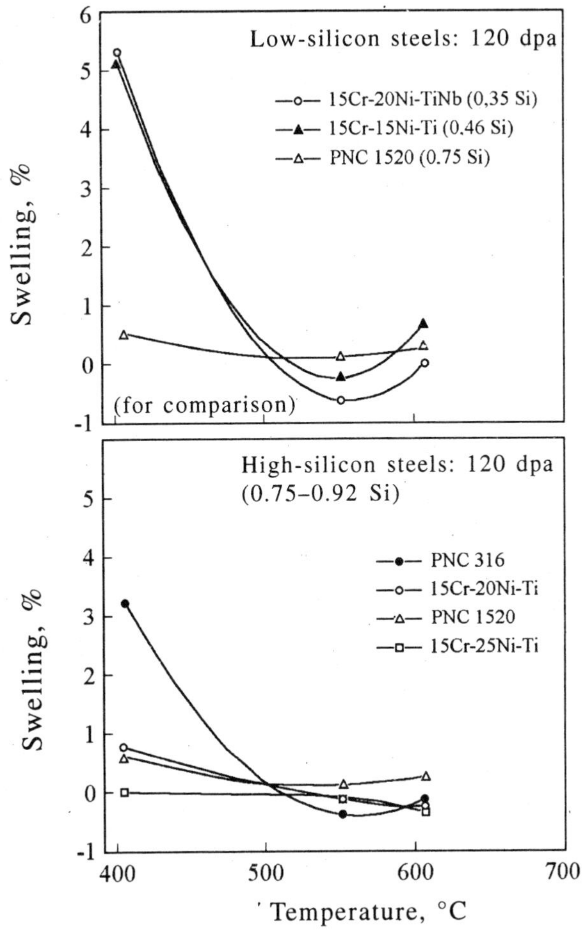

Fig. 6.63. Temperature dependences of swelling of several cold-worked austenitic chromium–nickel steels [274].

for 15Cr–20Ni–Ti, PNC1520 and 15Cr–25Ni–Ti steels. At a fluence corresponding to 150 dpa, the swelling of these steels did not exceed 2%. In contrast to these materials, identical steels with approximately 0.41–0.45% are characterised by a considerably higher swelling rate.

In section 4.6.3.1, we already presented a number of experimental data on the effect of Ti, P and cold plastic deformation on structural–phase changes in PCA-type steels, directly associated with their swelling resistance. We shall summarise other experimental results within the framework of the swelling process.

It is well known that the irradiation of chromium–nickel steels stimulates the segregation of Ni and Si with the formation of γ', G

Table 6.4. Composition of PNC316 steel and a number of advanced steels with higher nickel content [260]

Steels	Chemical composition, %										c/w, %
	Cr	Ni	C	Si	P	B	Mn	Mo	Ti	Nb	
PNC316	16.52	13.84	0.052	0.82	0.028	0.0031	1.84	2.49	0.08	0.079	20
15Cr–15Ni–Ti	15.10	15.61	0.060	0.46	0.029	0.0048	1.77	2.45	0.32	–	16
15Cr–20Ni–Ti	15.29	19.61	0.061	0.81	0.026	0.0043	1.57	2.46	0.18	–	16
15Cr–25Ni–Ti	15.09	24.41	0.060	0.85	0.020	0.0049	1.47	2.34	0.21	–	16
15Cr–20Ni–TiNb	14.89	19.66	0.057	0.41	0.026	0.0044	1.72	2.55	0.28	0.090	18
PNC1520	15.05	19.71	0.064	0.75	0.028	0.0031	1.92	2.56	0.25	0.110	16

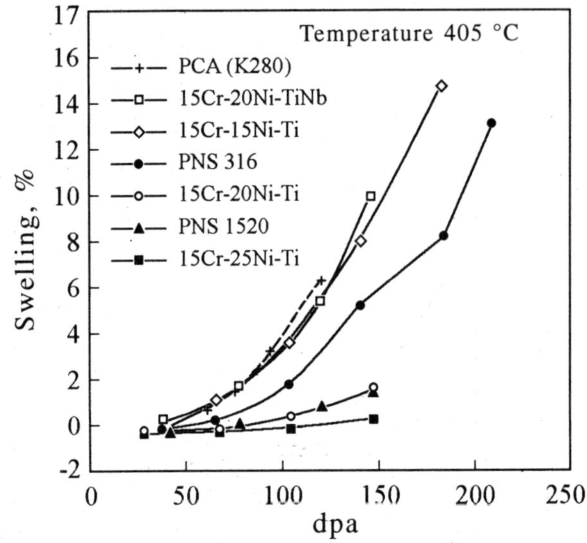

Fig. 6.64. Dose dependences of swelling of several cold-worked austenitic–chromium–nickel steels [274].

and η (M_6C) phases, enriched with these elements.

In this case, as shown in several investigations (see, for example, [271,281,282], the phases *G* and η are the centres of nucleation and preferential growth of voids. In particular, this void formation and swelling mechanism is especially intensive in annealed materials. An increase of the fluence in the annealed steels modified with titanium is characterised by a tendency for growth and dissolution of MC carbides (mainly TiC) [281] which are stabilizers of the dislocation network. Preliminary cold deformation stimulates the nucleation of the MC phase on the dislocations and suppresses the formation and growth of the *G*-phase, the η-phase, and swelling [274,277,281]. Additional alloying with boron and phosphorus of chromium–nickel steels, modified with titanium, results in the refining of MC carbides [283,284] and this also increases the stability of the dislocation network and swelling resistance.

The positive role of titanium in suppressing swelling is also reflected in the fact that titanium both in the solid solution and in the form of TiC carbides operates as a trap for gases suppressing also the growth of the gas bubbles as a result of their coalescence [97,278,283,285]. In this case, the additional alloying of chromium–nickel steels with boron and phosphorus also enhances the effect of suppression of the negative influence of helium on swelling [278,283,285]. Very suitable examples of both individual and combined effect of titanium and phosphorus on the density of voids and swelling in neutron-irradiated Fe–16Cr–17Ni alloys in relation to the value of the He/dpa ratio are presented in Fig. 6.55 and 6.66 [278].

The above experimental results show that phosphorus in chromium-nickel steels efficiently suppresses swelling both at elevated (in the vicinity of the low temperature peak of swelling at ~400 °C) and at high temperatures (≥500 °C).

On the basis of a number of experimental investigations it has been clearly established that the suppression of swelling at high temperatures is associated with the formation of phosphides (see, for example [273,274,277,286,287]). The main mechanisms of the effect of phosphides in suppressing swelling may be summarised as follows:

1. Like titanium carbides, phosphides also settle on the dislocations

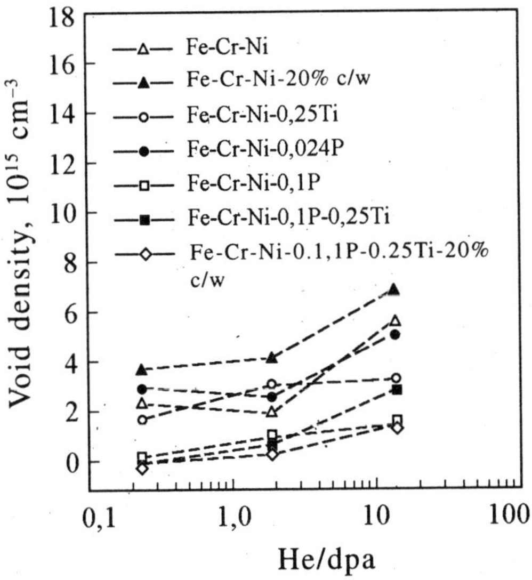

Fig. 6.65. Dependence of the density of voids in neutron-irradiated (at 415 °C) alloys Fe–Cr–Ni, Fe–Cr–Ni–Ti, Fe–Cr–Ni–P and Fe–Cr–Ni–Ti in annealed and cold-worked conditions on He/dpa ratio [278].

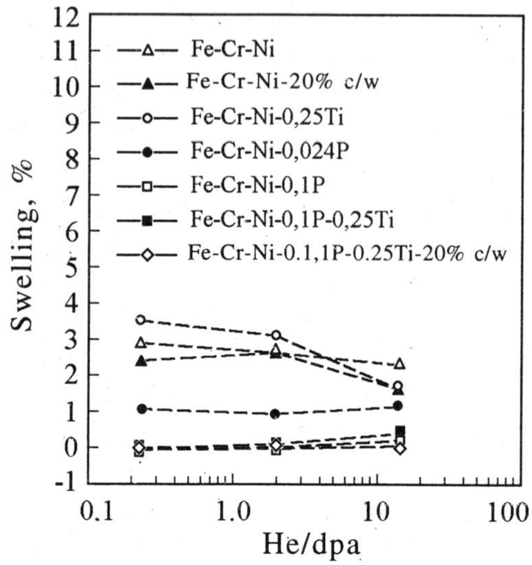

Fig. 6.66. Dependence of the degree of swelling in neutron-irradiated, at 415 °C, alloys: Fe–Cr–Ni, Fe–Cr–Ni–Ti, Fe–Cr–Ni–P and Fe–Cr–Ni–P–Ti in annealed and cold-worked conditions on the He/dpa ratio [278].

and stabilise the dislocation structure, suppressing the processes of recovery of this structure.

2. The formation of phosphides suppresses the nucleation and growth of radiation-stimulated precipitates, enriched with nickel and silicon (γ', G and η-phases).

3. The fine-dispersion phosphides, like the precipitates of titanium carbide, efficiently capture helium and, decreasing its concentration in the matrix, suppres the processes of the nucleation and growth of voids.

According to various data, the phosphides form in the chromium-nickel steels in the temperature rage ~470–520 °C which evidently depends on both the phosphorus content and other factors, such as the content of titanium and silicon, accelerating the formation and improving stability of the phosphides [287].

At low temperatures, the phosphorus atoms are found mainly in the solid solution. In this case, to analyse the effect of both phosphorus and other alloying elements on void formation, it is convenient to use considerations on vacancy supersaturation:

$$S_v^s = c_v / c_v^0 \tag{6.91}$$

In this equation, c_v and c_v are the concentrations of radiation and thermodynamically equilibrium vacancies, respectively.

The problems associated with the effect of vacancy supersaturation on the processes of void nucleation have been examined in great detail in section 3.8.1. Summarising some assumptions made in section 3.8.1, essential for further analysis, it is necessary to note that:

1. With a decrease of S_v^s, in accordance with equation (3.90), the activation barrier for the nucleation of voids increases, and with an increase of S_v^s it decreases.

2. The alloying elements which increase the effective coefficient of vacancy diffusion, decrease c_v in equation (6.91) and at a constant value of c_v^0 (equation (3.94)) suppress the void formation process. Otherwise, the rate of void formation increases.

At the same time, as shown in section 3.8.1, the process of void formation may also be suppressed in cases in which the effective coefficient of diffusion of the vacancies decreases (equation (6.2)) and the value of c_v in equation (6.91) increases. However, in this case, in accordance with equation (3.114), the thermodynamically equilibrium concentration of the vacancies in equation (6.91) increases and the overall effect of c_v and c_v^0 on the vacancy supersaturation evidently depends on their relationship. The combined effect of both processes on vacancy supersaturation was examined by the authors of this book on an example of numerical calculations for diluted nickel-based alloys (Fig. 3.20).

As indicated by Fig.3.20, the effect of the thermodynamically equilibrium concentration of vacancies at specific values of c_t and E_v^B is dominant and the rate of void nucleation consequently decreases. Similar results were also obtained in numerical calculations of steel 316.

As shown in [288,289], phosphorus in chromium–nickel steels has a very high energy of binding with vacancies (~0.4 eV). In this case, as mentioned previously, this results on the one hand, in a decrease of the effective coefficient of diffusion of the vacancies (equation 2.6) thus supporting void nucleation. On the other hand, the thermodynamically equilibrium concentration of vacancies increases in this case (equation 3.114) and the vacancy supersaturation decreases, with the void formation process suppressed.

As discussed in section 4.6.3.1, the undesirable segregation of phosphorus to the grain boundaries is suppressed in alloying chromium–nickel steels with titanium, stabilising the positive solid-solution effect of phosphorus on suppression of the void formation process. In addition to this, as mentioned previously, phosphorus, as an undersized atom in the iron lattice (ΔV = –13.6% [290] forms stable mixed dumbbells decreasing the diffusibility of interstitials in steels. Consequently, the degree of mutual recombination of interstitials and vacancies increases and this may also support the suppression of swelling as a result of decrease in the dynamic concentration of radiation point defects.

The experimental data, discussed above, show that silicon, like phosphorus, especially in concentrations of ~≥0.8%, efficiently suppresses swelling in cold-worked chromium–nickel steels (Figs. 6.63 and 6.64). The experimental results obtained for irradiated Fe–15Cr–20Ni steels show [291] that silicon has a strong effect in suppressing swelling in annealed chromium–nickel steels (Fig. 6.67).

Interpretation of the mechanism of the effect of silicon on suppressing the process of void formation is usually associated with the fact that silicon increases the effective coefficient of diffusion of vacancies (see, for example [274,292,293]). As mentioned previously, this should result in a decrease of vacancy supersaturation and, consequently, increase of the activation barrier to void nucleation.

However, as shown by analysis of the experiments carried out into the irradiation of 316 steel with a higher silicon content (~1.7%) in a high-voltage electron microscope at temperatures of 600–660 °C, the energy of binding of vacancies with silicon atoms is very high (~0.35 eV) [294]. On the basis of the considerations regarding the effective diffusion coefficient of point defects, this result leads to a decrease of the effective coefficient of diffusion of the vacancies (equation 2.6) and the mechanism of the effect of silicon

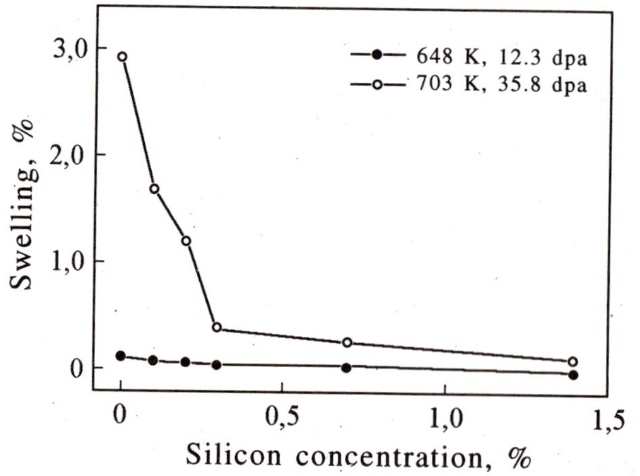

Fig. 6.67. Swelling of annealed Fe–15Cr–20Ni–Si alloys in relation to silicon concentrations [291].

on swelling evidently changes. In fact, this mechanism becomes identical with the mechanism of the effect of phosphorus on swelling, which has been examined in great detail.

In [294] it was also shown that the atoms of silicon, like atoms of phosphorus, are effective traps not only for vacancies but also for interstitials. The latter result is in complete agreement with theoretical considerations examined in section 1.3. Like phosphorus, silicon is an undersized atom in the iron lattice ($\Delta V = -7.88\%$ [290]) and forms stable mixed dumbbells decreasing the effective diffusibility of interstitials (section 2.2).

On the whole, the analysis results indicate that when silicon is in the solid solution, the mechanism of its effect on swelling is completely identical with the effect of phosphorus but this does not contradict the literature data. However, it should be mentioned that, on the whole (per concentration unit), phosphorus suppresses more efficiently swelling at both elevated and high temperatures and it does not cause the formation of phases unfavourable for swelling resistance, such as G and η, enriched with nickel and silicon. As already mentioned, these phases are the centres of nucleation and preferential growth of the voids.

Of the main matrix elements in both ternary Fe–Cr–Ni alloys and in chromium–nickel steels, nickel has the strongest effect on the suppression of swelling. In ternary Fe–Cr–Ni alloys, an increase of the nickel content results in a monotonic decrease of the degree of swelling to the minimum values at a nickel concentration of ~35–45% [91]. However, the resistance to swelling in both annealed and cold-deformed ternary Fe–Cr–Ni alloys decreases quite rapidly already at relatively small radiation doses.

In the first case, this is associated with the fact that the formation of voids in these alloys takes place at small radiation doses when a dislocation network is not yet formed. In the latter case, the dislocation density rapidly decreases during radiation in the absence of alloying elements–stabilisers of the dislocation structure [278] (for example, titanium, carbon, silicon, phosphorus).

The combination of these factors determines the real prospects for the development of chromium–nickel steels with increased nickel content, with additional alloying with appropriate elements. Evidently, titanium and phosphorus are the most promising of these alloying elements. A low concentration of boron in addition to phosphorus results, as shown in, for example [273], in an increase in the irradiation creep resistance of chromium–nickel steels.

As already mentioned, alloying with silicon, especially with an increase of nickel concentration, stimulates the processes of formation of G and η-phases, enriched with these elements, These processes reduce swelling resistance. In addition to this, steels with high concentrations of nickel and silicon show a tendency for a decrease of the creep resistance, supporting the formation of carbides [274].

As regards the mechanism of the effect of nickel on the suppression of swelling, this mechanism is associated, as in the case of silicon, with an increase in the effective coefficient of vacancy diffusion [274,292,293].

This interpretation is hardly accurate. As the oversized element in the iron lattice ($\Delta V = 4.6\%$ [290]), nickel in irradiated steel diffuses by the normal mechanism of direct exchange with vacancies and has the lowest diffusion coefficient in comparison with the main matrix elements ($D_{Cr}>D_{Fe}>D_{Ni}$ [295]). Enrichment of sinks in irradiated chromium–nickel steels with nickel and depletion in iron and chromium [296,297] also does not correspond to the considerations according to which nickel increases the effective diffusion coefficient of vacancies. In addition, as directly indicated by the analysis of the experimental data for annealing of radiation defects in irradiated Fe–Cr–Ni alloys [298,299], an increase in the nickel concentration either does not change or slightly decreases the effective diffusion coefficient of the vacancies. However, in this case, an increase of the nickel content results in a very large decrease of the energy of formation of the vacancies, i.e. the thermodynamically equilibrium concentration of vacancies in equation (6.91) increases.

As shown previously, the thermodynamically equilibrium concentration of vacancies has an almost controlling role in the variation of the extent of vacancy supersaturation S^s_v. This is especially important in this case in which the value of the effective diffusion coefficient of the vacancies is almost constant. Consequently, the effect of nickel on the suppression of swelling in the ternary Fe–Cr–Ni alloys and steels based on them should be associated in particular with the considerations indicating the increase of the thermodynamically equilibrium concentration of vacancies as a result of a decrease of their formation energy.

6.3.2. Austenitic chromium–manganese steels

One of the main factors causing swelling of the chromium–nickel steels is the large section of formation of helium on nickel, on both thermal and fast neutrons. Figure 6.68 shows the results of our in-

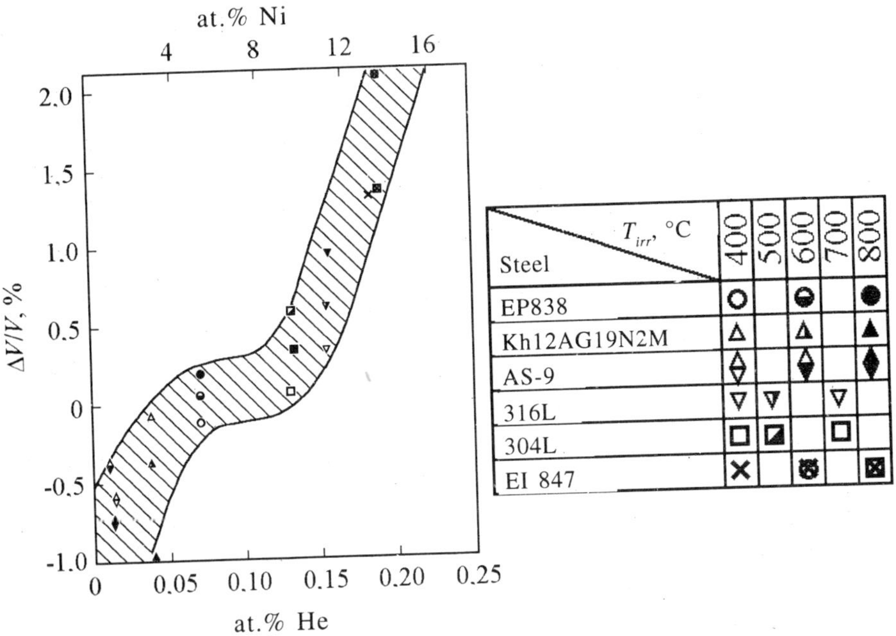

Fig. 6.68. Dependences of the swelling of austenitic steels on the content of nickel and transmutation helium [66, 127].

Table 6.5. Chemical composition of investigated steels

Steel grade	Composition of element, wt.%							
	C	Si	Mn	Cr	Ni	Mo	Al	Nb
04Kh12G14N4YuM (EP–838)	0.05	0.48	14.2	12.0	4.9	0.52	1.59	–
04Kh12AG19N2M	0.05	0.36	16.9	12.4	2.2	0.56	–	–
06Kh17G17DAMB (AS-9)	0.04	0.40	16.5	17.0	–	–	–	–
0Kh16N11M3 (316L)	0.03	0.35	1.67	16.4	11.9	2.6	–	–
0Kh18N9 (304L)	0.02	0.93	0.55	18.0	9.8	–	–	–
04Kh16N15M3B (EI–847)	0.05	0.56	0.27	15.6	14.6	3.3	–	0.53

vestigations on a series of austenitic steels with different nickel content irradiated in a CM-2 reactor [66,127]. The compositions of the steels are presented in Table 6.5.

The fluences of the thermal and fast (E>0.1 MeV) neutrons were 3.2×10^{22} cm^{-2} and 1.9×10^{22} cm^{-2}, respectively. The calculations of the concentration of transmutation helium were carried out in [300].

Figure 6.68 shows directly that the degree of swelling of steels increases with increasing concentration of nickel and transmutation helium which, in the present case, is formed by the two-stage reaction:

$$^{58}Ni(n,\gamma)^{59}Ni(n,\alpha)^{56}Fe \quad (6.92)$$

The austenitic steels in which nickel is fully or partially replaced by manganese show almost no swelling at the given irradiation parameters. Negative volume changes in 04Kh12AG19N2M and AS-9 chromium–manganese steels with a low nitrogen concentration (~0.2–0.3%) are associated evidently with the radiation-stimulating formation of nitride phases characterised by negative dilation volumes in comparison with the matrix.

On the whole, the results illustrate clearly the efficiency of restricting the nickel content of austenitic steels with a tendency for replacing nickel by manganese which has a considerably smaller formation section of helium and hydrogen in comparison with nickel. This fact is one of the most important primary reasons for starting investigations of the radiation resistance of steels based on Fe–Cr–Mn, as alternatives to chromium–nickel steels as materials for atomic and thermonuclear power engineering. In one of the first studies in this area [123], comparative investigations were carried out into the effect of irradiation at a temperature of ~100 °C on the mechanical properties of EP-838 steel developed at the Institute of Metallurgy and Materials Science of the Russian Academy of Sciences (IMET RAN), in comparison with the well-known 316 steel. The results of subsequent investigations, carried out at both high temperatures and irradiation fluences show that as regards the level of irradiation resistance and the variation of the mechanical properties, EP-838 steel is not inferior to the type 316 steel (section 6.2.1.4) and has a higher swelling resistance [128,301].

Further investigations in the development of chromium–manganese steels were stimulated by research into reduced-activation materials (see section 4.6.3.2, 6.2.1.4, and Chapter 5).

In this case, owing to the fact that nickel, in contrast to manganese, generates radionuclides with a considerably longer lifetime, the nickel concentration in austenitic steels with accelerated decrease of induced radioactivity should be, if possible, restricted at $\leq 1\times10^{-2}$ % (Table 5.1).

At the same time, it is also necessary to consider replacement or a large reduction in the concentration of a number of other conventional alloying elements of austenitic steels (Chapter 5).

As already reported in 6.2.1.4, the most promising compositions of chromium–manganese steels with accelerated decrease of induced radioactivity are at present steels of the type AMCR-0033 (one of the compositions of this steel is presented in Table 4.4) and steels based on the Fe–12Cr–20Mn–C system. The phase stability and variation of the mechanical properties of these materials under irradiation were investigated in sections 4.6.3.2 and 6.2.1.4, respectively.

Unfortunately, at the present time, the swelling of steels of the type Fe–12Cr–20Mn–W–C proposed by the IMET RAN and the Oak Ridge National Laboratory, has been studied only at relatively low temperatures (350 and 400 °C) and fluences (8–10 dpa) [132] (at least, according to the available literature data). According to the data published in [132], in irradiation of the 0.1C–12Cr–20Mn–1W steel (IMET RAN) in the annealed austenitised condition in a BOR-6 reactor at a temperature of 400 °C, at 10 dpa, the swelling of this steel does not exceed 1.4%. The swelling of the American steel based on the 0.25C–12Cr–20Mn–2W–Ti system and in the cold-worked condition (20%) in irradiation in the same conditions was only 0.6%.

Evidently, further investigations of the swelling resistance of steels should include examination of the effect of increasing temperature and radiation fluence. In this case, an additional factor of increasing the tendency of these steels for suppressing swelling is the optimum alloying of the steels with elements such as W, Ti, V, B and P. It should mentioned that, as shown in [134], the optimum combination of these elements resulted in a large improvement of the mechanical properties of the Fe–12Cr–20Mn–W–C steel, irradiated in the temperature range 420–600 °C with a fluence of up to 44 dpa (section 6.2.1.4).

The swelling of AMCR-0033 steels, developed by Euroatom, has been studied up to high fluences. According to the results in [302, 303], at a nickel content of this steel of 0.7%, a fluence of ~50 dpa and 520 °C, the swelling of the steel is ~4 and 1.6% in the annealed and cold-worked (20%) conditions, respectively. When the fluence is increased to 60 dpa, the swelling of the steel at 420 °C does not exceed ~1.2%, and at temperatures of 500 and 600 °C it is ~3%. However, at a fluence of >60 dpa, the swelling of the steel rapidly increases and at 80 dpa and a temperature of 520 °C it is already quite high (7%) [303].

In this case, it should be mentioned that the nickel concentration of the above steel was 0.7%, which is considerably higher than the maximum permissible concentration level for reduced-activation materials, mentioned previously ($\leq 1\times10^{-2}$%).

6.3.3. Ferritic steels

In contrast to austenitic steels, swelling is not the main problem of the application of martensitic and ferritic–martensitic steels in nuclear engineering. At relatively high fluences, the swelling of ferritic steels is very small and the value is usually an order of magnitude or more lower than in the case of the austenitic steels. For example, the swelling of EP-450 ferritic–martensitic steel (Fe–13Cr2MoVNbB) after irradiation at 400 °C with a fluence of 90 dpa, did not exceed 0.1%, and after irradiation with a fluence of 142 dpa it was 2% [304]. The extremely low values of swelling were recorded for martensitic and ferritic–martensitic steels Fe–9Cr1MoVNb and Fe–12Cr1MoVW (HT-9) (1.76% and 1.02%, respectively), irradiated at a temperature of 420 °C in an FFTF/MOTA reactor [305,306].

As already mentioned previously (section 6.2.3.1), the ferritic steels, at least up to temperatures of 500–550 °C, are superior to the austenitic steels in both the creep resistance and long-term strength.

As a result of the favourable combination of these properties, there are tendencies in both Russia and foreign atomic power engineering to develop ferritic steels as alternative structural materials for the active zone of fast reactors with liquid metal heat carriers. In contrast to the austenitic steels, the ferritic steels are not subjected to high-temperature helium embrittlement and this is also one of the advantages. In addition to this, at the present time, the ferritic steels are regarded as promising structural materials for the DEMO demonstration fusion reactor in which lithium or Li–Pb eutectic are to be used as heat carriers. In this case, the ferritic steels have sufficiently high corrosion resistance in these media [307], which is higher than that of the austenitic steels.

At the same time, one of the main disadvantages, typical of the ferritic steels, is their low-temperature embrittlement, especially in the conditions of higher concentrations of transmutation helium (section 6.2.1.3). However, it may be possible to reduce the effect of this factor by the application of ferritic steels with accelerated decrease of induced radioactivity. As shown in [308,309], reduced-activation steels, in which molybdenum is replaced by tungsten, and niobium by tantalum and vanadium, are characterised by lower ductile–brittle transition temperatures T_{DB} and by a small shift of these temperatures during irradiation.

Another important problem of the efficient replacement of austenitic steels in active zones by ferritic steels is the increase of their long-term strength at high temperatures. It is well known that the recovery of the martensitic structure and a decrease of the strength of ferritic steels during irradiation take place at temperatures of 500–550 °C. To investigate one of the promising directions, work is being carried out at the present time to develop the so-called dispersion-hardened martensitic and ferritic–martensitic steels (see, for example [306, 310–312]. The composition of these materials includes mainly particles of Y_2O_3 oxides whose task is to increase the stability of the martensitic structure and improve the creep and long-term strength parameters.

6.3.4. Vanadium-based alloys

As already mentioned in sections 4.6.4, 6.2.1.5 and in Chapter 5, in addition to the positive combination of the thermophysical, mechanical and corrosion properties, a number of reduced-activation vanadium-based alloys are also characterised by relatively higher swelling resistance. One of the main characteristic special features of alloying vanadium is that the undersized atoms (dilation volume <0)), i.e. iron, chromium and nickel, increase the degree of swelling of this material and the elements with positive dilation volume is (W, Mo and Ti) increase its swelling resistance [313].

Alloying of vanadium with titanium in the concentration up to ~20 at.% is especially effective in increasing the swelling resistance of vanadium [313].

Figures 6.69(*a*) and 6.69(*b*) show the dose dependences of the swelling of V–Ti binary alloys irradiated at temperatures of 420 and 600 °C, respectively [143,314]. The identical dependences for a number of alloys of the V–Ti–Cr system investigated extensively at the present time are shown in Fig. 6.70(*a*) and Fig. 6.70(*b*) [143,314]. The concentration of the interstitial impurities in the V–17.7 Ti alloy in Fig. 6.69 is considerably higher than in the V–20 Ti alloys.

Attention should be given to the anomalous nature of the variation of the degree of swelling with the radiation dose, i.e. the presence of maxima in Fig. 6.69 and 6.70. The formation of these maxima in the vanadium-based alloys is associated with the suppression of swelling after the application of specific fluences as a result of the radiation-stimulated formation of fine-dispersion precipitates of the Ti_5Si_3 phase, playing the role of additional sinks for vacancies and interstitials [314].

Figure 6.71 gives the estimated values (including extrapolated values) of the maximum possible swelling of a number of V–Ti and V–Ti–Cr alloys after irradiation with a fluence of ~120 dpa in the temperature range 420–600 °C, obtained on the basis of the experimental data presented in Fig. 6.69 and 6.70. They show that the alloys of the V–Ti and V–Ti–Cr systems are characterised by relatively high swelling resistance which, evidently, will not be the limiting factor in their application as materials for the first wall of the DEMO fusion reactor and, possibly, in reactors of a future generation. As indicated by the analysis of prospects for the development of materials for fusion reactors [315], the maximum number of atomic displacements in the materials of the first wall of the DEMO reactor will not exceed 80 dpa, and in the reactors of the next generation it will reach approximately 100–150 dpa.

For the vanadium-based alloys, like ferritic steels, one of the main critical factors of application of these materials in fusion reactors

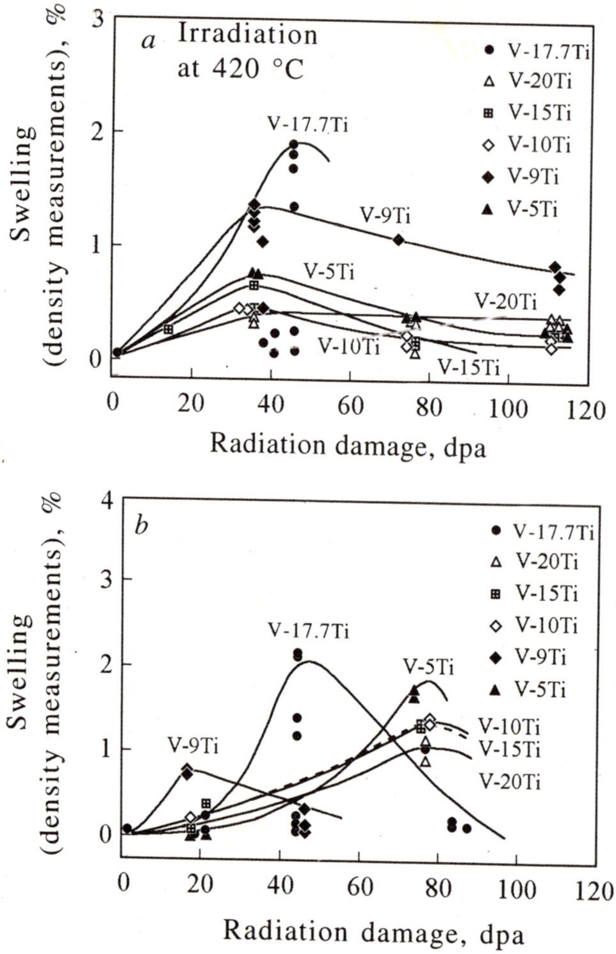

Fig. 6.69. Dose dependences of swelling in V–Ti alloys irradiated at temperatures of 420 °C (*a*) and 600 °C (*b*) [143, 314].

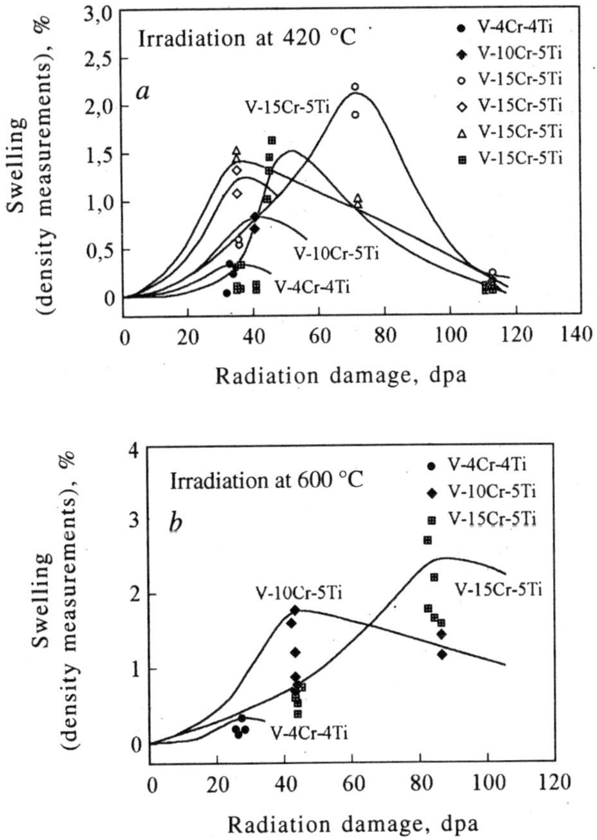

Fig. 6.70. Dose dependences of swelling in V–Ti–Cr alloys irradiated at 400 °C (*a*) and 600 °C (*b*) [143,314].

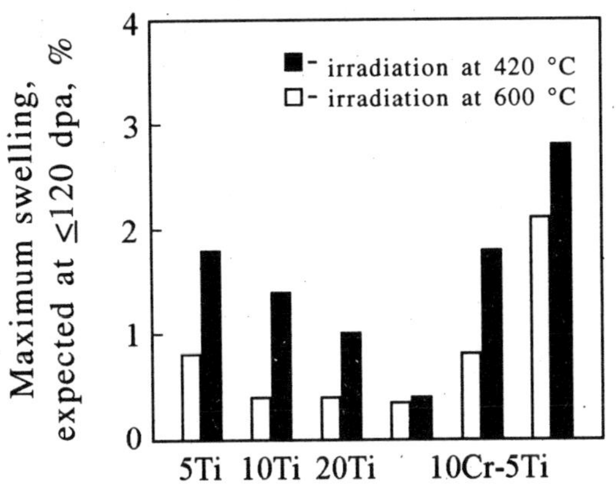

Fig. 6.71. Maximum values of swelling in a series of V–Ti and V–Ti–Cr alloys irradiated at temperatures of 420–600 °C with a fluence of ~120 dpa [143, 314].

will be the possibility of a large decrease of plasticity at both tensile stresses and under impact loading after irradiation at temperatures of ≤400 °C (see section 6.2.1.5). In addition to this, the most likely reasons for the embrittlement of these materials will be the arrival into vanadium alloys of elements such as oxygen, nitrogen and carbon from the contact media [315] and also the susceptibility of these alloys to high-temperature helium embrittlement at temperatures of ≥615 °C [312]. The extensive embrittlement of the vanadium alloys may also take place during the introduction of hydrogen into alloys (as a result of absorption from the contact media or the formation of hydrogen by nuclear reactions), especially in the presence of oxygen [316].

REFERENCES

1. Makin M.J., Blewitt T.H. Acta Met., 1962, v.10, p.241.
2. Dichl J., Shilling W. In: Proc. Third Int.Conf. on Peaceful Uses of Atomic Energy. N.Y.United Nations, 1965, v.9, p.62.
3. Koppenaal T.J. Phil. Mag., 1965, v.11, p.12
4. Makin M.J. In: Radiation Effects, N.Y., Gordon and Breach, 1967, p.627.
5. Ono K., Mifune T., Meshii M. Phil.Mag., 1968, v.17, no.146, p.235
6. Koppenaal T.J., Arsenault R.J. Metallurg.Rev.,1971, no.157, p.175.
7. Havancshak K., Szenes G., Skuratov V.A., Man Y.D., Malinovski W., Shchegolev V.A. Phys.Stat.Sol.(a), 1988, v.106, p.371.
8. Volkov M.G., Vondarenko G.G. and Kolokol'tsev V.N., In: Structural Materials for Fusion Reactors, 1983, p.82.
9. Farrell K. In: Dimensional Stability and Mechanical Behaviour of Irradiated Metals and Alloys. Nuclear Energy Society, London, 1983, p.73.
10. Wiffen F.W. Nucl.Met., 1973, v.18, p.176.
11. Tovtin V.I. Variation of the structure and properties of reduced-activation alloys based on Al and V in electron and neutron irradiation, Dissertation for the title of Candidate of Physical-Mathematical Sciences, Institute of Metallurgy, Russian Academy of Sciences, Moscow, 1992.
12. Schiraishi K., Fukaya K., Katano Y.J. J.Nucl.Mater., 1974, v.54, p.275.
13. Hiraoka Y., Okada M., Fujii T., Tanaka M., Hishinuma A. J.Nucl.Mater., 1986, v.141/143, p.837.
14. Arsenault R.J. Acta Met., 1967, v.15, p.1853.
15. Arsenault R.J., Pink E. Mater.Sci.Eng., 1971, v.8, p.141.
16. Bement A.L. Rev.Roum.Phys., 1972, v.17, no.3, p.361
17. Wechsler M.S. In: Fundamental Aspects of Radiation Damage in Metals. USERDA, 1976, v.II, p.991.
18. Platov Yu.M., Lazorenko V.M., Simakov S.V., Tovtin V.I., Ivanov V.V. Mater.Sci.Forum., 1992, v.97/99, p.253.
19. Shiraishi K., Fukaya K., Katano Y. In: Radiation Effects and Tritium Technology for Fusion Reactors, USERDA, 1976, v.II, p.122
20. Bradley E.R., Jones R.H. J.Nucl.Mater., 1986, v.141/143, p.686
21. Bajaj R., Wechsler M.S. In: Fundamental Aspects of Radiation Damage in

Metals. USERDA, 1976, v.II, p.1010.
22. Ivanov L.I., Kalinin V.G., Platov Yu.M., Pletnev M.N., Smirnov A.V., Tovtin V.I. and Shamardin V.K. Fiz. Khim. Obrab. Mater., 1983, No.6, p.3.
23. Ivanov L.I., Lazorenko V.M., Platov Yu.M. Kalinin V.G., Simakov S.V., Smirnov A.V., Tovtin V.I., Toropova L.S. and Shamardin V.K. Fiz. Khim. Obrab. Mater., 1985, No.4, p.16.
24. Zabolotny V.T., Ivanov L.I. Platov Yu.M. and Simakov S.V. Fiz. Khim. Obrab. Mater., 1985, No.1, p.124.
25. Farrell K., Richt A.E. ASTM. STP683, 1979, p.427.
26. Farrell K., King R.T. ASTM. STP683, 1979, p.440
27. Ivanov L.I., Lazorenko V.M., Platov Yu.M. and Tovtin V.I., Fiz. Khim. Obrab. Mater., 1990, No.3, p.5.
28. Ivanov L.I., Ivanov V.V., Lazorenko V.M., Platov Yu.M., Tovtin V.I. J.Nucl.Mater., 1992, v.191/194, p.928
29. Huang Y., Arsenault R.J. Rad.Effects, 1973, v.17, p.3.
30. Shiraishi K., Fukaya K., Katano Y. J.Nucl.Mater., 1972, v.44, p.228
31. Wechsler M.S. In: The Inhomogenety of Plastic Deformation. Ohio, ASTM, 1973, p.19.
32. Zelensky V.F., Kazachkovsky O.D., Reshetnikov F.G. and Tsikanov V.A. Seriya: Fiz. Radiats. Povrezhdenii i Materialovedenie, 1981, No.4(18),p.3.
33. Zelensky V.F., Kinryukhin N.M., Neklyudov I.M., Ozhigov L.S and Parkhomenko A.A. High-Temperature Radiation Embrittlement of Materials, Analytical review KhFTI 83-43, 1983, 47p.
34. Zelensky V.F. and Neklyudov I.M. Vopr. Atomn. Nauki i Tekhniki. Seriya: Fiz. Radiats. Poverzhdenii i Materialovedenie, 1984, No.1/2(29/30), p.46.
35. Atteridge D.G., Johnson A.B., Charlot L.A., Remark J.F. In: Radiation Effects and Tritium Technology for Fusion Reactors., 1976, v.II, p.307.
36. Votinov S.N., Sharov B.V., Tsykanov V.A., Shamardin V.K., Yarkovoi V.S., Sernyaev G.A., Raevsky V.M., Prokhorov V.I., Samsonov B.V., Lobanov G.P. and Pokrovsky A.S., in: Atomic Energetics, Fuel Cycles and Radiation Materials Science, 1971, v.3, p.426.
37. Pasechnik M.V., Kashuba I.E., Fedorov M.B. and Yakovenko T.I. Nuclei and Radiation Resistance of Structural Materials, Naukova dumka, Kiev, 1978, 312p.
38. Lijbrink E., van Grol H.J., Dekker F., van Witzenburg W. In: Effects of Radiation on Materials. Baltimore, ASTM, 1982, p.765.
39. INTOR Zero Phase Report. Vienna: IAEA, 1980.
40. Al'tovsky I.V., Votinov S.N., Dobatkin V.I., Drits M.E., Elagin V.I., Ivanov L.I., Orlov V.V., Platov Yu.M., Toropova L.S. and Filatov Yu.A. Structural Materials for Fusion Reactors, Nauka, Moscow,1983, p.40.
41. Shiller P., in: Structural Materials for Fusion Reactors, Nauka, Moscow, 1983, p.48.
42. Platov Yu.M., Votinov S.N., Drits M.E., Ivanov L.I., Kalinin V.G., Smirnov A.V., Toropova L.S. and Shamardin V.K. Fiz. Khim. Obrab. Mater., 1981, No.1, p.53.
43. Ivanov L.I., Krasnoselov V.A., Platov Yu.M., Tovtin V.I., Toropova L.S. and Shamardin V.K. Fiz. Khim. Obrab. Mater., 1990, No.4, p.5.
44. Ivanov L.I., Ivanov V.V., Lazorenko V.M., Platov Yu.M. and Tovtin V.I. Tekhnologiya Legkikh Splavov, 1990, No.12, p.46.
45. Ivanov L.I., Ivanov V.V., Lazorenko V.M., Platov Yu.M., Tovtin V.I., Toropova L.S. J.Nucl.Mater., 1991, v.191/194, p.1075

46. Ivanov L.I., Lazorenko V.M., Platov Yu.M. and Tovtin V.I. Fiz. Khim. Obrab. Mater., 1989, No.6, p.9
47. Balandin Yu.F., Gorynin I.V., Zvezdin Yu.I. and Markov V.G. Structural Materials for Atomic Power Stations, Energoizdat, Moscow, 1984, p.280.
48. Kolesova T.N., Lyadov G.D., Pecherin A.M. and Shamardin V.K. Effect of Parameters of Neutron Radiation on Embrittlement of Pressure Vessel Steels, Dimitrovgrad, NIIAR, 1984, p.38.
49. Effects of Radiation on Materials. Philadelphia, ASTM, 1982, 1225p.
50. Effects of Radiation on Materials. Baltimore, ASTM, 1985, 1267p.
51. Influence of Radiation on Materials Properties. Ann Arbor, ASTM, 1987, 800p.
52. Radiation Embrittlement of Nuclear Reactor Pressure Vessel Steels. Philadelphia, ASTM, 1993, 408p.
53. Amayev A.D., Kryukov A.M., Levit V.L., Sokolov M.A. In: Radiation Embrittlement of Nuclear Reactor Pressure Vessel Steels. Philadelphia, ASTM, 1993, p.9.
54. Vishkarev O.M., Zvezdin Yu.I., Shamardin V.K., Tulyakov G.A. In: Radiation Embrittlement of Nuclear Reactor Pressure Vessel Steels. Philadelphia, ASTM, 1993, p.218.
55. Bramovsky M., Pav T. In: Radiation Embrittlement of Nuclear Reactor Pressure Vessel Steels. Philadelphia, ASTM, 1993, p.57.
56. Howthorne J.R. In: Embrittlement of Structural Steels, Metallurgiya, Moscow, 1988, p.423.
57. Howthorne J.R. In: Optimizing Materials for Nuclear Applications. Warrendale, Metallurgical Society, 1985, p.167.
58. Havel R., Vacek M., Brumovsky M. In: Radiation Embrittlement of Nuclear Reactor Pressure Vessel Steels. Philadelphia, ASTM, 1993, p.163.
59. Simonen E.P. In: Optimizing Materials for Nuclear Applications. Warrendale, Metallurgical Society, 1985, p.225.
60. Amaev A.D., Gorynin I.V., IgnatovV.A., Nikolaev V.A., Rjazantzev E.P., Shamardin V.K. In: Proc.Int.Conf.on Radiation Materials Science, 1990, p.106.
61. Stahlkopf K.E., Odette G.R., Martson T.U. In: 4th Int.Conf. on Pressure Vessel Technology Material Fracture and Fatigue. London, 1980, p.265.
62. Steele L.E., Davies L.M., Ingham T., Brumovsky M. In: Effects of Radiation on Materials Properties. Baltimore, ASTM, 1985, p.863.
63. Odette G.R., Lombrozo P.M., Willaert R.A. Effects of Radiation on Materials. Baltimore, ASTM, 1985, p.840.
64. Lukas G.E., Odette G.R., Maiti R., Sheckherd J.W. In: Influence of Radiation on Materials Properties. Ann Arbor, ASTM, 1987, p.379.
65. Gelles D.S. ISIJ International, 1990, v.30, no.11, p.905.
66. Ivanov L.I., Platov Yu.M., Demina E.V. J.Advanced Mater., 1994, no.1(2), p.166. Ivanov L.I., Platov Yu.M. and Demina E.V. Perspektivnye Materialy, 1995, No.2, p.37.
67. Kai J.J., Klueh R.L. Fusion Materials. DOE/ER—0313/18, 1995, p.133.
68. Reduced Activation Materials for Fusion Reactors. Baltimore, ASTM, 1990, 253p.
69. Fusion Reactor Materials. Proc.4th Int.Conf. Kyoto, 1989, J.Nucl.Mater.,1991, v.179/181.
70. Fusion Reactor Materials. Proc.5th Int.Conf. Cleawater, 1991, J.Nucl.Mater.,1992, v.191/194.
71. Fusion Reactor Materials. Proc.6th Int.Conf. Stresa, 1993, J.Nucl.Mater.,1994,

v.212/215.
72. Gelles D.S. In: Fusion Materials. DOE/ER—0313/13, 1992, p.157.
73. Klueh R.L., Vitek J.M. J.Nucl.Mater., 1991, v.182, p.230.
74. Klueh R.L., J.Nucl.Mater., 1991, v.179/181, p.728.
75. Klueh R.L., Ehrlich K., Abe F. J.Nucl.Mater., 1991, v.179/181, p.116.
76. Klueh R.L., Kai J.J., Alexander D.J. In: Fusion Materials. DOE/ER—0313/16, 1994, p.128.
77. Little E.A., Stoter L.P. In: Effect Radiation on Materials. Baltimore, ASTM, 1982, p.207.
78. Klueh R.L., Maziasz P.J. J.Nucl.Mater., 1992, v.187, p.187.
79. Huang F.H., Hamilton M.L. . J.Nucl.Mater., 1992, v.187, p.278
80. Kimura A., Matsui H.J. J.Nucl.Mater., 1994, v.212/215, p.701.
81. Kohno Y., Kohyama A., Yoshino M., Asakura K. J.Nucl.Mater., 1994, v.212/215, p.707.
82. Kai J.J., Klueh R.L. In: Fusion Materials. DOE/ER—0313/18, 1995, p.133.
83. Kimura A., Narui M., Kayano H. In: Annual Progress Report for Manbusho—DOE, 1992, p.177.
84. Kohno Y., Gelles D.S., Kohyama A., Tamura M., Hishinuma A. J.Nucl.Mater., 1992, v.191/194, p.863.
85. Wassilew C., Ehrlich K. J.Nucl.Mater., 1992, v.191/194, p.850.
86. Hu W.L., Gelles D.S. In: Influence of Radiation on Materials Properties. Ann Arbor, ASTM, 1987, p.83.
87. Klueh R.L., Alexander D.J., Maziasz P.J. J.Nucl.Mater., 1992, v.186, p.185.
88. Kohyama A., Kohno Y., Asakura K., Kayama H. J.Nucl.Mater., 1994, v.212/215, p.684.
89. Zvezdin Yu.I., Vishkarev O.M., Tulyakov G.A., Magera Yu.G., Smirnov V.A., Al'tovski I.V., Grigoryan A.A., Shamardin V.K., Pecherin U.M. J.Nucl.Mater., 1992, v.191/194, p.855.
90. Klueh R.L., Maziasz P.J. J.Nucl.Mater., 1992, v.187, p.43.
91. Garner F.A. In: Materials Science and Technology. Nuclear Materials. VCH, Weinheim, 1994, Part 1, v.10A, ch.6, p.419.
92. Hosoi Y. ISIJ International, 1990, v.30, no.11, p.917.
93. Fish R.L. Straalsund J.L., Hunter C.W., Holmes J.J. In: Proc.Symp.Effects of Radiation on Structure and Mechanical Properties of Metals and Alloys. Philadelphia, ASTM, 1973, p.149.
94. Elen J.D., Fenici P. J.Nucl.Mater., 1992, v.191/194, p.766.
95. Odette G.R., Frey D.J. J.Nucl.Mater., 1979, v.85/86, p.817.
96. Odette G.R., Lucas G.E. J.Nucl.Mater., 1991, v.179/181, p.572.
97. Little E.A. J.Nucl.Mater., 1986, v.139, no.2, p.261.
98. Holmes I.J., Straalsund J.L. In: Radiation Effects in Breader Reactor Structural Materials. N.Y., AIME, 1977, p.53.
99. Rowcliffe A.F., Hishinuma A., Grossbeck M.L., Jitsukawa S. J.Nucl.Mater., 1991, v.179/181, p.125.
100. Kohyama A., Grossbeek M.L., Piatti G. J.Nucl.Mater., 1992, v.191/194, p.37.
101. Odette G.R., Lucas G.E. J.Nucl.Mater., 1992, v.191/194, p.50.
102. Grossbeck M.L., Mansur L.K. Mater.Sci.Forum, 1992, v.97/99, p.407.
103. Horsten M.G., deVries M.I. J.Nucl.Mater., 1994, v.212/215, p.514.
104. Mattas R.F., Smith D.L., Wu C.H., Karoda T., Shatalov G. J.Nucl.Mater., 1992, v.191/194, p.139.
105. Kalinin G., Gauster W., Matera R., Tovassoli A.F., Rowcliffe A., Fabritsiev S., Kawamara H. J.Nucl.Mater., 1996, v.333/337, p9.

106. Powel J.E., Alexander D.J., Grossbeck M.L., Rowcliffe A.F., Lucas G.E., Jitsukawa S., Hishinuma A., Shiba K. J.Nucl.Mater., 1993, v.212/215, p.442.
107. Shamardin V.K., Neustroev V.C., Golovanov V.N. and Povstyanko A.V. Radiatsionnoe Materialovedenie, Khar'kov, 1990, v.3, p.72.
108. Neustroev V.S., Povstyanko A.V., Shamardin V.K. Mater.Sci.Forum. 1992, v.97/99, p.329.
109. Shcherbak V.I. Preprint FEI-2308, Obninsk, 1993, p.17.
110. Lucas G.E. J.Nucl.Mater., 1993, v.206, p.287.
111. Fissolo A., Cauvin R., Hugot J.P., Levy V. In: Effects of Radiation on Materials. Philadelphia, ASTM, 1990, v.II, p.700.
112. Neustroev V.S. and Shamardin V.K. Atomnaya Energiya, 1991, v.71, p.345.
113. Chernikov V.N., Zakharov A.P. and Kazansky P.R. Vopr. Atomn. Nauki i Tekhn., Seriya: Fizika Rad. Povrezhdenii i Radiats. Materialoved., 1989, No.2,(49), p.83.
114. Miahara K., Itoh A., Sakamoto Y., Kayano H., Hosoi Y. J.Nucl.Mater., 1992, p.786.
115. Grossbeck M.L. J.Nucl.Matcr., 1991, v.179/181, p.568.
116. Votinov S.N., Prokhorov V.I. and Ostrovsky Z.E. Irradiated Stainless Steels, Nauka, Moscow, 1987, 128p.
117. Ulmaier H., Trinkaus H. Mater.Sci.Forum, 1992, v.97/99, p.451.
118. Schroeder H., Dai Y. J.Nucl.Mater., 1992, v.191/194, p.781.
119. Paulus M., Schroeder H., Deutz J. J.Nucl.Mater., 1992, v.191/194, p.798.
120. Bystrov L.N., Ivanov L.I., Platov Yu.M. and Ustinovshchikov V.M. In: Structural Materials for Fusion Reactors, Nauka, Moscow, 1983, p.5.
121. Bad'yanov A., Volkov M.G., Ivanov L.I. Lazorenko V.M. and Platov Yu.M. Fiz. Khim. Obrba. Mater., 1985, No.2. p.136.
122. Ivanov L.I., Bystrov L.N., Platov Yu.M., Lazorenko V.M., Bondarenko G.G., Kasjanov O.M., Ustinovshikov V.M. J.Nucl.Mater., 1992, v.191/194, p.733.
123. Votinov S.N., Zudin I.F., Ivanov L.I., Kalinin V.G., Platov Yu.M., Smirnov A.V. and Shamardin V.K., Fiz. Khim. Obrab. Mater., 1981, No.1, p.50.
124. Klueh M.L., Grossbeck M.L. J.Nucl.Mater., 1984, v.122/123, p.294.
125. Klueh R.L., Bloom E.E. In: Optimizing Materials for Nuclear Applications. Warrendale, AIME, 1985, p.73.
126. Shamardin V.K., Bulanova T.M., Neustroev V.S., Ostrovsky Z.E., Kosenkov V.M., Ivanov L.I., Demina E.V. and Prusakova M.D. Radiatsionnoe Materialovedenie. Khar'kov, 1990, v.3, p.81.
127. Shamardin V.K., Bulanova T.M., Neustroev V.S., Ivanov L.I., Djomina E.V., Platov Yu.M. J.Nucl.Mater., 1991, v.179/181, p.637.
128. Shamardin V.K., Bulanova T.M., Neustroev V.S., Ostrovsky Z.E., Kosenkov V.M., Ivanov L.I., Djomina E.V. J.Nucl.Mater., 1992, v.191/194, p.706.
129. Klueh R.L., Maziasz P.J. In: Reduced Activation Materials for Fusion Reactors. Philadelphia, ASTM, 1990, p.7.
130. Klueh R.L., Kenik E.A. J.Nucl.Mater., 1992, v.191/194, p.437.
131. Piatti G., Boerman D., Dos Santos Margues F. J.Nucl.Mater., 1992, v.191/194, p.662.
132. Bulanova T.M., Neustroev V.S., Shamardin V.K., Ostrovsky Z.E., Kosenkov V.M., Ivanov L.I., Dyomina E.V. In: Plasma Devices and Operation, 1994, v.3, p.331,
133. Maziasz P.J., Klueh R.L. In: Reduced Activation Materials for Fusion Reactors. Philadelphia, ASTM, 1990, p.56.
134. Klueh R.L. In: Fusion Materials. DOE/ER—0313/20, 1996, p.253

135. Gold R.E., Harrod D.L. J.Nucl.Mater., 1979, v.85/86, p.805.
136, Loomis B.A., Hull A.B., Smith D.L. J.Nucl.Mater., 1991, v.179/181, p.148.
137. Satou M., Abe K., Kayano H. J.Nucl.Mater., 1991, v.179/181, p.757.
138. Rasenwasser S.N., Miller P., Dolessandro J.A., Rawls J.M., Toffolo W.E., Chen W. J.Nucl.Mater., 1979, v.85/86, p.177.
139. Satou M., Abe K., Kayano H., Takahashi H. J.Nucl.Mater., 1992, v.191/194, p.956.
140. Loomis B.A., Nowicki L.J., Smith D.L. J.Nucl.Mater., 1994, v.212/215, p.790.
141. Satou M., Abe K., Kayano H. J.Nucl.Mater., 1994, v.212/215, p.794.
142. Smith D.L. Vanadium Alloy Development for ITER. Presented at US/RF Meeting Moscow, January 17-21, 1994.
143. Chung H.M., Loomis B.A., Smith D.L. J.Nucl.Mater., 1994, v.212/215, p.804.
144. Kurtz R.J., Abe K., Chernov V.M., Kazakov V.A., Lukas G.E., Matsui H., Muroga T., Odette G.R., Smith D.L., Zinkle S.J. J.Nucl.Mater., 2000, v.283/287, p.70.
145. Alexander D.J., Snead L.L., Zinkle S.J., Gubbi A.N., Rowcliffe A.F., Bloom E.E. In: Fusion Materials. DOE/ER—0313/20, 1996, p.87.
146. Lyakishev N.P., Dyomina E.V., Ivanov L.I., Kolotov V.P., Kashin V.I., Platov Yu.M., Savvateev N.N., Vinogradova N.A. J.Nucl.Mater., 1996, v.233/237, p.1516.
147. Ivanov L.I., Dedyrin A.I., Savvateev N.N., Borovitskaya I.V. J.Adv.Mater., 1994, no.1(5),p.435.
148. Forrest R.A., Endacott D.A.J. UKAE report, AERE/R—13402, 1989, p.117.
149. Ivanov L.I., Ivanov V.V., Kolotov V.P., Platov Yu.M. and Tovtin V.I. Metally, 1992, No.1, p.141.
150. Ivánov L.I., Platov Yu.M., Dedyurin A.I., Borovitskai I.V., Lazorenko V.M., Savvateev N.N., Sidorova V.V., Zakharova M.I., Artemov N.A. J.Nucl.Mater., 1996, v.233/237, p.395.
151. Response of Metals and Metallic Structure to Dynamic Loading. Publication NMAB—341, US National Academy of Sciences. Washington, D.C., 1978, 179p.
152. Orowan E. In: Symposium on Internal Stresses. Inst.Metals. London, 1947, p.451.
153. Seeger A. In: Proc. of Second Int.Conf. on Peaceful Uses of Atomic Energy. N.Y. United Nations. 1958, v.8, p.250.
154. Fleisher R. Acta Met., 1962, v.10, p.835.
155. Fleisher R. J.Appl.Phys. 1962, v.33, p.3504.
156. Arsenault R.J. Acta Met., 1967, v.15, p.1853.
157. Kayano H., Higashiguchi Y., Yajima S. J.Nucl.Sci. and Technol., 1977, v.14, no.2, p.117
158. Matta M.K., Batra I.S., Sharma B.D. Rad.Eff.Letters, 1979, v.43, p.223.
159. Corbett J.W., Smith R.B., Walker R.M. Phys.Rev., 1959, v.114, no.6, p.1452.
160. Piani C.S.B., Aspeling J.C. Rad.Effects, 1983, v.69, p.19.
161. Kessler G., Kulcinski G.L. In: Fusion and Fast Breeder Reactors. Int.Inst.Appl.Systems Analysis, Laxenburg, Austria, 1977, p.351.
162. Zinkle S.J., Kulcinski G.L. Effects of Radiation on Materials. Baltimore, ASTM, 1985, p.363.
163. Akishin A.I. Ion Bombardment in Vacuum, Gosenergoizdat, Moscow, 1963, 116p.
164. A Model of the Cosmic Space, Moscow University, Moscow,1983, v.3, 635p.
165. Gol'dshtein M.I., Litvinov V.S. and Bronfin B.M. Physics of Metals of High-

Strength Alloys, Metallurgiya, Moscow, 1986, p.311.

166. Simons R.L., Hilbert L.A. Effects of Radiation on Materials. Baltimore, ASTM, 1985, p.820.

167. Beskorovainyi N.M., Kalin B.A., Platonov P.A. and Chernov I.I. in: Structural Materials for Fusion Reactors, Energoizdat, Moscow, 1995, 704p.

168. Muroga T., Watanabe H., Yoshida N., Kirishita H., Hamilton M.L. J.Nucl.Mater., 1995, v.225, p.137.

169. Ruhle M. Phys.Stat.Sol., 1968, v.26, p.221.

170. Harries D.R. J.Nucl.Mater., 1979, v.82, p.2.

171. Ghoniem N.M., Alhajji J., Garner F. In: Effects of Radiation on Materials. Philadelphia, ASTM, 1982, p.1054.

172. Abe F., Garner F.A., Kayano H. J.Nucl.Mater., 1994, v.212/215, p.760.

173. Kronpa F., Hirsh P.B. In: Disc.Farad.Soc., 1964, no.38, p.49.

174. Kojiama S., Zinkle S.J., Heinisch H.L. J.Nucl.Mater., 1991, v.179/181, p.982.

175. Ohnuki S., Takahashi H., Kinoshita H., Machizuki S. J.Nucl.Mater., 1994, v.212/215, p.755.

176. Wyatt O. H. and Dew-Hughes D.D. Metals, Ceramics, Polymers, Atomizdat, Moscow, 1979, p.978.

177. Farrell K., Houston J.J. In: Rad. Effects and Tritium Technology for Fusion Reactors. USERDA, 1976, v.II, p.209.

178. Grossbeck M.L, Maziasz P.J., Rowcliffe A.F. J.Nucl.Mater., 1992, v.191/194, p.808.

179. Sikka V.K., Moteff J. J.Nucl.Technology, 1974, v.32, p.52.

180. Bradley E.R., Jones R.H. J.Nucl.Mater., 1985, v.133/134, p.671.

181. Hirsch P.B. In: Proc.Conf.Point Defects and Diffus.Process. London, 1977, p.95.

182. Kimura G. and Maddin R.V. in: Defects in Quenched Metals. Atomizdat, Moscow, 1969, p.188.

183. Haazen P. Physical Metallurgy, Mir, Moscow, 1968, v.3, p.248.

184. Inal O.T., Galligan J.M. Acta Met., 1970, v.18, p.557.

185. Cottrell A.H. Dislocations and Plastic Flow, Metallurgizdat, Moscow, 1958, p.267.

186. Kittel' Ch. Elementary Solid State Physics, Nauka, Moscow, 1965, p.366.

187. Massal'sky T.B. in: Physical Metallurgy, Mir, Moscow, 1967. v.1, p.144.

188. Ohr S.M., Tucker R.P., Wechsler M.S. Phys.Stat.Sol.(a), 1970, v.2, p.559.

189. Cottrell A.H. Trans.Metallurg.Soc. AIME, 1958, v.212, p.192.

190. Seeger A. In: Dislocations and Mechanical Properties of Crystals. N.Y., J.Wiley and Sons. 1957, p.243.

191. Rowcliffe A.F., Zinkle S.J., Stubbins J.F., Edwards D.J., Alexander D.J. J.Nucl.Mater., 1998, v.183/192, p.183.

192. Bad'yanov A.B., Volkov M.G. and Lazorenko V.M. In: Structural Materials for Fusion Reactors. Nauka, Moscow, 1988, p.178.

193. Ivanov L.I., Volkov M.G., Platov Yu.M., Sadykhov S.I.O. and Simakov S.V. Fiz. Khim. Obrab. Mater., 1988, No.1, p.28.

194. Cahn J.W. Acta Met., 1963, v.11, p.1275.

195 Saiko K. Phil.Mag., 1981, v.43, no.1, p.43.

196. Van Osch E.V., Horsten M.G., Vries M.I. J.Nucl.Mater., 1998, v.258/263, p.301.

197. Rowcliffe A.F., Robertson J.P., Klueh R.L., Schiba K., Alexander D.J., Grossbeck M.L., Jitsukawa S. J.Nucl.Mater., 1998, v.258/263, p.1275.

198. Trinkaus H. Rad.Effects. 1986, v.101, p.91.

199. Trinkaus H., Ulmaier H. J.Nucl.Mater., 1994, v.212/215, p.303.
200. Ibragimov Sh.Sh., Aitkhozhin E.S., Pyatiletov Yu.S. Influence of Radiation on Materials Ann Arbor, ASTM, 1987, p.5.
201. Platov Yu.M., Simakov S.V. Tsepelev A.B. Fiz. Khim. Obrab. Mater., 1989, No.1, p.11.
202. Bystrov L.N., Ivanov L.I, Martishin O.V. Rad.Eff., 1975, v.24, p.111.
203. Konobeevsky S.T. Effect of Radiation on Materials, Atomizdat, Moscow, 1967, 401p.
204. Kelly B. Radiation Damage in Solids, Atomizdat, Moscow, 1970, p.236.
205. Bystrov L.N., Ivanov L.I., Kaveshnikov V.G. and Platov Yu.M. Fiz. Khim. Obrab. Mater., 1972, No.4, p.15.
206. Ghoniem N.M., Kulcinski G.L. Rad.Effects, 1979, v.41, p.81.
207. Boilean F., Geffroy B., Paulin R. In: Positron Annihilation. Proc.6th Int.Conf. Amsterdam, 1982, p.509.
208. Bystrov L.N., Ivanov L.I., Tsepelev A.B. Rad.Effects, 1986, v.97, p.127.
209. Hansen H., Sehule W., Candy M.R. Fusion Technology, 1988, v.88, p.905.
210. Schule W., Hansen H. J.Nucl.Mater., 1994, v.212/215, p.388.
211. Grossbeck M.L., Gibson L.T., Jitsukawa S. J.Nucl.Mater., 1996, v.233/237, p.148.
212. Hudson J.A., Nelson R.S., McEroy R.C. J.Nucl.Mater., 1977, v.65, p.279.
213. Scholz R. J.Nucl.Mater., 1994, v.212/215, p.530.
214. Grossbeck M.L., Horak J.A. J.Nucl.Mater., 1988, v.155/157, p.1001.
215. Straalsund J.L. In: Radiation Effects in Breeder Reactors Structural Materials. N.Y., AIME, 1977, p.191.
216. Mosedale D., Harries D.R., Hudson J.A., Lewthwaite G.W., McElroy R.J. In: Radiation Effects in Breeder Reactors Structural Materials. N.Y., 1977, p.209.
217. Wassilev Ch., Ehrlich K., Bergmann H.J. In: Influence of Radiation on Materials. Ann Arbor, ASTM, 1987, p.30.
218. Puig R.J., Gilbert E.R., Chin B.A. In: Effects of Radiation on Materials. Philadelphia, ASTM, 1982, p.108.
219. Puig R.J., Gelles D.S. In: Fusion Reactor Materials. DOE/ER—0313/6, 1989, p.201.
220. Toloczko M.B., Garner F.A., Eiholzer C.R. J.Nucl.Mater., 1994, v.212/215, p.604.
221. Troyanov V.M., Bulkanov M.G., Kruglov A.S., Krjuchkov E.A., Nikulin M.P., Pevchykh J.M., Rusanov A.E., Smirnoff A.A., Votinov S.N. J.Nucl.Mater., 1996, v.233/237, p.381.
222. Garner F.A., Eiholzer C.R., Demina E.V., Ivanov L.I. In: Fusion Reactor Materials. DOE/ER—0313/14, 1993, p.221.
223. Herschbak K.K., Bergmann H.J. J.Nucl.Mater., 1990, v.175, p.143.
224. Vorobjev A.N., Budylkin N.I., Mironova E.G., Porollo S.I., Konobeev Yu.V., Garner F.A. J.Nucl.Mater., 1998, v.258/263, p.1618.
225. Neustroev V.S., Shamardin V.K. In: Effects of Radiation on Materials. Philadelphia, ASTM, 1993, p.816.
226. Garner F.A., Puigh R.J. J.Nucl.Mater., 1991, v.179/181, p.577.
227. Tolosko M.B., Garner F.A. J.Nucl.Mater., 1994, v.212/215, p.509.
228. Tolosko M.B., Garner F.A., Eiholser C.R. J.Nucl.Mater., 1998, v.258/263, p.1163.
229. Woo C.H., Garner F.A.. J.Nucl.Mater., 1992, v.191/194, p.1309.
230. Woo C.H., Garner F.A., Holt R.A. In: Effects of Radiation on Materials.

Philadelphia, ASTM, 1993, p.27.
231. Kohyama A., Kohno Y., Asakara K., Yoshino M., Nambe C., Eiholzer C.R. J.Nucl.Mater., 1994, v.212/215, p.751.
232. Kohyama A., Hishinuma A., Gelles D.S., Klueh R.L., Dietz W., Ehrlich K. J.Nucl.Mater., 1996, v.233/237, p.138.
233. Kohno Y., Kohyama A., Hirose T., Hamilton M.L., Nauri M. J.Nucl.Mater., 1998, v.271/272, p.145.
234. Hesketh R.V. Phil.Mag. 1962, v.7, p.1417.
235. Brailsford A.D., Bullough R. Phil.Mag., v.27, p.49.
236. Heald P.T., Speight M.V. Phil.Mag., 1974, v.30, p.869.
237. Bullough R., Hayns M.R. J.Nucl.Mater., 1975, v.57, p....
238. Wolfer W.G., Ashkin M. J.Appl.Phys., 1976, v.47, p.791.
239. Mansur L.K. Phil.Mag., 1979, v.30. p.497.
240. Savino E.J. Phil.Mag., 1977, v.36, p.323.
241. Saralidze Z.K. Fiz. Tverd. Tela, 1978, v.20, p.2716.
242. Woo C.N. J.Nucl.Mater., 1984, v.120, p.55.
243. Gittus J.N. Phil.Mag., 1972, v.25, p.345.
244. Heald P.T., Harbottle J.E. J.Nucl.Mater., 1977, v.67, p.229.
245. Heald P.T. In: Radiation Effects in Breeder Reactors Structural Materials. N.Y., AIME, 1977, p.781.
246. Nichols F.A. J.Nucl.Mater., 1978, v.69/70, p.451.
247. Bullough R., Wood M.H. J.Nucl.Mater., 1980, v.90, p.1.
248. Mansur L.K., Reiley T.C. J.Nucl.Mater., 1980, v.90, p.60.
249. Henager C.H., Simonen E.P. In: Influence of Radiation on Materials Properties. Ann Arbor, ASTM, 1987, p.75.
250. Bystrov L.N., Ivanov L.I. and Tsepelev A.B. 1985, Vopr. Atomn. Nauki i Tek. Seriya: Fizika Radiatsionnykh Povrezhdenii i Radiatsionnoe Materialovedenie, No.2(35), p.1.
251. Tsepelev A.B. in: Irradiated Stainless Steels, Nauka, Moscow, 1987, p.91.
252. Matthews J.R., Finnis M.W. J.Nucl.Mater., 1988, v.159, p.257.
253. Garner F.A., Gelles D.S. J.Nucl.Mater., 1988, v.159, p.286.
254. Borodin V.A. and Ryazanov A.I. In: Radiatsionnoe Materialovedenie. Khar'kov, 1990, v.3, p.3.
255. Saralidze Z.K. In: Structure and Radiation Damage in Structural Materials. Metallurgiya, Moscow, 1996, p.114.
256. Heald P.T., Speight M.V. Acta Met., 1975, v.23, p.1389.
257. Bullough R., Willis J.R. Phil.Mag., 1075, v.31, p.855.
258. Bullough R. In: Fundamental Aspects of Radiation Damage in Metals, USERDA, 1976, v.II, p.1261.
259. Woo C.N. J.Nucl.Mater., 1979, v.80, no.1, p.131.
260. Ghoniem N.M., Kulcinski G.L. Rad.Effects., 1979, v.41, p.81.
261. Dimitrov C., Benkaddour A., Dimitrov O., Corbel C., Moser P. Mater.Sci.Forum, 1987, v.15/18, p.1275.
262. Katoh Y., Miroga T., Kohyama A., Stoller R.E., Namba C. J.Nucl.Mater., 1996, v.233/237, p.1022.
263. Nichols F.A., J.Nucl.Mater., 1979, v.84, no.1/2, p.207.
264. Atkins T., McElroy R.J. In: Radiation Induced Changes in Microstructure. Philadelphia, ASTM, 1987, p.447.
265. Caillard D., Martin J.L., Jouffrey. Acta Met., 1980, v.28, p.1059.
266. Ulmaier H., Schilling W. Radiation Damage in Metallic Reactor Materials. In: Physics and Modern Materials. Vienna, IAEA, 1980, p.301.

267. Skiner B.C., Woo C.N. Phys.Rev.B., v.30, no.6, p.3084.
268. Woo C.N., Gosele U. J.Nucl.Mater., 1983, v.119, p.219.
269. Zelensky V.F., Neklyudov I.M. and Chernyaeva T.P. Radiation Defects and Swelling of Metals, Naukova dumka, Kiev, 1988, p.296.
270 Garner F.A., Wolfer W.G. In: Effects of Radiation on Materials, Baltimore, ASTM, 1982, p.1073.
271. Igata N., Irino T., Kohyama A. J.Nucl.Mater., 1992, v.191/194, p.1323.
272. Evans J.H. J.Nucl.Mater., 1980, v.88, p.31.
273. Hamilton M.L., Johnson G.D. Puigh R.J., Garner F.A., Maziasz P.J., Yang W.J. Abraham N. PNL-SA-15308, 1988, 53p.
274. Shibahara I., Akasaka N., Ohose S., Okada H., Ukai S. J.Nucl.Mater., 1994, v.212/215, p.487.
275. Uematsu K., Kodama T., Ishida Y., Suzuki K., Koyama M. In: Radiation Effects in Breeder Reactors Structural Materials N.Y. AIME, 1977, p.571.
276. Garner F.A.,Kumar A.S. In: Radiation-Induced Changes in Microstructure (Part I), Philadelphia, ASTM, 1987, p.289.
277. Katoh Y., Kohno Y., Kohyama A. J.Nucl.Mater., 1994, v.212/215, p.464.
278. Muroga T., Watanabe H., Yoshida N. J.Nucl.Mater., 1994, v.212/215, p.482.
279. Watanabe H., Muroga T., Yoshida Y. J.Nucl.Mater., 1994, v.212/215, p.503.
280. Sekimura N., Garner F.A., Newkirk J.W. J.Nucl.Mater., 1992, v.191/194, p.1234.
281. Hamada S., Suzuki M., Maziasz P.J., Hashinuma A. J.Nucl.Mater., 1991, v.179/181, p.515.
282. Lee E.H., Mansur L.K. Metallurg. Trans., 1992, v.23A, p.1977.
283. Jitsukawa S., Hojoa K., Hishinuma A. J.Nucl.Mater., 1991, v.179/181, p.538.
284. Marshall P. In: Austenitic Stainless Steels. London, Elsevier, 1984, p.254.
285. Kawanishi H., Ishina S. J.Nucl.Mater., 1991, v.179/181, p.534.
286. Maziosz P.J. J.Nucl.Mater., 1989, v.169, p.95.
287. Schibahara I., Ukai S., Onose S., Shikakura S. J.Nucl.Mater., 1993, v.204, p.131.
288. Azarian A. Contribution a l'Etude des Defauts Ponctuels dans les Aciers Inoxydables, 1974, CEA-R-4617, Doctoral Thesis, University of Paris.
289. Azarian A., Kheloufi K. J.Nucl.Mater., 1981, v.97 p.25.
290. King H.W. J.Mater.Science, 1966, v.1, p.79.
291. Watanabe N., Garner F.A., Muroga T, Yoshida N. J.Nucl.Mater., 1995, v.255, p.76.
292. Esmailzadeh B., Kumar A.S. In: Effect of Radiation on Materials. Baltimore, ASTM, 1985, p.468.
293. Garner F.A. In: Phase Stability Durring Irradiation. Warrendale, TMS-AIME, 1981, p.165.
294. Fisher S.B., Madden P.K., Miller K.M. Phys.Stat.Sol. (*a*) 1980, v.57, p.281.
295. Rothman S.J., Nowicki L.J., Murch G.E. J.Phys.F.Metal.Phys., 1980, v.10, p.383.
296. Kenik E.A. J.Nucl.Mater., 1992, v.187, p.239.
297. Kenik E.A., Hojou K. J.Nucl.Mater., 1992, v.191/194, p.1331.
298. Dimitrov O., Dimitrov C. In: Point Defects and Defect Interaction in Metals. University of Tokyo Press, 1982, p.656.
299. Dimitrov C., Benkaddour A., Dimitrov O., Corbel C., Mozer P. Mater.Sci.Forum., 1987, v.15/18, p.1275.
300. Pimenov V., Polyakov Yu. and Slushaev V. Preprint. RIAR-9 (417), Institute of Atomic Reactors (1980).

301. Shamardin V.K., Bulanova T.M., Golovanov V.N., Neustroev V.S., Povstyanko A.V., Ostrovsky Z.E. J.Nucl.Mater., 1996, v.233/237, p.162.
302. Garner F.A., Brager H.R., Gelles D.S., McCarthy J.M. J.Nucl.Mater., 1987, v.148, p.294.
303. Garner F.A., Brager H.R., Takahashi H.J. J.Nucl.Mater., 1991, v.179/181, p.633.
304. Solonin M.I. J.Nucl.Mater., 1988, v.258/263, p.30.
305. Gelles D.S. J.Nucl.Mater., 1994, v.212/215, p.714.
306. Gelles D.S. J.Nucl.Mater., 1996, v.232/237, p.293.
307. Ioltukhovsky A.G., Kondrat¢ev V.P., Loont'eva-Smirnova M.V., Votinov S.N., Schamardin V.K., Povstyanko A.V., Bulanova T.M. J.Nucl.Mater., 1996, v.233/237, p.299.
308. Klueh R.L., Kai Ji-Jung, Alexander D.J. In: Fusion Materials. 1994, DOE/ER-0313/16, p.128.
309. Rieth M., Dofferner B., Rohring H.D. J.Nucl.Mater., 1996, v.233/237, p.351.
310. Hishinuma A., Kohyama A., Klueh R.L., Gelles D.S., Dietz W., Ehrlich K. J.Nucl.Mater., 1998, v.258/263, p.193.
311. Saito J., Suda T., Yamashita S.,Ohnuko S., Takahashi H., Akasaka N., Nishida M., Ukai S. J.Nucl.Mater., 1998, v.258/263, p.1264.
312. Bloom E. J.Nucl.Mater., 1998, v.258/263, p.7.
313. Loomis B.A., Smith D.L. J.Nucl.Mater., 1992, v.191/194, p.84.
314. Chung H.M., Loomis B.A., Nowicki L., Gazda J., Smith D.L. In: Fusion Reactor Materials. 1993, DOE/ER-0313/15, p.223.
315. Ehrlich K., Bloom E.E., Kondo T. J.Nucl.Mater., 2000, v.283/287, p.79.
316. Rohring H.D., DiStefano J.R., Chitwood L.D. J.Nucl.Mater., 1998, v.258/263, p.1356.

Index

N

O

P

Q

R

S

T

U

V

W

Z